D1566020

SCIENCE AND CONTROVERSY

Lockyer in middle age.
(Photograph by courtesy of Capt. Lockyer)

SCIENCE AND CONTROVERSY

A biography of Sir Norman Lockyer

by

A. J. MEADOWS

Professor of Astronomy, University of Leicester

THE MIT PRESS
Massachusetts Institute of Technology
Cambridge, Massachusetts

First published in the United States of America in 1972

ISBN 0 262 13 079 3
Library of Congress catalog card number: 72-4536

Printed in Great Britain

PREFACE

One very good reason for the current rapid expansion of nineteenth-century studies is our growing understanding that many of the problems facing the Victorians still face us today. In a very immediate sense, what happened in Victorian times has a bearing on what is happening in Britain now. However, in studying the development of science, and the attitudes of the general community towards it, not all decades of Queen Victoria's reign are of equal interest. It is customary to consider the years round the 1860s as a watershed, dividing the early-Victorian period from the late-Victorian. In these terms, the most important questions concerning science and its cultural context were posed in the later period. Yet the later decades of the nineteenth century have, for the most part, been relatively less well studied by historians of science than the earlier decades. There are several good reasons for this, including the increasing complication and specialisation of science towards the end of the century, and the superabundance of relevant printed and manuscript material available. However, so important an epoch cannot be ignored indefinitely. There are, indeed, now signs that it is beginning to receive the greater attention it deserves. The present study may, perhaps, help in the process of clarifying the internal and external relations of science at this time.

Sir Norman Lockyer is a key figure of this later period. It is some measure of his stature that, amongst Victorian scientists, he is most nearly comparable with T. H. Huxley in breadth of interest. Lockyer, like Huxley, thrived on controversy; but, whereas Huxley generally came out the winner, Lockyer, in his scientific controversies at least, generally came out the loser. Yet this can make Lockyer the more interesting of the two. Standing at a slight angle to the world view of his scientific peers, he seems to hold up for us a mirror to their beliefs. Throughout it all he retained that self-confidence which is often claimed as a Victorian characteristic. It has been said that, of all the decades in our history, a wise man would choose the eighteen-fifties to be young in. Lockyer was young in that decade, and he carried some of its ebullience with him throughout his life.

Whilst writing this book I have frequently had the sensation that I was describing not events of fifty or a hundred years ago, but things happening today. The danger, and futility, of trying to find parallels in history to present-day events is well known. But the great questions of science constantly recur, though maybe in different guises, and so it seems do the great questions of science policy and education. I have drawn no analogies; but I shall be surprised if my readers are similarly self-denying.

One biography of Lockyer, now out of print, already exists: that by Professor Dingle in a volume of essays on Lockyer's life and work published by Macmillan in 1928. As one would expect with such an author, the biography is a model of accuracy, so far as it goes. The volume was, however, edited by Lockyer's second wife and by one of his daughters. It was natural, therefore, that the controversial side of Lockyer's nature should be toned down. But it is in the controversies that both the man and his time are most clearly revealed, and I have endeavoured to re-emphasize this in the present work.

I am always pleasantly surprised, when writing a book, at the kindness and helpfulness of everyone with whom it brings me into contact. It is a genuine pleasure for me to thank the many people who have helped in the preparation of this volume. First of all, I would wish to thank Professor Conn and the Norman Lockyer Observatory for permission to use and quote their Lockyer archives; and to Professor Conn, himself, for the hospitable welcome I received at Sidmouth. Then my thanks are due to Mrs. Pingree for advising me concerning the archives of Imperial College, and to the College for permission to use them. I am indebted to Mrs. Pingree, in particular, for showing me both a manuscript copy of the history of the Physics Department at Imperial College, and her own notes on the history of the buildings. I am extremely grateful to Captain Lockyer and his wife for entertaining me so generously whilst I examined the manuscript material on Lockyer which they have retained; and to Captain Lockyer for permission to quote from it. I am indebted to Dr. Dewhirst both for introducing me to the archives of the Cambridge Observatories, and for entertaining me during my visit to Cambridge. I would also like to express my thanks to Professor Redman for permission to quote from the Cambridge Observatories' archives. My thanks are due to Mr. Laurie for his help in using the archives at the Royal Greenwich Observatory, and to Sir Richard Woolley for permission to quote from them; to Mr. Kaye, the Librarian of the Royal Society, for permission to examine relevant archival material belonging to the Society, and to the Royal Society, itself, for permission to use some of this material in the present book. I am also most grateful to Dr. MacLeod for providing me with copies of the correspondence between Stokes and Lockyer now held by Trinity College, Cambridge, and to Dr. Smeaton for providing me with a copy of Meldola's diary for the period 1875–6. Whilst I was looking for illustrations for this book, Professor Hide drew my attention to a set of photographs which had been deposited at the Meteorological Office by the widow of Dr. W. J. S. Lockyer. I would like to express my thanks to Professor Hide and to Mr. Lock, Librarian at the Meteorological Office, for their aid with this material. I am indebted to my colleagues at Leicester University—particularly those who are members of the Victorian Studies Centre—for many lively discussions; and to the Research Board for the provision of funds to help towards the cost of travel and maintenance during the writing of this book and for the cost of copying much of the manuscript material used. I am greatly indebted to Dr. W. H. Brock for undertaking the chore of reading through the whole manuscript. His comments throughout have been invaluable. Dr. D. H. Menzel also read through the entire manuscript and made many detailed comments on both style and content. I am extremely grateful to him for his apposite advice.

CONTENTS

I	The Militant Civil Servant	1
II	The Man of Letters	16
III	The Man of Science	39
IV	The Devonshire Commission	75
V	South Kensington and Meteorology	113
VI	What is an Atom?	135
VII	The Philosopher's Stone	175
VIII	Family and Friends	209
IX	A New Orientation	238
X	Education and National Progress	258
XI	The Final Push	280
	Epilogue	308
	Bibliography	310
	References	316
	Index	327

I

THE MILITANT CIVIL SERVANT

In May, 1859, *The Times* published a sonorous call to arms.

> 'Storm, Storm, Riflemen form!
> Ready, be ready against the storm!
> Riflemen, Riflemen, Riflemen form!'

Alfred Tennyson was celebrating the end of his first decade as Poet Laureate by expressing his countrymen's fears of war.

We tend to think of Victorian England as a country generally at peace, but in the late 1850s contemporary commentators believed that a major conflict was almost inevitable. The Crimean War lay in the past, but new, threatening tensions had arisen. Most of the British army was a long way away. Some units were tied up in the aftermath of the Indian mutiny, some were involved in the second—and third—Chinese wars, and some had been dispatched to Canada, for the resident British minister in the United States had been dismissed, and there was a consequent possibility of war. Just at this point in time, with the British Isles very feebly defended, a new quarrel flared up with France.

At the beginning of 1858, a determined attempt to assassinate Napoleon III was made by Orsini, an Italian, who believed that Napoleon was hindering the cause of Italian unity. It was discovered that Orsini had planned his attempt in London, and had used bombs bearing that familiar Victorian designation—'made in Birmingham'. Napoleon's Foreign Minister protested violently, and a group of French colonels proposed that an attack should be made on London. But when Palmerston tried to introduce a Bill increasing the punishment for conspiracy to murder, he was strongly opposed, and was accused of weakness towards the French. His Bill failed, and he was, himself, forced to resign. War seemed very close, and with nearly all the regular units abroad, the outlook was unpleasant.

There was a considerable unanimity of opinion as to the immediate need—a volunteer army must be formed and armed at once. The unanimity did not initially extend to the Army heads of staff, who were noticeably unenthusiastic. Popular pressure was too much for them, and, three days after the appearance of Tennyson's poem, permission was given for corps of volunteers to be raised. It was not intended that the British working classes should be armed : it was required, at least to begin with, that members of the new corps should buy their own equipment and defray all personal expenses. As a result, the recruits were predominantly men from commerce or the professions.

One of the earliest volunteers at Wimbledon (the corps being raised mainly on a territorial basis) was an enthusiastic civil servant, Joseph Norman Lockyer. Every morning he rose at 5.30 in order to drill with his neighbours, from six to seven o'clock, for Queen and country. Wimbledon was a good place for military training; near the village the long common developed into a smooth expanse of grass which was ideal for drilling. There, earlier in the century, many of the great military reviews had been held, and now under similar circumstances these were renewed. Most of the volunteers were to be trained as riflemen, so, towards the end of 1859, a National Rifle Association was set up making provision for shooting practice for the volunteers as one of its main aims. Its first national competitive meeting was held on Wimbledon Common in the summer of 1860. This event was of more than passing interest to Lockyer. Not only was it on his own territory, but he was, himself, a keen marksman, good enough to become an instructor. Moreover, Sidney Herbert was the President of the Association and Minister at the War Office, where Lockyer was a clerk.

In those days a Civil Servant was not expected to start work until ten o'clock at the earliest, but Lockyer worked to a much more rigid schedule. After drill he would walk down to his house on the hillside, between the Common and the railway station, take a quick breakfast, and then catch a train into town. He usually travelled via the London and South Western Railway to Waterloo, and then, negotiating the incipient chaos which was later to make Waterloo the most ill-reputed terminus in London, made his way northwards to the British Museum. At this point Lockyer clearly distinguished himself from his colleagues at the War Office. Some of them certainly called in at the British Museum Reading Room from time to time, but none of them went there to read science books. Yet such was Lockyer's purpose.

In the late 1850s scientific training in England was extremely limited. Where interest in science existed, it had often been aroused outside the normal school and university system. As a consequence, many important contributors to science at this time had had little or no formal training. The Victorian scientific scene, and Lockyer's subsequent contribution to it, can only be properly understood if we realize the remarkable diversity of paths by which a man could then become a scientist, and the essential role which the amateur played in science. Lockyer's progress into science is a case in point.

Lockyer's education was that of a typical member of the mid-Victorian middle class: it contained a certain amount of elementary mathematics, but no science at all. During his teens he had studied at Kenilworth school in Warwickshire, and had acted as a student-teacher both there and at a similar establishment in Weston-super-Mare. In 1856, the master of Kenilworth school wrote to the main landowner in the area, Lord Leigh of Stoneleigh Abbey, on Lockyer's behalf, asking for his assistance in getting Lockyer a Government appointment. He painted the following picture of his pupil's accomplishments.

> Mr Lockyer is 20 years of age, and has lately been Tutor at a School in Somersetshire: he had previously resided in my School, as Student and Teacher, six years.
>
> His stature is about 5ft 6, and his health uniformly good.
>
> He is a sound English Scholar, and a Classic: a good Accountant and Mathematician, and has a knowledge of French. In a word, his abilities are of the highest order, and his punctuality and perseverance indefatigable.[1]

Lockyer at this stage was mainly interested in languages. Before taking up a temporary appointment at the War Office in 1857—for he succeeded in obtaining Lord Leigh's patronage—he spent some months in Switzerland studying French and German, stopping on the way out and back to attend lectures at the Sorbonne. Indeed, his other major excursion from his native Midlands, to the school at Weston-super-Mare, had been undertaken in the hope of improving his Greek.

The West Country, however, was also the area from which his family had originally hailed, not so long before. His grandfather had moved to Kensington to practise as a surgeon-apothecary, and there Joseph Hooley Lockyer, Norman's father, had been born. The name Hooley came from his mother, Rebecca Hooley, who was a Midlander from a village just outside Leicester. She died in 1823, when J. H. Lockyer was still in his teens, to be followed only five years later by his father. He and the rest of the family were left with very little money, and prospects would have been grim had they not possessed one very influential relative. This was William Howley (an alternative spelling of Hooley), Bishop of London, who about this time was transferred to the See of Canterbury. Howley had a well marked reputation both for sanctity and for promoting the interests of his relatives. As uncle and godfather to J. H. Lockyer, he was prepared to help him in every possible way. At first, it was thought that this could best be done by paying for an education at Oxford; but, eventually, it was decided J. H. Lockyer should follow his father's old profession and qualify as a surgeon-apothecary.

In the early nineteenth century three kinds of doctor, in the modern sense of the word, were recognized—physicians, surgeons and apothecaries.

Of these the physicians formed the most privileged and respected group; they were regarded as gentlemen, university educated, who worked with their heads rather than their hands. The surgeons belonged to a lower class of practitioner using manual techniques, not only for internal surgical operations, but also for the external application of salves or plasters. The third group of medical practitioners, the apothecaries, figured even lower on the social scale. Basically their task was to compound and dispense drugs, although the tradition had grown up that they could also, with some restrictions, visit and prescribe for their patients as well. Because they derived their livelihood mainly fom the sale of drugs, they were classified as tradesmen and despised accordingly.

The physicians ministered predominantly to the needs of the gentry and the upper-middle class, and when in the early years of the nineteenth century a demand for better medical attention arose from the rapidly increasing middle classes, they were slow to respond. Instead, the gap was filled by surgeon-apothecaries; men who had qualified for membership of both the Royal College of Surgeons and the Society of Apothecaries, and were therefore entitled to administer to their patients both externally and internally. At least, this was the theory: in practice, by no means all who called themselves surgeon-apothecaries were legally entitled to do so. In the 1820s when J. H. Lockyer reached his decision to become a surgeon-apothecary, the old medical hierarchy still stood, but it was becoming increasingly evident that it did not truly reflect the relative merits of the different practitioners. A surgeon-apothecary qualified for his title by a long apprenticeship of five years or more, receiving a training which was always practical, and often quite efficient. An English physician, on the other hand, qualified via a university education which contained little that was practical, and was seldom efficient. Even the class distinction which had previously differentiated physicians and surgeon-apothecaries was becoming blurred. The aspiring physician at this time was actually more likely to have come from a lower class home than the average trainee surgeon-apothecary. In fact, it was men of J. H. Lockyer's generation, with his type of training and background, who began to break down the old distinctions between the different types of doctor, and to substitute the modern concept of a general practitioner.

From our point of view, these 'new men' in the medical world had one attribute of immediate importance: they were much more concerned with a scientific approach to medicine than any of their predecessors. J. H. Lockyer was in this respect typical. After qualifying in the 1830s, he moved to the Midlands to practise, probably through the influence of his mother's side of the family. He settled first in Rugby, and very quickly became involved there in the formation of a Literary and Scientific

Institution. The appearance of such societies was a common feature in the early nineteenth century, and, in this case, reflected especially the growing interest in the world round about. J. H. Lockyer lectured extensively on science to the new institution, concentrating first on chemistry and a little later on electro-magnetism. This was the period when Michael Faraday's work on electro-magnetism at the Royal Institution in London was creating a great stir. It appears that J. H. Lockyer was repeating Faraday's experiments for the benefit of his Midland audience, and may even, though this is less certain, have carried out some original demonstrations of his own. Through his participation in the Literary and Scientific Institution, J. H. Lockyer became acquainted with Thomas Arnold, who was another keen member, and so developed an interest in Rugby School, where Arnold had been headmaster since the latter part of the 1820s. Many years later his son was also to become involved in the fortunes of science at Rugby, but there seems to have been no direct connection with this earlier interest of his father.

In 1835, shortly after establishing his practice at Rugby, J. H. Lockyer married Ann Norman, the daughter of one of the local gentry : her father was the squire of a small village nearby. Eleven months later their first child, Joseph Norman Lockyer, was born. His Christian names were, of course, taken from his father's first name and his mother's surname; to avoid confusion with his father, however, he was always known simply as Norman Lockyer. After a few years in Rugby the family moved to Leicester, and there, in 1845, Mrs Lockyer died, leaving J. H. Lockyer with the task of bringing up Norman and one other child, a daughter three years younger. Norman remained with his father in Leicester until 1849, then, partly because of ill health, he was sent to stay with some of his mother's relatives back in Warwickshire. This was the reason why he became a pupil at Kenilworth School. J. H. Lockyer died in 1855, and so we find his son seeking Lord Leigh's patronage in the following year.

Norman always retained a vivid remembrance of his father's interest in science. In so far as his inclination towards science derived from this impetus, Lockyer was in good company. The largest single source of scientists in the mid-nineteenth century was probably medical men and their families. However, it is doubtful if his father's example by itself would have been sufficient to have turned Lockyer's interest so definitely towards science. After all, during his teens, when he might first have become interested in science, he was separated from his father for much of the time. Something further was needed to arouse Lockyer's latent interest : the spark was to be found amongst the new friends he made when he went to live in Wimbledon.

Lockyer was from the beginning a great organiser, another talent, perhaps, which derived from his father. Wimbledon, at the time he went there, was just becoming a popular, and growing, commuter suburb. One sign of this growth was the formation of a village club on the Ridgeway, not far from Lockyer's house. Lockyer immediately became involved in the new club as a committee member, and had as one of his fellow members on the committee George Pollock, a barrister whose work took him frequently to the North of England. Pollock was acquainted from his travels with Thomas Cooke who had become by the late 1850s one of the leading English instrument makers. Pollock purchased a refracting telescope from Cooke, set it up in his garden at Wimbledon, and invited his friends to look through it. Lockyer did so, was fascinated, and in 1861 ordered an identical instrument to be installed in his own garden. His serious study of science had begun.

Lockyer then, sitting at his place in the British Museum Reading Room, was following a natural development of interest from his own point of view, though an unusual one for a clerk at the War Office. Yet even here, there was a certain logic at work. A striking characteristic of Victorian science is the amount of research which was carried out by officers of the Army and Navy. For example, the President of the Royal Society, when Lockyer first began his scientific studies, was Major-General Sabine of the Royal Engineers, famous for his correlation of magnetic activity on Earth with spot activity on the Sun. Hence, at least amongst the military with whom Lockyer came into contact in the course of his work, there was a tradition of sympathy for scientific research. But the intensity of activity which Lockyer brought to his new hobby was peculiarly his own.

It may equally be said, on the other side, that there can have been few places in Victorian England which proffered greater time for self-improvement than the War Office. In the 1860s the Civil Service (a term only introduced in the preceding decade) ambled along at a very sedate pace. Hours of work were customarily from ten in the morning to four in the afternoon. Even during these hours the atmosphere was noticeably relaxed; the War Office, in particular, seldom over-worked its clerks. The flavour of life on Horse Guards at this time has been recorded by one of Lockyer's colleagues.

> There were too many clerks and too little work. I came as my brother's *locum tenens* at a time when his place could easily have been allowed to remain unoccupied. I reported myself to my chief, a most delightful gentleman some twenty or thirty years my senior. He introduced me to all his colleagues as 'Coco's brother'. 'Coco' was my brother's nickname. Everyone was as kind and as civil as could be. There were five desks, and one was allotted to me. Any amount of stationery and all sorts of official books of reference, a most cheerful apartment, with a number of specimen-glasses, which later on contained flower

button-holes to be worn in the hour allowed for lunch, say 12 noon in Piccadilly or Rotten Row. I was most courteously invited to read the *Times*. I got through the *Times* fairly rapidly while my colleagues had a friendly chat about the doings of the day, or, rather, night. Someone had gone to such a ball where the supper was only 'tol lol', and the play of yesterday evening was, in spite of what the papers said of it, awfully dull. My colleagues were the most charming of companions, but I must confess that Dickens' sketch of the Circumlocution Office was scarcely a caricature.

After reading the *Times* through and finding the conversation of my courteous colleague a little above me, as at the age of seventeen I had scarcely plunged into the vortex of polite London society—the word 'smart' in its modern sense had not been invented—I ventured to address my chief.

'Can I do anything?' I asked, as I stood beside him at his desk. 'Is there anything for me to do?'

He seemed a little perplexed. The other denizens of the room paused for a moment in their conversation to hear his reply. It seemed to me that they appeared to be amused. My chief looked at me and then at the papers in front of him.

'Ah!' said he at last, with a sigh of relief, 'are you fond of indexing?' I replied I was fond of anything and everything that could be of the slightest service to my country. If those were not the exact words I used, that was the spirit of my answer.

'I see a glutton for work,' observed my chief, with a smile that found reflection on the faces of my other colleagues. 'Well, à Beckett, just index this pile of circulars.'

I seized upon the bundle and returned to my desk. Oh, how I worked at those circulars. There were hundreds of them, and I docketed them with the greatest care and entered their purport into a book. From time to time my official chief, so to speak, looked in upon me to see how I was getting on.

'I say,' said he, 'there is no need to be in such a desperate hurry. I am not in immediate need of the index. You can take your time, you know. Wouldn't you like a stroll in the Park? Most of us have a little walk during the day. We none of us stand upon ceremony, and are quite a happy family.' But no, I stuck to my indexing, and, after some three days of fairly hard work, found my labours done. I took up the bundle of circulars, now in apple-pie order, and laid them on my chief's desk.

'I say, à Beckett,' said he, 'this won't do. You are too good a fellow to be allowed to cut your own throat, and for your brother's sake I will give you a tip. Don't do more than you are asked to do. Now I gave you those circulars to index because you would bother me for work. I didn't want the index. Now it's done it's not the least bit of use to me. Of course it may come in useful some day, but I scarcely see how it can, as the lot are out-of-date, but of course it may.'[2]

Each Government department at this time had its own conditions and terms of appointment, although the recently formed Civil Service Commission was trying to eradicate some of the major differences. It was a uniform requirement, however, that a candidate for a Government appointment had to have his name put forward by some appropriate person. Lord Leigh, as head of one of the most influential Warwickshire families, was thus capable of intervening decisively on Lockyer's behalf. Lockyer started as a temporary clerk, a position in which many of his colleagues remained for several years; but he was allowed to take the qualifying examination for a permanent post within a few months of his arrival at the War Office. He then became a third-class clerk at an initial salary of £100 a year (with annual increments of £10) and a leave allowance of 48 days per year. The latter was generous, even by modern standards, though it must be remembered that Saturdays were counted as working days.

Of all the Government departments, the War Office when Lockyer joined it was the most over-crowded. Its establishment had been allowed to rise rapidly during the Crimean War, the assumption naturally being that it would fall again after the emergency had passed. But the same causes that led to the formation of the Volunteers also made the Cabinet reluctant to reduce the War Office establishment. As a result, Lockyer joined a vast number of other third-class clerks, and found himself not only with little work to do, but also with remote prospects of promotion.

Following the Northcote-Trevelyan report of 1853 there were continuing attempts to reform the Civil Service, and the War Office was one of the most obvious targets for would-be reformers. In 1863, Gladstone, as Chancellor of the Exchequer, began to press strongly for economy at the War Office. This led in 1864–5 to an inquiry into the running of the War Office, which resulted in a recommendation that the number of staff should be immediately cut. '. . . the character of the War Office is such that, if the clerks did their work with diligence 10 per cent of their numbers might be reduced.'[3]

The clerks at the War Office were dismayed, but not Lockyer. For him the inquiry marked the beginning of his rise in the Civil Service, since another reform—the formation of a new section at the War Office—had been decided on at the same time. A memorandum concerning this was circulated in December 1865. 'The Secretary of State for War has established a Branch to codify and revise Warrants and Regulations issued from this Office, and has appointed Mr J. Norman Lockyer to the charge of the Branch.'[4]

To understand how Lockyer had managed to single himself out for promotion from the numerous other third-class clerks at the War Office, we must return again to the village club at Wimbledon. The impetus for the formation of this club had been specifically Christian. Pollock was a churchwarden of the parish church; Lockyer in his early years and for some time afterwards was a keen Anglican. But the two most prominent people on the club committee were Tom Hughes and J. M. Ludlow, leading figures in the Christian Socialist movement. By the time that Lockyer met them the early heyday of that movement was already over, but it was still a potent intellectual influence within and without the Church of England. In particular, the Christian Socialists, as broad churchmen, were playing a significant part in the dialogue between churchmen and scientists, a dialogue which, as a consequence of the appearance of Darwin's *Origin of Species* in 1859, was elsewhere under some strain. Hughes and Ludlow had had a house specially built for their joint occupation on the Ridgeway at Wimbledon. There from 1853 onwards they formed the nucleus of one of the most important intellectual networks of the period; a network into which Lockyer was gradually to be introduced.

Lockyer often travelled up to town with both Ludlow and Hughes, but it was with the latter that he struck up the closer acquaintance. Hughes had been at Rugby during the period when Lockyer's father had been practising there. Like Lockyer he was an enthusiastic supporter of the Volunteer movement, and had, indeed, raised his own corps from the members of the Working Men's College which the Christian Socialists had founded in London. Although Hughes was by profession a barrister at Lincoln's Inn, his varied interests had brought him a wide range of friends, and one of these was Earl de Grey who, besides being the man mainly responsible for the formation of the Volunteers, became, in 1863, Secretary for War and therefore Lockyer's ultimate chief.

We have seen that War Office reconstruction became a major issue during the 1860s. This was, in fact, only a particular aspect of a more general demand for reform of the Army, sparked off initially by the military muddle revealed during the Crimean War. In the early 1850s the Colonial Office, the Treasury, the Home Office, the Secretary at War and the Commander-in-Chief were all responsible for some aspect of Army administration, as were a variety of other, smaller departments. Some amalgamation of responsibilities had subsequently occurred, but throughout the 1850s a combination of pressures from the Tory party, the Court and the Permanent Under-Secretary at the War Office, Sir Benjamin Hawes, prevented any major reforms. The Palmerston-Russell administration of 1859–66, however, decided to tackle the problem vigorously. First Sidney Herbert, who had been pressing strongly for Army reform, became Secretary for War with de Grey as Under-Secretary, and then, from 1863–6, de Grey was promoted and took over the battle. He was appointed by Palmerston against Gladstone's wishes as the result of pressure brought by Florence Nightingale. The death of Sir Benjamin Hawes in 1862 removed the last major obstacle, and de Grey immediately began to push ahead with plans for organisational reform. He had realised that the main problem was the lack of any clearly defined spheres of operation and of permitted responsibilities within the Army. He therefore insisted that a vital first requirement was the codification and coordination of the vastly ramified system of Army regulations. The best thing to do, he decided, would be to appoint someone with a legal background to consider the revision of the regulations. He naturally turned to his friend Tom Hughes who was qualified both by his training and by his military enthusiasm. Some members of the Tory party immediately raised the cry of Whig jobbery, but Hughes went ahead with his task and, requiring a member of the War Office to assist him in his work, equally naturally chose Lockyer. When Hughes retired from the work, de Grey decided that a permanent Army Regulation Branch should be set up with Lockyer as its head, on Hughes' recommendation.

From the financial viewpoint Lockyer's promotion was a major advance (there are indications that he was finding it very difficult to make ends meet in the early 1860s) for he was granted an additional remuneration of £300 per annum. On the other hand, the work was distinctly onerous for he had not only to codify and coordinate existing regulations, but also to examine new decisions and enquire whether they accorded with precedent. As a result, he had to curtail his outside activities, and he now often overstayed the nominal office hours of ten till four.

It is evident that Lockyer happily seized this opportunity to exercise his own initiative. He claimed later that he had been at one time or another a member of five War Office committees connected with regulations, and these on occasion required his attention for as much as five days a week. It appears that he himself was responsible for much of the organization. Take, for example, the following letter from the Office of the Quarter-Master-General.

> In compliance with your request, I have just spoken to General Forster on the subject of a joint committee of representatives of the various Departments at the War Office and Horse Guards with authority to arrange questions that may arise during the revision of the R. Warrant and Army Regulations.
>
> The General thinks that His Royal Highness would offer no sort of objection to the appointment of such a Committee but, on the contrary, is of the opinion that this would be the most effectual method of expediting the process of codification and of authoritatively settling the various points of detail, that must constantly arise as the work progresses.[5]

Despite the statement by his Kenilworth schoolmaster that Lockyer's health was good, Lockyer was, in fact, very liable to breakdowns from overwork in his early years. He had a short illness at the end of 1864, followed by another early in 1867, and a more prolonged one in the middle of 1868. Fortunately, War Office regulations for sick-leave were distinctly generous : anything up to six months could be taken off on full pay, so Lockyer did not suffer financially. His last breakdown was sufficiently serious for him to decide to recuperate in Switzerland, a country which had been in use as a health resort by Englishmen for the previous two or three decades.

Lockyer's idea of recuperation was a strenuous walking tour. He was at this time a small, slim, physically very active man, though by middle age he became distinctly stout and had to watch his weight very carefully. He decided to go with a few companions who were equally keen walkers. Noticeable amongst these was Edward Whymper. Whymper was a few years younger than Lockyer; he had been drawn into the growing English craze for alpine mountaineering from the start, and had already obtained a permanent claim to fame in this activity by making the first ascent of the Matterhorn in the summer of 1865. In the process, however, four members of the team had been killed in an accident, and the English press, which had previously ignored mountaineering, happily took this opportunity to condemn it. Thus, in the latter half of the sixties, Whymper possessed an unenviable notoriety. He seems, indeed, to have been greatly shaken by the accident, and during the period of Lockyer's continental tour was doing hardly any serious climbing: instead he was writing up his experiences in one of the classic mountaineering books—*Scrambles in the Alps.* He was also becoming increasingly interested in the scientific side of mountaineering, which perhaps partly accounts for his friendship with Lockyer. For Lockyer's growing scientific interests included the dispute which was raging in the 1860s over the importance of glaciation as a force shaping mountains. People attracted into mountaineering at that time could be divided into two camps, scientists and non-scientists, who tended to regard each other with contempt. Lockyer's friendships were, as might be expected, mainly with the scientists.

Lockyer's breakdown in 1868 may also have been connected with increasing personal tensions at the War Office, for his post as head of the Regulation Branch was now in question. Lockyer had accepted the position on the understanding that, if he gave satisfaction, the post would be made permanent, and he would be correspondingly promoted. In May 1868, however, it was proposed that Lockyer should lose his extra allowance and should be placed under the control of a first-class clerk. Lockyer immediately sought an interview with the Secretary of State and persuaded him that a Committee of Inquiry should be set up. He then began to solicit the help of any of his acquaintances whom he thought might influence the committee's decision. He had by this time built up, via his literary and scientific work, a wide circle of friends, and although most of these could not help directly, they sometimes had invaluable contacts of their own. For example, John Forbes-Robertson, art critic and father of the actor, accompanied Lockyer and Whymper on their 1868 walking tour. Lockyer now persuaded him to approach one of his friends at the Treasury, Lord Stansfield, to exert pressure on the War

Office. This particular attempt seems to have borne little fruit, but there can be no doubt that Lockyer soon managed to build up significant outside backing. Most of his influential friends, however, were on the Liberal side in politics, whereas the reigning government was Conservative; so some of the pressure exerted on his behalf may have backfired.

When the Committee of Inquiry finally met, its investigations gave Lockyer little cause for joy: indeed, he accused it of failing to stick to its brief, for it reported back to the Secretary of State, now Sir John Pakington, that Lockyer's demotion should be confirmed. As Lockyer bitterly pointed out at the end of 1868, he was still doing the same amount of work, but now in a subordinate position and with his salary reduced by over fifty per cent.

A main factor in his demotion was probably the continuing demand for economy at the War Office. A contributing factor may well have been latent antagonism to Lockyer within the department. There can be no doubt that some such antagonism existed: partly due to Lockyer's rapid advance, but mainly in reaction against his attempts to increase efficiency, Lockyer was not always very tactful in trying to get colleagues to mend their ways. Whatever the cause, at the end of the 1860s Lockyer found himself, so far as his War Office career was concerned, not very much further on than the point at which he had started.

During these years of progress and regress in his work at the War Office, Lockyer's social life had attained a high level of activity. Very soon after his nomination to a clerkship at the War Office, Lockyer married a girl whom he had known in Warwickshire, Winifred James. His wife, although typical of Victorian wives in that she tends to be hidden in her husband's shadow, was a strong personality in her own right. She was a daughter of William James, one of the important figures in the early history of railways. James was a Warwickshire man who became agent and solicitor to Lord Warwick, a position in which he accumulated considerable wealth. He owned the Snowford estate (of about 1000 acres) in Warwickshire and the Trebenshun[6] estate (of about 700 acres) in South Wales, during the early years of the nineteenth century. James, however, got bitten very early on by railway mania and as early as the end of the eighteenth century he proposed plans for the laying out of railways. During 1821–2 he drew up the preliminary report for a Liverpool and Manchester Railway, and, at about the same time, W. H. James, one of his sons, took out a patent for a tubular steam boiler. Both father and son became involved with George Stephenson, and it was firmly believed within the family that Stephenson cheated both of them of their rightful recognition as pioneers in the development of railways and of the locomotive. James became so much concerned with railway projection that he lost most of the money and estates he had previously accumulated, and was at one stage declared bankrupt.

Not long before his death William James married a second time, and Winifred James was a daughter of this second marriage.[7] It is evident that there was considerable enmity between the wife and children of the second marriage and the children of the first marriage. The latter, having started life with a wealthy father who had gradually become poorer, watched with chagrin whilst the second wife received not only the monies remaining in her husband's will, but also, in due course, a substantial amount under the wills of James' brother and also of his brother-in-law. According to one of the daughters of the first marriage, the cruellest cut of all was that the second Mrs James obtained £300 from the Liverpool and Manchester Railway in recognition of her husband's work and told her step-children nothing of it, even though the eldest step-daughter was dying at the time.[8]

Whatever the truth of these accusations, there seems to have been sufficient money to provide Winifred with an adequate all-round education. Like Lockyer she was a proficient French linguist, and when he started to take an active interest in astronomy she began to study the subject too. In the mid-1860s a translator was required for a French book on astronomy, *Le Ciel* by Guillemin, and she agreed to undertake the task. The translation was so successful that she was immediately accepted in the publishing world as a leading translator of French scientific texts. In 1868, she published another translation entitled *Volcanoes and Earthquakes* and in the early 1870s another work by Guillemin on *The Forces of Nature*. This success was, of course, very helpful in balancing the Lockyers' budget during financially difficult years. At the same time, Winifred Lockyer was fully occupied with the typical task of a Victorian wife in producing children. In the first fifteen years after their marriage she gave birth to nine children, and, unlike the average Victorian household, only one of these died in infancy. In looking after them, she had little assistance from her husband, who was occupied most weekends and evenings, often till two o'clock in the morning, with his scientific and literary work.

In early 1865, the Lockyers moved from Wimbledon closer into London, taking a house in West Hampstead. They had evidently been thinking of moving throughout 1864, for a letter from one of Lockyer's new scientific acquaintances, William Huggins, suggested that they might like him to look for an old cottage near his own house on Tulse Hill.[9] Wimbledon was then a village well separated from London, though during lunch-time walks in Green Park, Lockyer could see back to Wimbledon Hill where he lived, and, despite the growing number of commuters, still had a social life of its own. Hampstead, on the other hand, was already on the outskirts of London; although parts were well wooded and hilly, and in Hampstead Heath there was a tract of open country for walks of the sort the Lockyers appreciated. When they moved to Hampstead, they actually found considerable local anxiety about encroachment on this heath, as the Lord of the Manor was granting portions of it for building purposes. Not long after their arrival, the local residents banded together to form a Heath Protection Fund, and started legal proceedings against the Lord of the Manor, which led eventually to the Metropolitan Board of Works buying him out. Nevertheless, despite this show of local spirit, the social life of Hampstead was much more identified with that of the metropolis than had been the case at Wimbledon. In particular, Hampstead at this time had a certain attraction for artists and men of letters. Alfred Tennyson's married sister lived there with his mother, and the poet was a frequent visitor to the neighbourhood. In December 1865, Tennyson came to Hampstead to stay with his sister, his mother having died earlier in the year at about the time the Lockyers moved to the area. Whilst there he was inducted as a Fellow of the Royal Society, a gesture to acknowledge his great interest in science. During this visit, he was introduced to Lockyer by Thomas Woolner, a mutual friend. Woolner, a leading sculptor of the day, was almost certainly one of the friends whom Lockyer made via his literary work. Tennyson and Lockyer found that they had several friends in common; indeed they had both independently become acquainted with Tom Hughes at almost the same time. Tennyson was reckoned to be one of the Broad Church group: he was a great admirer of the master of the Christian Socialists, F. D. Maurice, and was a member of the group which had formed round Alexander Macmillan, the publisher, to which Hughes also belonged.

The Lockyers' social life now began to expand and diversify rapidly, and acquired the characteristics which, in retrospect, can be seen as typical of liberal-intellectual circles in London in the 1860s. For example, this was a period of great interest in spiritualism, and so we find A. R. Wallace, one of the great scientific converts to spiritualism, inviting them along to his private seances.[10] It seems that Lockyer was sceptical of the validity of spiritualist phenomena, though many years later he confessed to Oliver Lodge—like Wallace, a scientist who became a convinced spiritualist—that he had experienced an apparently psychical happening.[11] Wallace, incidentally, had returned to London from his travels in 1862 and so entered the scientific world of the metropolis at about the same time as Lockyer.

Not long after the move to Hampstead, perhaps because of the greater ease of social contact with his friends, perhaps to help offset the problems of his work at the War Office, Lockyer instituted his 'smokers' or informal weekly 'at homes', usually on a Wednesday evening, to which any of his closer friends were welcome to come. Many of them were so regular in attendance that individual clay pipes, labelled with their names, were kept in a special rack in the smoking room. Although Tennyson only came when he was in London, there was a clay pipe kept specially for him. The guests who attended came from a wide range of backgrounds, from Winwood Reade, the author of *The Martyrdom of Man* and a leading free thinker, on the one hand, to Thomas Baines, the African explorer and artist, on the other.

Lockyer was an entertaining conversationalist and keen on music and in his early days he enjoyed singing and playing the flute. These activities all figured in his 'at homes', but always the emphasis was on conversation. He also entertained his guests on clear nights with views through his telescope which was by this time a larger refractor, again made by Thomas Cooke. Lockyer was a very gregarious man, and his personality, which in most of his public dealings was from the beginning noticeably aggressive, became much milder on ordinary social occasions. As examples are accumulated of Lockyer's public controversies, these must be balanced against the evident delight which a wide variety of eminent men found in personal contact with him. Relaxed amongst his peers Lockyer was an excellent host.

II

THE MAN OF LETTERS

Literary ambition was a common prerogative of Civil Servants when Lockyer joined the War Office. It was commonly said at the time that all the leading writers of the period were recruited either from the Bar or from the Civil Service. When War Office reform was under consideration in 1865 the main object of the inquiry was said to be, 'to improve the efficiency of the Establishment and getting the clerks to understand that they are paid for work and not for literary distinction'.[1] There was an especial reason for increased literary activity in the sixties—the removal of certain restrictions during the 1850s, such as the abolition of newspaper duty, led to a sudden burgeoning of periodical literature and a consequent demand for competent writers. It is therefore not surprising that Lockyer should have tried his hand in this direction very soon after settling at Wimbledon. His literary interests were, however, certainly reinforced by his friendship with Tom Hughes.

Hughes had won national fame not long before Lockyer met him with the publication by Macmillan of *Tom Brown's Schooldays.* Hughes had naturally gone to the Macmillan brothers with his book, for they were strong supporters of the Christian Socialist movement. Alexander Macmillan, in particular, was always prepared to consider works by members of the movement for publication. About the time that Lockyer first came to London, Alexander Macmillan began to hold regular weekly meetings of a small group of his friends including Woolner, Tennyson, Tom Hughes, Kingsley, Holman Hunt, Huxley, and Spencer. Apart from the usual conviviality, the group also discussed with Macmillan possible future projects which he had under consideration. Thus, one result of these meetings was the initiation of *Macmillan's Magazine*, a liberal monthly journal with David Masson, a Christian Socialist and another member of Macmillan's weekly meetings, as editor.

Ludlow was rather disappointed by this decision as he had hoped to have the appointment himself. Instead, he decided to join with Tom Hughes in publishing a new weekly periodical of their own which would carry on a free and uninhibited discussion of contemporary controversial topics in science, religion and the arts. They believed, with some justice, that no contemporary journal really permitted this. They particularly emphasized the need for having signed articles, an attribute in which their journal became a pioneer. If it appears odd that this should be considered an important characteristic, it is worth noting that it was named later in the century by the historian, Lecky, as one of the major advances in increasing liberty of thought. One reason presumably why the journal was published independently by the proprietors, and not by Macmillan, was because a major item in the journal, as in most of its contemporaries, was the reviewing of books; and they desired to make it completely manifest that there would be no bias in these reviews in favour of a particular publisher. This again was something that could not be said of all the other journals then appearing. They naturally envisaged a publication which would be both Broad Church and politically liberal in outlook. The first number of their new journal, which they called the *Reader*, appeared on 3 January 1863.

Lockyer's early literary work really grew out of his scientific interests. His first articles were simply notes on astronomical topics, sometimes including descriptions of the observations that he himself was carrying out. Some of these appeared in the *London Review*: a journal edited by one of the founders of the *Illustrated London News*. But he occasionally contributed elsewhere, e.g. to the *Spectator*. It was not surprising therefore that Ludlow and Hughes should ask Lockyer to contribute regularly to their new journal; though now he was required to cover the whole area of science, not just those parts connected with his astronomical interests. Some years later Lockyer wrote to Huxley about these early days, but he remembered them in the following form.

> With regard to the *Reader* I can tell you this about its early stages. I believe it was started by Macmillan and Stevens the printer (Family Herald). T. Hughes and J. M. Ludlow had most to do with it at starting and as I knew them both they asked me to look after the science.[2]

Despite the apparent inevitability of Lockyer's involvement in the *Reader*, the effect it had on his future life, not only in the literary sphere, but also on his scientific work and his Civil Service career, cannot be overestimated. A major reason was that it brought him into close contact with Thomas Huxley. Huxley was eleven years older than Lockyer, and in the 1860s had reached the peak of his form; highly influential in scientific circles, and well known to many outside those circles by his forthright and controversial advocacy of Charles Darwin's ideas on evolution. Huxley had realised for some time the need to establish a new journal where he and his friends might have an *entrée*, and this for several reasons. There was a general feeling amongst English biologists that communication between themselves was poor; Huxley believed this deficiency could be repaired if some journal could be pressed into service as a kind of scientific newspaper. Moreover, such a journal could present the scientific viewpoint to the general public. In this latter respect, he had come to feel the need for a journal that would treat science impartially with particular intensity as a result of popular attacks on the *Origin of Species*, and on evolutionary ideas in general, during the early 1860s.

Maskelyne, one of the leading British mineralogists and another member of the Christian Socialist group, had tried, and failed, during the fifties to float a general science journal on his own account. In 1858, he held discussions with the editor of the *Saturday Review*, which had begun publication three years before, and to which he already contributed, in order to see whether that journal could initiate a fortnightly science column. The editor asked Maskelyne to get together a small group of scientists to provide material, and Maskelyne approached various potential contributors, including Huxley. Unfortunately, despite considerable interest, the scheme foundered. Huxley, undeterred by this lack of success on Maskelyne's part, made the next move himself: he agreed to take over the editorship of the *Natural History Review*, hoping he could mould it into the sort of journal he judged to be needed. By the middle of 1863, however, he had found the task too great in combination with all his other commitments, and had been forced to resign. The journal ceased publication a couple of years later. Huxley was disappointed by the rather poor response of the scientific public to the *Natural History Review*, and concluded that a quarterly publication, such as the *Review*, appeared too infrequently to act as an efficient medium for scientific communication. When therefore the *Reader* began weekly publication in 1863 with the avowed intention of publishing scientific material, Huxley immediately declared his interest; more particularly since he was on very friendly terms with the Christian Socialist group.

It was probably due to Huxley's influence that Lockyer towards the end of 1863 began to look for support in increasing the emphasis on science in the *Reader*. On the one hand, he tried to obtain new contributors; recruiting, for example, James Glaisher, an astronomer and meteorologist who pioneered the use of balloons in meteorology. (Balloonists came to be divided, like mountaineers, into the scientists and the non-scientists with a tendency towards a similar mutual feeling of contempt.) On the other hand, Lockyer tried to extend the scope of his scientific reporting. Thus, he obtained permission from Sabine, the President of the Royal Society, to report meetings of the Royal Society in the *Reader*. Sabine, a military man himself, seems to have liked the War Office clerk, and helped to facilitate Lockyer's first close contacts with the Royal Society.

By this time, however, the *Reader* had already undergone various reorganisations. The first editor, Ludlow, resigned after three months, and was replaced by David Masson, the man who had been preferred to him as editor of *Macmillan's*. Lockyer thus came into close contact with Masson, and struck up a friendship which lasted for many years until the latter's death. Despite these rapid changes, the *Reader* retained the name it had immediately gained for literary excellence. Besides the Christian Socialists, Ludlow, Tom Hughes, and Charles Kingsley, other regular contributors included W. M. Rossetti (brother of the pre-Raphaelite) and Tom Taylor who, although little remembered now, was one of the best known poets of the period. He was another Civil Servant, and a neighbour of Lockyer at Wimbledon. Unfortunately, this excellence was not rewarded by the financial return : the 1860s besides being a prolific time for the foundation of journals were also a time when many foundered. The competition was strong; journals with similar aims were often started by different publishers almost simultaneously; for example, the *Cornhill* was founded at almost the same time as *Macmillan's*. Similarly the *Reader* overlapped in time and content with the *Athenaeum*, and also, on the purely scientific side with other journals. For example, William Crookes, then a young chemist rapidly rising to fame, helped initiate the *Quarterly Journal of Science* at the beginning of 1864, in addition to editing the more specialized *Chemical News*. The *Quarterly Journal* is worth noting especially, as it was explicitly intended to aid communication amongst scientists, and, moreover, Lockyer was named as one of its supporters. It became evident during 1864 that the Christian Socialists would be financially unable to continue with the *Reader* for much longer. However, both Lockyer and, more importantly, Huxley were unwilling that the journal should be allowed to fail without a struggle, for they felt that it had proved its usefulness in the dissemination of scientific information.

Lockyer's main task on the *Reader*, apart from editorial work, had been supplying summaries of scientific matters of current interest, and providing long descriptions of the annual British Association meetings of 1863 and 1864. The British Association at this time played a very important part in the scientific life of the country. The addresses, particularly by the Presidents of the Sections, often led to considerable controversy. For example, the debate on evolution was often developed at British Association meetings. Huxley was therefore a frequent attender. Lockyer too attended frequently during the sixties, both initially as a reporter for the *Reader* and increasingly as a scientist in his own right. He became keenly interested at this time in the new and, from the point of view of contemporary religion, dubious field of anthropology, and went to some of the sessions with Huxley. It seems likely that Lockyer and Huxley used the 1864 meeting to discuss the state of the *Reader*, and this may have spurred them both on to greater efforts in its support.

Huxley and a small number of his closest scientific friends had been meeting fairly regularly for some time: in the latter part of 1864 they put these meetings onto a regular basis by forming themselves into the X Club. Some of the other members of this club, especially Frankland, were to figure prominently in Lockyer's subsequent career though he, himself, never became a member. The first meeting of the club was early in November 1864, and one of the main topics of discussion was the future of the *Reader*. In the next few days, half the members of the X Club agreed to contribute towards the purchase of the *Reader*, turning it into limited liability company. At the same time Lockyer had been proselytizing successfully on his own account for more support.

Having attended to the financial side, Huxley, who had taken charge of proceedings, was now faced with the problem of finding someone to do the general routine work of editing. Towards the end of November, Thomas Hughes, who remained as one of the proprietor's of the paper after Huxley's takeover, wrote to him

> My strong belief is that at present Lockyer can do all the general editing and will be the best man for us. He knows the machinery having been there from the first, has been in constant relations with such men as Ludlow etc. who must be had for writers on general subjects not yet occupied, has the science already in the right grooves, and is not above taking advice, is a real good worker and above all has his heart in the business.
>
> He will do the work too gladly at a lower figure than any other competent man, a consideration to be regarded at the present until we get our capital and know where we are. We have bought a pig in a poke and shall want careful management, and above all plenty of give and take and pulling together to pull our pig out of the poke and make him a comely grunter fit to run alone. With care the *Reader* may be made a first class paper and a good property, but a little bolting and jibbing will have us all in the ditch amidst the jubilations of newspapers.[3]

It is worth noting that Lockyer was prepared to accept a very low salary for the work since in 1864 he was still in considerable financial straits. Huxley agreed with Hughes, and within three days he had invited Lockyer to superintend the general editing of the *Reader*, saying 'If you can find the time requisite—there is no one I would so gladly see looking after matters at the office of the *Reader*.'[4] In the event, however, Lockyer was not finally appointed general editor. Various other people were approached instead, including Leslie Stephen, whose only apparent connection with the group of scientists was a blatantly non-scientific interest in mountaineering.

Lockyer, nevertheless, found himself involved with the entire gamut of the *Reader*'s contents on occasion. He was thus led to correspond with many figures of the time who were quite unconnected with science. For example, Tom Hood wrote to him asking him for 'any bad trashy novel' to review,[5] and the two were evidently on excellent terms. Indeed, the change in the fortunes of the journal obviously cheered Lockyer considerably, even if he was not appointed to the highest position. A letter from one of his friends commented 'I saw Ludlow from whom I learnt that the *Reader* was casting off its slimy old skin and coming out renewed in better style. I hope it is so for your sake—you were very despondent about it when I saw you.'[6]

Despite the sound advice of the editor of the *London Review*, who warned him that although the *Reader* was an excellent journal taking shares in it would be a gamble, Lockyer plunged some of his small reserves of money into the venture and became one of the proprietors. Fellow proprietors among the scientists, besides Huxley, Lubbock, Spencer, Spottiswoode, and Tyndall of the X Club, included Charles Darwin and his half-cousin Francis Galton. It is some indication of the personal way in which shareholders were recruited that Galton was persuaded to join by Spencer, who had in turn been approached by both Huxley and Lockyer. Among the non-scientific proprietors appeared the name of J. S. Mill, whose *On Liberty* had been published in that literary *annus mirabilis*, 1859, along with Darwin's *Origin of Species*. This was no bad company for a young man not yet thirty, who aspired to scientific and literary recognition. Moreover, Huxley obtained the declared support for the journal of some seventy-five of the leading British scientists of the day, including many who figured largely in Lockyer's subsequent career. One of these, J. D. Hooker, referred to the scientific proprietors of the journal, Huxley, Galton, Lockyer, etc., as the 'Young Guard of Science'[7] : a name that stuck to them long after their mutual relationship in the *Reader* had lapsed.

The *Reader* was thus able to call upon a most distinguished circle of supporters, though more of Huxley's generation than of his seniors. Nevertheless, the circulation of the journal continued to be a source of anxiety, and the *Reader* staggered a little uncertainly into 1865. By the summer of that year it was evident that the financial situation of the journal had deteriorated even further, though an attempt had been made to increase the circulation by lowering the price from 4d. to 2d. Moreover, there were continuing difficulties in the day-to-day running of the paper. Galton later recalled

> It was an amusing experience, owing to Mr B's insistence [John Bohm was the paid sub-editor], from a commercial point of view, about the necessity of obtaining advertisements by all sorts of ingenious means, but some of which, in our opinions, were not quite above-board. Then it was brought home to us that, as our venture was one of limited liability, whatever we bought must be paid

> for at once, while what we were to receive would not be paid for many months. We were like children in the hands of Mr B., who knew all the ins and outs of the commercial conditions of success, concerning which we were almost childishly ignorant. The newspaper proved dull, notwithstanding some really good articles. The management was naturally too amateurish; promised articles were delayed, and the time of the committee was too much wasted in frequent discussions about first principles, upon which Spencer loved to dilate.[8]

Ultimately, to the regret of the Young Guard, it became necessary to get rid of the journal in order to avoid further financial loss, and towards the autumn of 1865 it was sold to Thomas Bendyshe, a Fellow of King's College, Cambridge, and a violent atheist. Huxley fought to retain control until the last. Writing to Lockyer in August 1865, he remarked, 'Mr Bendyshe is one of the lights of the Anthropological (Society) and I should be sorry to see the *Reader* fall [?] into his hands.'[9] Lockyer, as a member of the Anthropological Society, may have been acquainted with Bendyshe. Under Bendyshe, the journal lingered on for another couple of years before finally expiring, but he changed both the policy and the contributors. Effectively therefore, the main scientific interest of the *Reader* lapsed after 1865.

Lockyer's association with the *Reader*, if not financially very advantageous, was of great importance to him for the close contacts he established with the leading British scientists, and for the friendships which he made with people in the fields of literature and the arts. The extensiveness of his social life after the move into London partly reflects this. But the experience of writing for the *Reader* was important in another way: the need to fill a weekly scientific column encouraged Lockyer to publish his own early speculations on several of the major scientific problems which came before him. He discussed, for example, the importance of meteorites, the study of spectral lines and the need for state support of science; all topics which were to engage him deeply in later years. Yet if Lockyer's work led him to speculate profitably for his personal scientific future, it is equally important to remember that his writings may themselves have played some part in determining the course of science, for they were read with interest by many of the younger scientists. The *Reader* was esteemed as an important journal during its brief life, and was therefore used as a mouthpiece by more than one eminent scientist. For example, one of the leading British geologists, Roderick Murchison, used the *Reader* during 1864 to launch a violent attack on the theory that extensive glaciers, which have since disappeared, were an important influence in shaping the Earth's surface features. This idea is today generally accepted, but at that time the more general belief was that glaciers were of minor importance compared with the role of flowing water.

Lockyer's literary activities occupied a major part of his time during the period of his connection with the *Reader*. It may be added that the *Reader* was not alone in consuming his literary energies; in 1864, for example, he had also briefly edited the *Electrician* while its normal editor was away. It seems, indeed, that even his burgeoning astronomical interests were curtailed at this time. After he ceased working for the *Reader*, his literary activity decreased considerably because of the increasing responsibility of his work at the War Office. When he took over the Regulation Branch, he noted that

> The additional sum of £300 was fixed upon as a special remuneration for the special character and onerous nature of the work, and on the condition of my giving up literary and scientific work which at that time afforded me an income of larger amount.[10]

As a result, although Lockyer seems to have had in mind the possibility of a new general scientific journal to follow the *Reader* throughout the latter part of the sixties, it was some time before he made any move to implement the idea. This does not mean that actually he gave up all his literary work. He continued to contribute occasional articles to the *Athenaeum*, the *Saturday Review* and the *Spectator*, and he began to consider the possibility of writing a book.

By the mid-sixties, Lockyer was becoming recognised as knowledgeable concerning the Sun; the suggestion arose therefore that he might team up with Balfour Stewart, one of the leading British authorities on solar physics at that time and a recent acquaintance of Lockyer, to produce a definitive book on the subject. The negotiations were broken off when Balfour Stewart decided that he had insufficient time to participate, and the idea had to be abandoned. But the possibility of publishing a book attracted Lockyer, and he started to write one on his own account, this time a simple introductory text on astronomy intended for people with no previous knowledge of the subject. This book, *Elementary Lessons in Astronomy*, appeared in mid-1868 and was well received. As we shall see, the ancient science of astronomy was undergoing a renewal in the 1860s, and there was a widespread interest in the subject. Lord Farrer, one of Lockyer's influential supporters, told him, 'We are delighted with the simplicity and clearness of your elementary lessons and especially with the absence of controversy and hypothesis'.[11] Few of Lockyer's subsequent books were to receive this last commendation. The *Elementary Lessons*, which went on selling for many years in various revisions, formed a useful additional source of finance for Lockyer at a difficult period.

Later Victorian times were good for best-selling authors, and Lockyer had from the start an understanding of the value of publicity and proper business management in literary endeavours. His belief that the *Reader* had foundered through clumsy business management only enhanced his wish to manage such a journal himself. In the case of this first book, he wrote to the Astronomer Royal, Sir George Airy, with whom he had become acquainted, and asked if he could dedicate the book to him. Sir George, however, knew the significance of his office well enough by this time, as he had been Astronomer Royal since the year before Lockyer was born, and he replied, 'My responsibility as a man in office makes me very nervous about any measure which might seem liable to be interpreted as an ostentatious recommendation of one work in preference to others.'[12]

The projected book with Balfour Stewart had been suggested by Alexander Macmillan, and *Elementary Lessons* was published by him. It was almost inevitable that Macmillan should become Lockyer's publisher. As we have seen, many of Lockyer's early literary and scientific acquaintances in London were closely connected with the firm, particularly with Alexander Macmillan, and wrote for it. Alexander Macmillan shared the Broad Church view that science and religion were allies, even if the controversy over the *Origin* had obscured the fact. He therefore felt it a religious duty to produce elementary science books which would introduce the general reader to the essential nature of contemporary sciences. He was also, of course, guided by the question of commercial interest, for it was evident to him in the 1860s that science was going to become of increasing importance as a school subject, and there would therefore be a growing demand for elementary texts in schools. Macmillan's particular contribution to the question of textbooks was to decide that they should be written, not by the schoolteachers who were actually involved in teaching at an elementary level, but by 'the recognized masters in each branch'.[13] Apart from Balfour Stewart and Lockyer, he had in mind such experts as Roscoe to write about chemistry, Jevons to write about political economy, then considered a science as is indicated by its inclusion as a section of the British Association, and Huxley to write on biology.

By the time that Macmillan's idea came to its final fruition, Lockyer was already highly involved in the workings of the firm. In the middle sixties there was something of a crisis following Alexander Macmillan's serious illness. Up to that time, he had virtually handled most of the forward planning of the firm alone. Now he gradually began to shift some of the planning responsibilities to the shoulders of other people whom he used as advisers. In this category, Lockyer eventually came to be regarded as his 'consulting physician with regard to scientific books and schemes.'[14]

Lockyer and Macmillan seem to have been generally on the best of terms, but on at least one occasion, probably during 1870, they quarrelled violently over Lockyer's relationship with the firm. There is a letter, unfortunately undated, to Lockyer from Archibald Geikie, one of the leading geologists of the day, who was evidently acting as a mediator between Lockyer and Macmillan.

I have just seen Mac. I did not tell him I had seen you nor did I allude to you until he asked me what mood you were in. I told him you were naturally indignant at his conduct as I myself had been but that I knew you were quite ready to listen to any honourable compromise. He then told me what he purposed. I don't know whether I am breaking confidence in retailing it to you. But forgive me if I am and of course receive this proposal as if you had never heard anything about it before.

He said that I had put things in a new light to him and that he was deeply grieved for what had happened and was anxious to make amends to you. He proposes a series of popular books of which he gave me the subjects to be either written by you or under your direction. He proposes to offer you a retaining fee of £250 per annum . . .'[15]

The series of books mentioned is presumably Macmillan's *Nature* series of general scientific texts, to which Lockyer himself contributed *The Spectroscope and its Applications.*

This mention of *Nature* brings us back again to Lockyer's desire to found a new scientific periodical. By the end of 1868 his expectations at the War Office were in eclipse, and he no longer felt himself bound to limit his literary activities in deference to his ordinary work. In the summer of 1868 he had already tried to persuade one of the daily newspapers, the *Daily News,* to run a scientific column. He remained, however, chiefly concerned with the need for a special weekly journal, like the *Reader*, but restricted entirely to science. Lockyer now discussed the possibility of such a journal with Macmillan, and the latter, still wanting to expand his scientific publications, was keenly interested. The basic policies of the new journal were determined by their joint experience, and by the previous rather dismal experiences of other scientific journals. When Lockyer was writing some years later to Huxley, he remarked, 'I don't know anything special about the death of the N.H.R. [*Natural History Review*]. . . . I only know that its circulation was 700 and that is why when *Nature* was started we had to appeal to a larger and not only a special public.'[16] Macmillan immediately set about acquiring support amongst his scientist friends for the new journal. One person to whom he turned was J. D. Hooker, the director of Kew. Hooker's reply pledged support, but was couched in pessimistic terms as regards the proposed journal, and was obliquely questioning as regards Lockyer.

By all means make public my good will to Mr Lockyer's periodical. I fear however that scientific support is a broken reed: and that it will be very difficult to supplant the *Athenaeum* bad as it is—the failure of scientific periodicals patronized by men of mark have been dismal . . .

I do not see how a really scientific man can find time to conduct a periodical scientifically—or brain to go over the mass of trash that is communicated to it and requires expurgation.[17]

Macmillan, however, was not deterred, and Lockyer, for his part, drummed up considerable additional support in the scientific community by means of a circular which he sent round in June, 1869. All the replies seem to have been favourable, including that from Crookes, who was still a potential competitor. Huxley was once again optimistic.

One of Lockyer's most pressing tasks was to recruit editorial assistants from people with some scientific background. Here again his scientific acquaintances were helpful, particularly Michael Foster, who advised him on possible people to engage. Foster, who was born in the same year as Lockyer, was a physiologist in London at this time, though he moved to Cambridge in 1870. Huxley, Sharpey and he were currently leading an attempt in London to develop the use of laboratory methods in biology teaching. The Exeter meeting of the B.A. occurred at a convenient time for Lockyer to interview some of Foster's recommended candidates. A few already had some of the necessary expertise. Thus one was John Brough, who, amongst other editorial experience, had been briefly editor of a *Nature*-like journal called *Laboratory* in 1867, before it went bankrupt. Foster also advised Lockyer on general matters concerning the proposed journal, 'I have come to the conclusion that 24pp. is quite enough to start with and that 4d. is quite high enough price.'[18]

An early decision was required on the title of the new journal. Both Lockyer and Macmillan considered the question, but the final decision may have been taken by Macmillan. Huxley wrote to Lockyer, in July 1869, 'Macmillan told me yesterday that he had nailed his colours to the mast—and was going in for "Nature" pure and simple. I am inclined to think it is the best plan on the whole.'[19] So *Nature* the new journal became, and soon a motto from Wordsworth was added to the title, '. . . To the solid ground / Of nature trusts the Mind that builds for aye'.* But, significantly, the transcription of this in the journal capitalized 'nature', but spelt 'Mind' with a small 'm'.[20]

The proposed name went down quite well with the scientific community. 'Nature' (with, or without, the initial capital) was in much greater use as a word amongst Victorian scientists than it is amongst contemporary scientists. Few, however, were quite as enthralled by it as J. J. Sylvester, the mathematician.

> What a glorious title, *Nature*—a veritable stroke of genius to have hit upon. It is more than Cosmos, more than Universe. It includes the seen as well as the unseen, the possible as well as the actual, Nature and Nature's God, mind and matter. I am lost in admiration of the effulgent blaze of the ideas it calls forth.[21]

* This was taken from one of his miscellaneous sonnets, and the quotation continues, 'Convinced that there, there only, she can lay / Secure foundations.'

But Sylvester was well known for the extremity of his enthusiasms, and Lockyer doubtless took this with the necessary grain of salt.

Lockyer pushed forward as rapidly as possible with plans for the first issue of *Nature* to appear in the autumn of 1869. He was now receiving a large number of enquiries from the scientific public about the new journal and suggestions for possible articles. In the middle of October, for example, Sylvester wrote to him again.

> When is *Nature* to reveal herself to the expectant world? I am often asked the question and can only answer that I am not in her secrets. I am desirous to get an idea of the tone of the articles. . . . Would you entertain a preliminary objection to anything relating to war as breaking in upon the feeling of quiet and rapt contemplation which the name of Nature suggests?[22]

The first number of *Nature* was finally scheduled to appear on the 4th November. Indeed, Lockyer fixed not only the day of publication, but even the time. As Macmillan explained to a friend,

> *Nature* is to be published on Thursday in London at 2.30. . . . Lockyer was peremptory that our publication day should indicate the point to which our information is brought up. The fallacy of a Saturday publication with a Thursday actual information he does not think right. . . . We start with 18pp. of advertisements. . . . I think it will look nice.[23]

Lockyer was, of course, concerned with the possibility of competing journals, but it seemed at first that there was little need for worry on this score. Late in 1868, a new weekly science journal, *Scientific Opinion*, had begun publication, but it was known in 1869 to be losing money and, in any case, had nothing like the massive backing, both financial and scientific, that *Nature* would have. (It actually ceased publication in 1870.) Lockyer therefore could afford to ignore it. But rumours were circulating during the summer of 1869 of a potentially much more damaging competitor. As Hooker wrote to Macmillan in July: 'There is *between ourselves*, a talk of one being established in Oxford.'[24] And, indeed, just a month before *Nature* finally appeared, another journal, the *Academy*, was published, claiming, amongst other things, that it intended to provide a forum for science. The editor was C. E. Appleton of St John's College, Oxford, a friend of Lockyer. Perhaps this circumstance, together with the fact that the new journal was a monthly, calmed some of Lockyer's fears. Nevertheless, he evidently sought reassurance from those of his acquaintances who knew something of the editorial policy of the *Academy* that there would be no great overlap of readership between the two journals. Michael Foster who was engaged to write for both journals told him that Appleton had already said there would be no interference between the two.[25] So it seems to have proved in practice, though the *Academy* initially reviewed *Nature* rather harshly, and, perhaps more significantly, converted itself into a fortnightly in 1871 and a weekly in 1874.

The first issue of *Nature* when it appeared was found to begin with a declamation on nature by Goethe, translated by Huxley. Some readers supposed that it had actually been written by Huxley. Huxley was the obvious choice to provide the opening fanfare, for no one had tried harder to establish a general scientific weekly in England, but his choice of source is interesting. German *naturphilosophie* held its own seductions for the generation of Huxley and Lockyer, in so far as they, too, believed that man and nature formed a unity which could be studied and understood by the proper application of science. Moreover, Germany had already become the Mecca of research scientists, and sympathy for at least some German ideas was strong amongst them.

Apart from Huxley's introduction, the first issue of *Nature* tried to set before its readers some sample of the type of material that it proposed to include in the future. In the circular which Lockyer had sent round to solicit support for the new venture, he had claimed that the two basic aims of the journal would be,

> First, to place before the general public the grand results of Scientific Work and Scientific Discovery, and to urge the claims of Science to a more general recognition in Education and in Daily Life; and
>
> Secondly, to aid Scientific men themselves, by giving early information of all advances made in any branch of Natural Knowledge throughout the world, and by affording them an opportunity of discussing the various Scientific questions which arise from time to time.[26]

These general aims were to be implemented by having eminent scientists write on the political, social and cultural aspects of science and by including articles on science in schools, as well as by the more usual reports of scientific discoveries. Lockyer also attached particular importance to the book reviews and correspondence. Although the letters were signed, the book reviews were either signed with initials or left unsigned. Lockyer called on many of his old friends and acquaintances for these reviews. For example, he immediately turned to Charles Kingsley for reviews of books on marine biology. Kingsley wrote a few days after the first issue appeared, enclosing a review and saying that he had received the first number of *Nature*. He added, 'I am exceedingly desirous that your paper should succeed.'[27]

Three years later, we find Kingsley writing again concerning his overall impression of the journal. 'I trust that Macmillan did not say that I had a *"bad"* opinion of *Nature*. On the contrary, I have the highest respect for it, and I wish I were wise enough to understand more of it. But I fear its circulation must be more limited than you would wish.'[28] Thus, although it had been one avowed purpose of *Nature* to present the results of science to the general public, it is evident that, if even so able an amateur as Kingsley could not keep up, the journal had soon become primarily a mode of communication between professional scientists.

Macmillan and Lockyer were only too acutely aware of the limitation on circulation to which Kingsley had pointed. In 1871, whilst Lockyer was away on a solar eclipse expedition, he received a letter from Macmillan, saying:

> But above all I am very anxious about *Nature*. I cannot help feeling that a very little more of *something* would make it a success, and if so of course it would be a permanent benefit to you. I have been thinking of many things. At present we are endeavouring to get it more widely taken at schools, and if we succeed in this will go into some other line.[30]

Lockyer, himself, claimed that *Nature* in the first year of its publication reached nearly 5,000 subscribers, and it was, of course, certainly read widely by many non-subscribers.[30] Unfortunately, detailed records are not available, but it is generally believed that Lockyer's estimate was a considerable exaggeration. It has been suggested, in fact, that the initial number of subscribers may have been only between 100 and 200.[31] One point of interest is that *Nature* almost certainly had a larger circulation abroad than in the United Kingdom.

Macmillan, although, of course, bearing all the cost of *Nature*, including Lockyer's salary, which was probably larger than his basic War Office salary, interfered very little in any aspect of the running of the journal. This was partly a matter of policy on his part, but was also due to his continuing ill-health. As he remarked in the letter to Lockyer quoted above, 'I have only seen Mrs Lockyer once since you left. I would have been up to see her but I have not been sleeping well and am troubled with my old swimming in the head, and trying what quiet at home and early hours will do.' The brickbats and bouquets therefore all accumulated to Lockyer. In the early years of *Nature*, there were a fair number of both. In the former category came the inevitable printer's errors. Huxley complained, 'Your printers are abominable. They make me say that "Tyndall did not see the *drift* of my statement", when I wrote "draft" as plainly as possible.'[32] A brief acquaintance with Huxley's handwriting is sufficient to load one's sympathy in favour of the printers in this particular case.

Then there were the violent objections of contributors whose articles had been cut.

> I have taken a good deal of pains for some years now to give you accurate accounts of the monthly meetings of the R.A.S. [Royal Astronomical Society] in a very concentrated form. They cannot be so condensed, and yet give a true representation, without extremely careful writing. I have taken pleasure in doing it both for the sake of 'Nature' and the society. It is no advantage to me: I don't even get a copy of the paper.

Now, if after I have taken all this trouble, you cut out some of my most careful paragraphs, as you did last report, for the sake of a trumpery advertisement, you must dispense with your report and keep your advertiser; and I must conclude that his three and sixpence is more important to 'Nature' than correct scientific information. You had better go back to little Ranyard: he will quite approve of the principle, and the scissors are not likely to do any harm. If you don't like little Ran, try Lord Lindsay or any other member of the advertising sandwich men.

You are bound to have a little truth sometimes: I only hope it was done by some of your understrappers.[33]

Brett, the author of this letter, was a minor Pre-Raphaelite painter and a keen amateur astronomer. It will become apparent in the sequel that, when Brett wrote this letter in the late seventies, Lockyer had become involved in a major controversy with a group including Ranyard and Lord Lindsay. This group centred its activities on the Royal Astronomical Society where Brett acted as one of Lockyer's defenders.

Another difficulty was the need for unusual illustrations, this at a time when woodcuts were still the normal process. For example, Francis Galton was experimenting in the seventies with attempts to determine whether certain physical traits were common either to different people of apparently similar personality, or to different members of a single family. In order to try and trace these similarities, Galton formed composite pictures which fitted together the facial features of all the people he was interested in, and so emphasized the characteristics they had in common. He wrote hopefully to Lockyer: 'I suppose it would not be possible to reproduce in "Nature" one of the composite portraits by any process of wood-photography?'[35]

A problem which all publishers in this country faced was the lack of a copyright agreement with the U.S.A. An example of the complications to which this could lead appears in the story of the creation of a new American journal, the *Popular Science Monthly*. Lockyer was naturally interested in this journal as a possible competitor to *Nature* since perhaps a quarter of *Nature*'s circulation was in the United States. When he tried to obtain a copy, E. L. Youmans, who was Herbert Spencer's chief American disciple, and founder of the *Popular Science Monthly*, wrote to him apologetically concerning the journal.

It is a piratical concern of which I am duly ashamed although if it ever pays us, my intention is to pay for foreign selections.

But I wish you to understand the way it was started. Having with great labor got Mr Spencer's consent to prepare a volume for the International Series, I next got his consent to publish it first in the periodicals, and I promised to arrange for them here. I accordingly sold them to the *Galaxy*, at high rates, but the first article came too late for simultaneous insertion, and I saw at once that if we did not in some way secure it it would be pounced upon by the Boston publishers, and that Spencer would get nothing. There was not a moment

> to lose . . . We accordingly resolved to print it ourselves with a lot of other stuff in a periodical form, and from the time that we decided to do this it was just twelve days to the publication of the *Popular Science Monthly*.[36]

Youmans went on to say that the resulting journal, based on the reproduction of original British and European articles, was much to his surprise flourishing, but he had only plucked up enough courage to send it to two people in England, namely Spencer and Tyndall.

Lockyer ran into difficulties with his editorial staff too. At the end of 1872, Huxley wrote to him on the subject, 'I have been trying to think of somebody who would do for your subediting place—but to no effect. Michael Foster generally knows of a lot of loose-ended clever fellows—why not try him for a recommendation.'[36]

This letter was written in the aftermath of a dispute over what might be called the Bennett affair. J. D. Hooker had been instrumental in persuading Macmillan to have a young man called Bennett appointed as sub-editor of *Nature* to deal with botanical matters. It appears that Lockyer was glad enough to have him. But as Hooker explained to Huxley during the controversy, Lockyer was not at all happy with Hooker's attitude towards the journal. 'It was through my interference that a Botanical Sub-Editor was appointed, and I have long thought that Lockyer owed me a grudge for the way I pitched into *Nature* to Macmillan and ridiculed the astounding blunders and ignorance its early botanical notices displayed.'[37]

At about this time, Hooker was engaged in a long drawn-out altercation with A. S. Ayrton, Commissioner of Works in Gladstone's first administration. Ayrton wished to convert Kew from a botanical garden into an ordinary park, and to that end he harried Hooker, the director of Kew, in the hope that he would feel forced to resign. Hooker's scientific friends rallied round, and through Sir John Lubbock, who, besides being a member of the X Club, was also a Member of Parliament, presented a petition to the House of Commons. To strengthen his own hand, Ayrton then prevailed on Sir Richard Owen to write a report on Kew and its management, which he proceeded to publish without showing it to Hooker. Owen and Hooker were violently antagonistic, partly due to opposing scientific views, but more particularly, there was an unresolved clash in the activities of the two. Owen was in charge of the Natural History departments of the British Museum, which had as part of their duties the collection of plant specimens, but this was also a duty of the herbarium at Kew. Owen's report, when it appeared, was found to allege that there had been gross mismanagement at Kew.

It is at this point that *Nature* became involved in the quarrel. Hooker felt that Lockyer did not support him against Ayrton in the pages of

Nature as he should have done. The last straw came when Lockyer published a letter from Owen on the 7th November 1872, vigorously attacking Hooker.

Hooker began fairly moderately, writing to Huxley, 'I well know that Lockyer has no ill-feeling—but . . . he is regarded by those who know something [?] of the circumstances as using *Nature* most unfairly by giving piecemeal scraps of evidence—and often this not in favour of Kew.'[38] Two days later, however, the tone hardened, 'My suspicions are strong against Lockyer of whom I have heard much that I do not like.'[39]

Bennett told Huxley that Lockyer had authorised the publication of Owen's letter, and Huxley passed this information on to Hooker. Lockyer, when tackled by Huxley insisted that he had nothing to do with the matter and said, 'You know that I am so busy with the Countries affairs now [i.e. his work on the Devonshire Commission], that Bennett manages *Nature* practically and this matter I have especially left to him.'[40]

Hooker then drafted a reply to Owen which Lockyer published in *Nature* on the 21st November. Huxley wrote to Lockyer about this letter, saying :

> It is all that it ought to be in my judgment and seeing that Owen's second attack has turned out to be, like the first, a mass of lies and suppressions—it seems to me that a grave question arises for you as Editor whether 'Nature' ought to be any longer made the vehicle of attacks which are simply refuted as fast as they are made.
>
> It is one thing to give a man fair play and another to afford him the opportunity of publishing a set of scurrilous libels in the hope that some of the mud he throws will stick. I only put this to you—for your consideration—of course the responsibility is yours and you must decide for yourself whether there are any considerations outweighing these.[41]

Huxley was as opposed to Owen as Hooker was. Nevertheless, he tried to act as a mediator between Hooker and Lockyer, and eventually the matter was smoothed over, though it is dubious whether Hooker was ever entirely reconciled to Lockyer. The upshot was that Bennett left the *Nature* offices.

Finally, from the point of view of administrative problems, Lockyer was faced with that continual difficulty of all editors of periodicals, issues going astray and consequent irate subscribers. Some of these complaints came from very far afield. For example, B. A. Gould wrote from Argentina to complain of the non-arrival of some copies of *Nature*.

> If ever you find yourself 500 miles from the coast or from any large town, in a part of the world where neither English, French nor German is spoken, not a scientific book to be found excepting your own store, and not the primary rudiments of science known to the population—you may then be able to picture to yourself my present condition . . . when as before intimated, I have the good fortune to receive the weekly issue of 'Nature' once more, I shall not feel quite so isolated.[42]

But this complaint also suggests the reason why Lockyer, and, more importantly, Macmillan, was prepared to go ahead with *Nature* despite its continuing inability to make a profit throughout its early years. From the beginning *Nature* was recognised by scientists all over the world as one of the significant organs of scientific opinion. This appears clearly from the editorial correspondence, both published and unpublished. Here are three examples of the influence attributed to *Nature* during the 1870s in the restricted field of meteorology alone.

In 1874, relations within the meteorological service of France were distinctly strained. Leverrier, the director of the Paris Observatory, was under fire from his meteorologist compatriots. In the nineteenth century, astronomy and meteorology remained fairly closely linked, as we shall see, and the main observatories also recorded meteorological data. Leverrier, although a major theoretical astronomer, was cordially detested by his colleagues as an administrator and person, but Lockyer seems to have got on with him quite well. R. H. Scott, who was superintendent of the Meteorological Office for the last thirty years of the nineteenth century, wrote privately to Lockyer on the approach he felt *Nature* should take.

> Matters as regards meteorology in France are so ticklish that I dare not write to your sub-editor about them . . . the enclosed circular . . . I am privately informed is meant for a 'coup' against Le Verrier.
>
> Such being the case I dare not go near Paris, as I have to be strictly neutral.[43]

Lockyer handled the situation in such a way as not to involve British meteorologists.

Three years later Geikie was writing to Lockyer from Edinburgh urging him to use the influence of *Nature* to press the significance of the Meteorological Report on public opinion. 'This Meteorological Report business is so melancholy that it seems to me no effort should be spared to rouse public attention. You did good service last week in *Nature* . . . but the authorities need a great deal of battering.'[44] This report was the work of a government committee, and proposed a thorough reorganisation of British meteorology. In general, the proposals were acceptable, but Scottish meteorologists felt they had been ignored, and Lockyer agreed with them. Partly as a result of comment in *Nature* the Treasury relented and allowed funds to be diverted to Scotland.

Finally, at the end of the seventies, the support of Lockyer in *Nature* was influential in obtaining a grant for an important meteorological observatory of the period, which was established on Ben Nevis.[45]

A part of the freedom with which the articles and reviews were written depended on Lockyer's preparedness to publish unsigned articles and reviews, although this clashed with the policy that the *Reader* had stood for originally. For example, we find P. G. Tait, Professor of Natural Philosophy at Edinburgh for the last forty years of the nineteenth century, telling Lockyer, "Dewar is to send you, by this post, an account of his and my paper, drawn up by Scott-Lang my Assistant. He will put it right, and you will take care to head it *From a Correspondent.* Otherwise it might be looked upon as an infringement of the rights of the Royal Society of Edinburgh . . .'[46] Not all contributors were equally pleased, however, and some complained that the anonymity worked against them.

> You are very hard on your poor Reviewers, in allowing criticised authors to pitch into them in your column; particularly as I conceive it would be *infra dig* for a reviewer to justify his opinions in the publication which professes to adopt them. It seems impossible to criticize freely and justly under this condition.
>
> I think in future I shall ask you to let me *sign* my articles, and to use the pronoun I instead of the editorial WE.[47]

But the part of *Nature* in which Lockyer was most personally and controversially involved was, of course, the correspondence. Britain in the 1870s was prolific with scientists who were convinced of the importance and accuracy of their own views, and were correspondingly certain of the triviality, even dishonesty, of the views of many of their fellow-scientists. Sometimes these scientists played a lone hand; at others they formed small groups for mutual offence and defence. The composition of these cliques was often quite fluid. The importance of *Nature* as a scientific forum meant that the correspondence columns were often used as a battleground between the different groups, and Lockyer, however impartial he might try to be, was frequently accused of favouritism and high-handedness. On occasion, the dispute spilled outside the limits of the British Isles, and eminent foreign scientists might become involved. For example, Lockyer approached Helmholtz in 1874 to ask him to review some of Tyndall's work for *Nature.* Helmholtz replied:

> I would not like to meddle purposely and actively into these discussions. I have had much friendly intercourse with Tyndall, Sir W. Thomson and some also with Professor Tait. I must say, I have a strong feeling that my Scotch friends have been misled by a kind of national jealousy, but it is not my calling to stand up publicly and to declare it into face, although I have defended very often Tyndall against them in private conversation.[48]

As this letter implies, there was a long-continuing dispute between some Scottish scientists, especially Tait, and Tyndall.

Tyndall was, in fact, one of the storm centres of *Nature* correspondence in the seventies. In 1873, he was involved in a virulent clash with Tait in *Nature.* The immediate point at issue was the importance of the ideas concerning glaciers put forward by the eminent Scottish scientist, J. D. Forbes, who had died a few years before. Tyndall implied not only that

Forbes' ideas were fundamentally unsound, but also that he had derived them from other people. Tait immediately sprang to the defence of his fellow-countryman, and attacked Tyndall's scientific knowledge, saying that he had now become merely a popularizer of science, and could no longer be considered a scientist. Lockyer was ultimately drawn into the maelstrom: in one of his letters to *Nature*, Tyndall reproved him as editor. '. . . if I might venture a suggestion, you would wisely use your undoubted editorial rights, and consult the interests of science by putting a stop to proceedings which dishonour it.'[49] This hit Lockyer on a tender spot and he retorted immediately. '. . . if the Editor were to assume the power and responsibility that Professor Tyndall suggests, *Nature* might easily fall from the position of absolute justice and impartiality in all scientific matters which it now occupies and become the mere mouthpiece of a clique.'[50]

Lockyer was, indeed, always exceedingly anxious that his journal should speak for all scientists and not just for some specific group. He carried this policy out to such an extent that, when in the 1890s he and Huggins were in the middle of one of their spasmodic disputes, he printed Huggins' attack in *Nature*, but did not print his own reply. Not that Lockyer was particularly averse to controversial correspondence. It was, after all, a certain means of keeping interest in the journal alive. The insults in the Tyndall-Tait controversy became so extreme, however, that Lockyer eventually had to take Tyndall's advice and close the correspondence. This immediately led to a violent complaint from Tait.

> The fact is that your impartiality as Editor has all along *told against me*. . . . you allow T[yndall], under pretext of withdrawing them [the insults], to *reprint* two of the low things he said. (Enough, however, remains unretracted to make it impossible for me to meet him except with the tip of my toe.)[51]

Two years later Tyndall was again the subject of attack, this time by H. C. Bastian, Professor of Anatomy at University College, London. Bastian was a firm believer in the possibility of spontaneous generation, and, indeed, thought that he had managed to demonstrate it experimentally. This had led him in 1870 into a controversy with Huxley, who had chosen to discuss the question of spontaneous generation in an address at the British Association in that year. Bastian decided that Huxley had slighted his work, so he proceeded to take Huxley's address to pieces in the pages of *Nature*. Huxley wrote to Lockyer,

> I have been obliged much against my will to take notice of Bastian's 'Reply' —What was his reason [?] for going out of his way to be so offensive?
>
> He knew exactly what I thought about his work and therefore must have known that in my judgment the kindest thing I could do was to be silent about him.[52]

Huxley's reply was, indeed, devastating, and effectively concluded that argument.

Bastian's dispute with Tyndall was an extension of that with Huxley. Tyndall had been propounding the idea that diseases were due to the action of germs which were carried about in the air. Bastian reiterated that some of the evidence for germs was really proof of spontaneous generation. With supreme confidence he wrote to Lockyer, '[Tyndall is] to show me some of his results and in turn to look at mine. This was not accorded without some hesitation—and I hope he won't draw his head out of the noose too soon!'[53] In fairness to Bastian, it should be said that the remarkable heat-resisting ability of bacterial spores had not been fully realised at the time. However, Tyndall brought in Pasteur to write in his defence, and the academic community judged that Bastian had once again been squashed.

A consideration of the early volumes of *Nature* shows that the leading members of the X Club were constantly involved in disputations in the pages of *Nature*, and constantly complaining to the Editor of *Nature* over his handling of contentious matters. After the correspondence with Hooker, Tyndall, and Huxley, we might end by way of comparison with a complaint from another member of the Club, Spencer. Although his prose style was somewhat more grave and official than theirs, his meaning was evidently the same.

> I believe the parliamentary usage is that one whose motion is attacked has a right to reply, and that there the matter ends; and this, I think, is the example habitually followed in oral controversies. Were it customary in oral controversies to permit a rejoinder, I presume that, in pursuance of the same principle, a re-rejoinder from the person attacked would end the matter. . . .
>
> I would suggest that the same principle holds in controversy carried on in print. Both in the *British Quarterly* and in *Nature* I am the one attacked and in the alternation of attack and defence there is not, up to last week, even an equality of opportunities.[54]

This reproof was delivered in the course of a dispute between Spencer and the irrepressible Tait. Spencer had claimed that some scientific knowledge was *a priori*, whereas Tait believed that it was all based on experimental evidence. Despite his editorial impartiality, Lockyer found Tait easier to get on with than either of the latter's *bêtes noires*, Tyndall and Spencer, as is reflected in the following piece of versification that Tait sent him about this time.

> Your printers have made but one curious blunder,
> Correct it instanter, and then for the thunder!
> We'll see in a jiffy if this Mr S[pencer]
> Has the ghost of a claim to be thought a good fencer.
> To my vision his merits have still seemed to dwindle,
> Since I found him allied with the great Dr T[yndall]

While I have, for my part, grown cockier and cockier,
Since I found an ally in yourself, Mr L[ockyer]
And am always, in consequence, thoroughly willin',
To perform in the pages of *Nature* (M[acmillan]).[55]

Lockyer's considerable degree of impartiality was presumably one of the features which finally enabled *Nature* to win through. But as editor he was obviously in an exposed position for earning his fellow-scientists' dislike, even had he been more diplomatic than he was. His editorial work, combined with his opinions concerning the advancement of science, earned him several enemies by the end of the 1870s. In 1879, Lockyer was proposed for election to the Athenaeum. C. W. Siemens, the engineer, who was one of his proposers, when writing to tell Lockyer that he had been elected, remarked, 'This ought to have been a mere matter of course but you have aroused the jealousy or enmity of some persons which made it necessary for your friends to be on the lookout for mishaps and it is on this account that I have much pleasure to congratulate you upon your victory!'[56]

But Lockyer's editorship of *Nature* also, needless to say, brought him many friends all over the world. For example, Edison, who was in many ways a kindred spirit, after promising to try and let *Nature* have news of his work first, went on, 'I hope you will come over here again (after you have become well smoked up in London) with several other deep and mighty intellects we will take to the Mountains for a grand hunt.'[57]

As we have seen, *Nature* was not the only publishing work in which Lockyer was involved in the 1870s, for he was also advising Macmillan on science books, especially the 'Nature' series of texts on science. These —like *Nature* itself—were not doing too well in the mid-seventies and required a fair amount of attention from Lockyer. However, his worries in this respect were somewhat lifted in the summer of 1876 when he received a letter from the firm saying, 'Macmillan is all for going on with the Nature series, believing that a first rate book or two would lift it tremendously all round.'[58] In fact, some of the books in the series were to become best-sellers. Thus Lockyer engaged Archibald Geikie to write a primer on physical geography. This was published in 1873, and during the next fifty years sold well over half a million copies.

Unfortunately, Lockyer's contribution to the series, *The Spectroscope and its Applications*, was financially far from successful. Another letter from the firm in 1877 pointed out with some feeling, 'We have a debt of £1,400 against the book and since midsummer we have sold four and had seven returned from agents. We know, of course, that you took a lot of trouble about the book, but a loss like this on one book is enough to make one's hair stand on end . . .'[59]

Lockyer wrote various other books which were published by Macmillan during this period: all of them were astronomical. *Solar Physics* published in 1873 was a reprint of articles and lectures which he had given in previous years, and so required little effort from him. *Studies in Spectrum Analysis*, although also based on his own work, was specially written, as was an elementary book, *Primer of Astronomy* (both published in 1872). Another introductory text, *Star-gazing, Past and Present*, which he wrote with G. M. Seabroke, a friend from Rugby, was also based on a series of lectures he had given. Lockyer, in fact, typically intermixed oral and written work. He used his lectures, of which he gave a large number in the 1870s, to perfect the presentation, and then wrote them up as books or articles. In general, however, he was too occupied with other affairs to be able to produce any very long original book during the seventies. In any case, his astronomical ideas were in too great a state of flux for him to write to a coherent theme, and this deterred him, for he always wrote out of his own experience and wanted his scientific books to have a logical sequence of development, like the plot of a novel.

Although Lockyer's main ventures with Macmillan were unprofitable during their early years together, their mutual relationship was remarkably cordial. An example of Macmillan's personal generosity to his friend can be found in 1877, when Lockyer was once again ill through overwork and depression. Macmillan wrote to him, 'Brunton tells Craik that he thinks you should have a little quiet rest on the continent. If the enclosed cheque will help to make this easier for you the firm desires your acceptance of it with the love of all the members.'[60] (Launder Brunton was a mutual medical friend of Lockyer and Macmillan; he was the editor of another of Macmillan's new journals—the *Practitioner*—founded the year before *Nature*.)

Lockyer was mainly a Macmillan author, but, of course, he contributed shorter articles elsewhere. Thus towards the end of the 1870s, there was a curious foretaste of things to come almost a century later. The editor of *The Times* got in touch with Lockyer as editor of *Nature*. 'I called on you the other day with the purpose of asking you, as the most eminent literary exponent of science with whom I am acquainted, to contribute occasionally to the columns of the Times . . .'[61] In effect, Lockyer was being asked to set up a *Times–Nature* news service, and, indeed, for several years he provided the daily with occasional articles on scientific topics.

III

THE MAN OF SCIENCE

The 1860s formed a crucial period in the development of astronomy. Since the scientific revolution of the seventeenth century, in which astronomy had played a major role, astronomers had been trying to build up a picture of the world based mainly on the accurate measurement of position and of changes in position. Newton had suggested that one force, gravitation, ruled the universe, and throughout the eighteenth century astronomers had been concerned with substantiating his claim and extending it to as many cases of bodies in motion as possible. Since the planets and their satellites constituted the most obvious examples of motions amongst the celestial bodies, attention had at first been mainly confined to the solar system. By the nineteenth century, the most obvious difficulties in the reconciliation of Newtonian theory with the observations had been overcome; the prospect for astronomy seemed to be mainly one of increasing the accuracy of both observation and theory. In this respect, the state of astronomy in the first half of the nineteenth century might be compared with that of physics in the second half. However, just when the heroic age of classical astronomy seemed to be drawing to a close, a new aspect of astronomy began to blossom, the study of celestial bodies as physical objects rather than simply as dynamical ones. Of course, there had always been some work of this type in astronomy. When Galileo first used a telescope for astronomical observation, two of his earliest discoveries were the rugged nature of the Moon's surface and the existence of dark spots on the Sun's face. But this sort of observation had come to take a subordinate place to the main interest in positional work.

The first major signs of a coming change of attitude appeared at the end of the eighteenth century. In Britain, Sir William Herschel was the first major astronomer whose work was not oriented predominantly towards positional measurement. Gradually, during the first half of the nineteenth century, the new concern with the physical nature of astronomical objects gathered momentum, to break through finally in full force during the 1860s. It is to this period that the modern growth of astrophysics may be best dated. At the time, however, many of the leading astronomers, whose careers had been entirely devoted to positional astronomy, found the new developments distasteful. For example, Admiral Smyth, a well-known amateur astronomer (his *Cycle of Celestial Objects* is still used by amateurs), was a close friend of William Huggins, one of the leading proponents of the new type of astronomy, yet he nevertheless felt forced to declare,

> One word in conclusion. With all my admiration of the marvellous and extensive power of Chemistry in disintegrating the nature and properties of the elements of matter, I really trust it will not be exerted among the Celestials to the disservice or detriment of measuring agency; and this I hope for the absolute maintenance of GEOMETRY, DYNAMICS, and pure ASTRONOMY.[1]

Lockyer, himself, writing in 1873 remarked, 'in England, though happily not abroad, many professional astronomers and physicists regard it [the spectroscopic examination of the Sun], as a rule, as a matter of tenth-rate importance.'[2]

The breakthrough in the 1860s stemmed basically from the development of astronomical spectroscopy, to be later supplemented by the growth of astronomical photography. At the same time the direct visual study of objects through a telescope had been increasing throughout the century, and by the sixties was occupying a place of major importance in astronomical work. Just looking through a telescope was obviously an ideal approach to astronomy for amateurs. It did not require an extensive mathematical background, such as was required for theoretical work in gravitational astronomy, nor did it need a very accurately calibrated telescope to be used on a tedious series of observations often lasting many years—the sort of work still emphasized by the professional observers. All that was needed was the patience to watch, or to search, over a reasonable period of time. It is not surprising therefore that when Lockyer first became interested in astronomy it was to the visual study of the Moon and the planets that he turned.

As we have seen, Lockyer's friendship with Pollock on the committee of the Wimbledon village club, led him to buy a replica of Pollock's 3¾-inch refractor, and to set it up in his own garden. By modern standards, this was a very small instrument; but the era of large telescopes as a necessity for research had still to dawn in the 1860s. It was during this decade, for example, that the German astronomer Argelander completed the most famous star atlas of the century, the *Bonner Durchmusterung*, which was compiled with a telescope of only 3½-inches aperture.

During his first months of observing, Lockyer received considerable advice from the maker of his telescope, Thomas Cooke, on how to go about the business of astronomical observation. Cooke also read the proofs of Lockyer's earliest articles on astronomy, and, in due course, when Lockyer felt the need for a longer telescope, lent him a 6¼-inch refractor, which Lockyer later bought. It was with this latter instrument that Lockyer made those observations during the sixties which were to turn him into one of the acknowledged leaders of the new astronomy. Lockyer was always extremely grateful to Cooke for his help, and, where the opportunity arose, tried to repay the kindness by recommending his products.

Lockyer's earliest detailed observations were of the Moon. He had become acquainted with W. R. Birt, an enthusiastic amateur, who concentrated on lunar studies and who encouraged him to do the same. Writing to Lockyer in May 1861, Birt explained his reason.

> By far the best mode for a student to adopt is to follow the bent of his inclination provided he is quite satisfied he is not doing work that has been well and ably done by others before him. The latter portion of this remark cannot apply to the Moon although it has been under observation so long. . . . Changes may have taken place since the date of the great map [of Beer and Mädler]. Most . . . differences are in minute objects and when we consider the smallest object clearly discernible a second of arc being about equal to a mile the action must be of *considerable energy* to produce any thing that we *can* behold and further it requires close watching on our parts to detect the changes as they arise otherwise they will be *lost* being referred to inaccuracy of detail in former observers.[3]

To understand why there was this heightened interest in lunar studies in the 1860s it is necessary to understand the situation which then existed with regard to lunar maps. Telescopic observations of the lunar surface had been made ever since Galileo sketched its appearance in the early seventeenth century, but no reasonably detailed map of the Moon became generally available until the nineteenth century. Then, in the 1830s, two Germans, Mädler and Beer (the latter a brother of the composer Meyerbeer*), produced after several years of intensive work the first moderately accurate Moon chart. Astronomers at last had a fairly reliable means of checking their lunar observations. The map was of sufficient value to be still in general use when Lockyer began to study the Moon. Indeed, it may have been that the initial result of the publication of the map was to dampen interest in charting the Moon's surface, as it hardly seemed worth repeating what had already been done so well. But, as time passed, there came to be a growing feeling that the details of Beer and Mädler's map needed careful checking. In particular, they had assumed that the Moon was a dead world, whereas there continued to be an undercurrent of feeling amongst astronomers that changes in minor features might actually occur. By the early 1860s, a certain number of lunar observers, especially in the British Isles, were looking for signs of change on the Moon. As a result of Birt's encouragement Lockyer became a

* The composer changed his name after a wealthy relative called Meyer had left him a legacy.

member of this British group. In 1861 Lockyer claimed that he had himself discovered a significant change in one small feature of the lunar surface, and Birt supported him.[4]

Interest in the nature of the Moon's surface grew sufficiently for the British Association at its meeting at Bath in 1864 to recommend that a committee should be set up for the purpose of constructing a map to four times the scale employed by Beer and Mädler. It was reasoned that small lunar features could then be recorded, and possible changes detected. Both Birt and Lockyer were made members of the British Association committee. Lockyer had only joined the British Association in 1863, by which time he had moved to Hampstead and set up his larger $6\frac{1}{4}$-inch telescope in the garden there. But as a result of his membership of the committee Lockyer soon found himself in the company of several influential scientists; for example, Sir John Herschel (the son of Sir William Herschel) who, now in his seventies, was regarded in many ways as the senior spokesman for science in Britain, and Lord Rosse, a member of a wealthy and distinguished family which produced more than one outstanding scientist. Other members included the Reverend W. R. Dawes who had been in communication with Lockyer on astronomical matters ever since the latter's first publication on astronomy in the *London Review*; Warren De la Rue, a wealthy paper manufacturer with whom Lockyer came to be closely connected during the next few years; John Phillips, Professor of Geology at Oxford, who became President of the British Association in 1865; and James Nasmyth, the engineer, inventor of the steam hammer, who in the latter part of his life came to be an enthusiastic, if somewhat controversial, astronomer.

One aspect of lunar work which received especial attention in Britain was the photography of the Moon. All the major nineteenth-century lunar charts were produced by visual mapping, but there was a continuing hope that this tedious approach might eventually be replaced by photographic recording. In 1852, De la Rue had obtained some of the first good images of the Moon using a new type of photographic process proposed the year before in France. This was the wet collodion process. Collodion was obtained by brewing up gun-cotton with sulphuric acid and potassium nitrate, and dissolving the residue with various iodides and bromides. The resulting liquid was evaporated to leave a thin layer of collodion on a glass plate. This was subsequently sensitized for exposure in the telescope by dipping in a solution of silver nitrate and silver iodide. The plate had to be exposed whilst wet, which restricted the maximum exposure time to about a quarter of an hour. Nevertheless the new method was much more sensitive to light than the old daguerreotype process, the original photographic process introduced by Daguerre and

Niepce in France at the end of the 1830s. So, despite its drawbacks, the wet collodion process remained the standard means for astronomical photography until the end of the 1870s, at which point something like the modern dry plate was introduced. Owing to the limited exposure time, however, photography during this period could only be used for bright astronomical objects such as the Moon.

Shortly after De la Rue had obtained his first photographs, Phillips at Oxford was also successful in obtaining photographic images of the Moon. He noted as one reason for the work that, 'if photography can ever succeed in portraying as much of the moon as the eye can see and discriminate, we shall be able to leave to future times monuments by which the secular changes of the moon's physical aspect may be determined.'[5] Since this was one of Lockyer's chief concerns, he was soon in touch with the leading British exponents of lunar photography, De la Rue, Phillips, and William Crookes. (As we shall see, Crookes' interests often overlapped with Lockyer's in these and later years.) Lockyer soon found that the most skilful and persistent exponent was De la Rue. By the end of the 1850s the latter had produced sharp lunar images showing quite considerable detail. He continued with this work into the 1860s in the hope that a comparison might provide definite evidence for surface changes.

In 1866, the Lunar Committee submitted a long report to the British Association on its first year's deliberations, and also coopted some new members. These included the Reverend Charles Pritchard, then President of the Royal Astronomical Society, who was to be closely connected with Lockyer for some years, and William Huggins, an amateur astronomer who was already recognised as one of the leading English astrophysicists. This proved to be the most exciting period of all for the study of changes on the lunar surface. In the autumn, Julius Schmidt, a German astonomer for many years director of the Athens Observatory and one of the greatest living lunar observers, announced that a small, but quite well-known crater called Linné had disappeared. (Lunar craters, ever since the seventeenth century, have been called after scientists and philosophers, especially, though not exclusively, those interested in astronomy. Linné is an example of one named after a non-astronomer: in this case, Karl von Linné, the famous Swedish botanist, better known by the latinized version of his name—Linnaeus.) Linné had been observed earlier as a definite crater by Beer and Mädler, and also by their contemporary selenographer, Lohrmann. Schmidt himself had previously noted it as a crater, but now he saw it simply as a white patch on the Moon's surface. Shortly afterwards he reported that he had found a small mountain at the centre of this white patch. Then Angelo Secchi, a Jesuit priest and the leading Italian astrophysicist, announced early in 1867 that the 'mountain' was really a small crater, much smaller than the original dimensions allotted to Linné. Schmidt's observation led to a renewed flurry of lunar observation, and still constitutes an interesting problem. By this time, however, Lockyer had become almost entirely involved in solar astrophysics, and his lunar observations, which had been sporadic for some time, seem to have been entirely discontinued. In 1868, he ceased to be a member of the British Association Lunar Committee. This never produced a lunar map, though the work was carried on for some time, mainly by Birt, and in the 1870s the upsurge of interest in lunar changes died away.

Not long after he began studying the Moon, Lockyer also started observing Mars. He had joined the Royal Astronomical Society in March 1862, when there was considerable interest among the Fellows of the Society over the opposition of Mars due to occur in the autumn. Mars and the Earth orbit the Sun at different speeds. Hence they are sometimes close together on the same side of the Sun, and sometimes they are far apart on opposite sides. They are closest together when the Sun–Earth–Mars axis is a straight line. At this point Mars is said to be in opposition. However, whereas the Earth has a fairly circular orbit round the Sun, Mars has an elongated orbit. The exact distance between the Earth and Mars at opposition therefore depends on the position of the latter planet in its orbit. The most favourable oppositions, when the Earth and Mars are closest together, occur about every fifteen years. Beer and Mädler, besides mapping the Moon, had also spent a good deal of time observing Mars during the 1830s. The prevailing opinion at the time was that no permanent features could be discerned on Mars, the features that were detected being simply evanescent cloud formations as on Jupiter. Beer and Mädler showed, however, that certain markings were stable, i.e. they could be seen at successive oppositions at the same place on the martian surface, although they were frequently blurred by overlying clouds. This seemed satisfactory evidence for the existence of permanent markings, with the implication, of course, that one could see down to the solid planetary surface. But the controversy over their nature, after dying down for a while, flared up once more at the end of the 1850s, when Mars again became well placed for observation. In 1862, the year of a very favourable opposition, Lockyer, together with several other observers in England and elsewhere, therefore turned his attention to observing Mars. By now he was observing entirely with his larger instrument. The $3\frac{3}{4}$-inch refractor was probably sold about this time.

Lockyer was already in close touch with the other Mars observers in Britain, especially with Dawes and Phillips, but he now also began to correspond with some of the British theoretical astronomers, in order to obtain predictions of the tilt of the Martian rotational axis during the periods when he was observing the planet. He encountered in this way J. C. Adams, the doyen of theoretical astronomers in the British Isles, who many years before had predicted simultaneously with the Frenchman Leverrier the existence of the planet, Neptune. Phillips encouraged Lockyer to publish the series of drawings he made of Mars during the opposition, telling him that they were outstanding representations of the phenomena. Lockyer, gratified, took his advice, and went on to point out that there was a remarkable agreement between the surface markings recorded on his own map, and those depicted some thirty years before by Beer and Mädler. This seemed to be positive proof of the permanency over long periods of time of the Martian surface features. However, prior to the appearance of Lockyer's drawings, Secchi in Italy had produced what were agreed to be excellent drawings of Mars at the opposition of 1858, and he had asserted that a comparison of his own results with the details recorded by Beer and Mädler indicated that they were totally irreconcilable. Some observers in this country agreed with him; as a result the first major piece of research by Lockyer led—prophetically—to a short, but sharp, dispute. This ended in the general acceptance of Lockyer's views, but the dispute is significant as it marks Lockyer's first clash with Secchi, an encounter that was to be repeated later.

Phillips referred to Mars in his Presidential address to the British Association at Birmingham in 1865, but he noted that Lockyer had by this time turned to solar studies. Lockyer, in fact, never again returned to serious lunar or planetary observing after the early 1860s. His last connection with Mars was actually by proxy: another English amateur astronomer, R. A. Proctor, drew up a new map of Mars in 1867, in which for the first time names were allotted to the major permanent features. One marking near the South pole of Mars he called 'Lockyer Land'. Proctor's map was extremely parochial in its nomenclature, emphasizing British observers at the expense of others (Phillips had two features named after him, and Dawes four), and it was not accepted on the Continent. These names were used for some years in this country, but eventually they gave way to the neutral, latinized names which are still employed today. It is ironical that this was the only occasion when Lockyer had an astronomical object named after him, and it was done by a man who very shortly afterwards became his greatest enemy in the scientific world.

Lockyer's growing interest in solar spectroscopy in the mid-1860s can doubtless be ascribed to a variety of causes. Two specific stimuli can, however, he pointed out. In the first place, he was undoubtedly stimulated by William Huggins' initial researches in astronomical spectroscopy. Huggins was an acquaintance with whom Lockyer came to be on close terms in the mid-sixties. In 1864, Huggins described to Lockyer the results that he was just beginning to obtain from spectroscopic studies of Jupiter, and suggested that Lockyer should take particular note of them. It may well be significant that Lockyer, shortly after this, began negotiating for a spectroscope to attach to his $6\frac{1}{4}$-inch telescope. At the same time, Lockyer's attention must have been turned specifically to the Sun by the controversy which was then raging over the appearance of the solar surface.

The fact that the Sun had a mottled appearance when seen through a telescope had been known for some time. However, in 1862–3 Nasmyth claimed that a more detailed examination of the Sun's surface showed it to be covered with bright objects of definite size and shape, resembling a willow leaf in appearance, which were scattered fairly randomly over the surface. This claim initiated what came to be called the 'willow-leaf' controversy. Nasmyth was supported by some observers including Sir John Herschel, who even speculated that the willow-leaves might be some exotic form of organism, of remarkable size, since an individual 'willow leaf' was alleged to be 1000 miles long and 200 to 300 miles wide. But he was strongly opposed by others, especially by Lockyer's friend, Dawes, who claimed that the granulation (his name for it) was completely irregular in appearance, except near sunspots. The controversy bubbled on, for it must be remembered that all the work was being done visually, and Lockyer, now fascinated, became involved. He agreed basically with Dawes, and in April 1865 obtained evidence which he believed supported Dawes' position. He observed the formation of some willow-leaf shaped objects in the penumbra at the edge of a sunspot, and comparing them with the surrounding bright solar photosphere felt able to declare that no willow-leaf objects were to be observed on the latter. The ultimate majority verdict amongst astronomers was in favour of the view that Dawes and Lockyer held, thus further enhancing the latter's reputation as an observer.

In making this set of solar observations, Lockyer had noted something that was to lead him on to much more important work. He had seen motions in the penumbra of spots which seemed to suggest a downward current of material into the spots. In 1865, a new controversy arose over the nature of sunspots, between the French astronomer, Faye, on the one hand, and a group of observers at Kew—De la Rue, Balfour Stewart and Loewy—on the other. Kew contained at this time the most important British centre for solar research. Faye believed that spots were regions of the Sun's surface where there was an upflow of material, and that they were dark in appearance because the spot material was too hot to radiate light properly. The Kew observers believed the exact opposite, namely that spots were regions of downflow, and that they were darker than their surroundings because they were cooler. Since Lockyer was on friendly terms with De la Rue and Stewart he was gratified that his observations seemed to support their position, and was encouraged to go on and examine the other point at issue, the temperature of sunspots. By this time Lockyer had acquainted himself with the use of a spectroscope, and with the main results that had so far been obtained from spectroscopic investigations. This enabled him to realise that the question of sunspot temperature was one that might be amenable to a spectroscopic analysis. In order to follow his line of argument, we must first consider briefly the state of astronomical spectroscopy in the mid-1860s.

The study of the solar spectrum dates back, in principle, to Newton's investigations of the nature of light in the latter part of the seventeenth century. However, real development did not occur until after the beginning of the nineteenth century. The vital observation, that the solar spectrum was not a complete range of colours, but had certain narrow bands of colour missing, was made by a British scientist, W. H. Wollaston, in 1802. However, Wollaston did not appreciate the significance of his discovery. The real origin of solar spectroscopy must therefore rather be ascribed to Josef Fraunhofer, the great Bavarian optician, who a decade later rediscovered the dark divisions in the Sun's spectrum and examined them in some detail. Because the light from the Sun was passed through a narrow slit before being dispersed by the prism, the spectrum produced took the form of a narrow band of light crossed by dark lines. These were subsequently designated as 'Fraunhofer lines'. The basic problem of solar spectroscopy was therefore posed immediately—what caused the dark lines?

During the first half of the nineteenth century, most of the spectroscopic advances were in the laboratory, but the spectra examined there proved generally to be unlike the Sun, in that they showed bright lines only. It came to be realised that there was often a connection between the colour of these bright lines and the particular substance which was being examined. Nevertheless, the connection did not appear to be unique, for many of the substances tested seemed to have lines in common. In particular, there were two yellow lines close together which seemed to be present in every spectrum. It was found that these two bright lines observed in the laboratory coincided in wavelength (that is in colour) with two dark lines in the solar spectrum which Fraunhofer had labelled with the letter D.* Ultimately it was realised that these two lines were characteristic of the element sodium, and that they appeared everywhere because

*Fraunhofer had attached Roman capitals (and some lower-case letters) to the most obvious dark lines he observed in the solar spectrum, starting with A at the red end and working down towards the violet.

sodium chloride (common salt) was always present in the atmosphere. Unless special precautions were taken, therefore, it would always occur in any laboratory material as a contaminant.

By the 1850s, the stage was set for a basic advance in understanding how spectra could be produced, whether in the laboratory or in the Sun. Indeed, during that decade, scientists in various countries did suggest many of the major points involved, but it was not until the very end of the fifties that a reasonably comprehensive picture was evolved. Its acceptance then was due mainly to the researches of Gustav Kirchhoff, a German physicist, working initially with Robert Bunsen, a chemist. Their work, it must be emphasized, marks a turning point in science not because of its uniqueness, but because they were responsible for finally convincing their fellow-scientists of the validity of the new outlook.

A joint assault *via* laboratory research and direct observation of the solar spectrum finally enabled Kirchhoff to show that a continuous spectrum could be produced by a sufficiently hot liquid or solid; a bright line, or band, spectrum by a heated gas; and a Fraunhofer-type spectrum, with dark lines, by passing a continuous spectrum through a somewhat cooler gas. It thus appeared that the dark-line spectrum of the Sun was simply the reversal of the bright-line spectra of various elements. Kirchhoff and Bunsen further showed that each element had a characteristic bright-line laboratory spectrum by which it could be identified. Combining these two results together they were able to determine, from a comparison of the Fraunhofer spectrum with laboratory spectra, that some nine terrestrial elements were certainly present above the surface of the Sun. This represented the first use of spectrum analysis to determine chemical composition in astronomy. (August Comte, the French philosopher, had declared categorically a few years before that the determination of the chemical composition of celestial bodies was an example of something that could never be done.)

In order to express the exact colour of a line, Kirchhoff found it necessary to introduce an arbitrary scale of numbers, running from small numbers at the red end of the spectrum to large numbers at the blue. However, this 'Kirchhoff scale' was not easily reproducible in other laboratories, so, at the end of the 1860s, a new method of calibrating the positions of spectral lines was introduced by the Norwegian, Ångström. This depended on the fact that light is a wave-like phenomenon, and the distance between the waves, the wavelength, varied with the region of the spectrum under consideration. Ångström therefore labelled any spectral line in terms of its corresponding wavelength, expressed in units of one ten-millionth of a millimetre. This unit was subsequently called an 'Ångström' unit after him. Ångström's system came into general use in astronomy in the 1870s, although particular lines were long referred to by their Kirchhoff numbers, and the Fraunhofer designations continued to be used for the stronger solar lines.

Kirchhoff, besides explaining why spectra were sometimes bright-line and sometimes dark-line, also used his spectroscopic ideas to propose the first model of the Sun based firmly on physical reasoning. He suggested that the main body of the Sun was a hot molten ball, whose surface, the photosphere, produced a continuous spectrum. It had to be fluid because of its observed differential motions. Above this was an extensive atmosphere which he identified with the hazy corona seen round the Sun at eclipses, in which he supposed the Fraunhofer lines were formed. The sunspots, he believed, were dark clouds floating just above the Sun's surface.

This theory was a distinct advance on the previous most popular concept of the Sun, proposed originally by Sir William Herschel, which had supposed that there was a cool, dark solar core surrounded by a layer of bright, luminous clouds. But Kirchhoff's model did not immediately replace that of Herschel. The reason was not simply innate conservatism on the part of astronomers, though there was doubtless an element of this, but was related to the fact that Kirchhoff's new picture had its own observational improbabilities. Kirchhoff was not an astronomer, so he tended to underrate these difficulties. In particular, there was the contrary evidence of the Wilson effect. Alexander Wilson, a Professor of Astronomy at Glasgow in the latter part of the eighteenth century, had observed a particularly large spot crossing the Sun's disc, due to the rotation of the Sun, and had noted that it formed a saucer-shaped depression in the solar surface. William Herschel had taken this observation as part of the evidence for his solar model: a spot was a region where the luminous clouds had parted so that one could see the dark nucleus of the Sun below. The point was that spots were thought necessarily to be concavities, and not projections as Kirchhoff supposed.

A new investigation of sunspots by De la Rue, at the beginning of the 1860s, seemed to provide conclusive evidence for the general occurrence of the Wilson effect, and therefore to disprove Kirchhoff's theory of spots. For this reason, most observers continued to favour the Herschellian concept of spots and, hence, of the Sun. Dawes was still using it without question in his correspondence with Lockyer in the mid-1860s. Nevertheless, Kirchhoff's picture of the formation of the Fraunhofer lines was so compelling, that it became of major interest to try and reconcile his mechanism with a model for the Sun which better fitted observation. In 1864, Secchi and Sir John Herschel proposed that the old picture of the photosphere as a layer of cloud should be retained, but the idea of a solid solar core should be dropped; instead it should be supposed that the Sun's interior consisted of high-density gas. (It had been found that hot, dense gas could give a continuous spectrum just like a liquid or a solid.)

This then was the new picture of the Sun which was rapidly gaining ground when Lockyer first began to study sunspots. He realised that Faye's idea of the gas in a spot radiating feebly because it was much hotter than the background, would result in a different type of spectrum from the Kew observers' opposite assumption that the gas radiated feebly because it was much cooler than the background surface. Applying Kirchhoff's concepts to these two contrary models, Lockyer argued that the first case should lead to the spectrum of a spot having bright lines, whereas the second should lead to dark lines—darker, indeed, than those from the Sun's surface as a whole, in proportion to the lower heat content of the spot material. Lockyer was now treading completely new ground. Previous work on the solar spectrum had examined light coming from the entire solar disc, whereas he was preparing to isolate light from the spot alone. What he did was very simple: he projected the image of the Sun formed by his telescope onto a screen, and drilled in this a fine slit through which light from a part of the solar image could be admitted to the spectroscope. The screen was then moved about until the slit lay across both a sunspot and a small part of the photosphere on either side. Lockyer found that no bright lines were visible; the Fraunhofer lines simply appeared to be widened in the spots. He could therefore claim that he had provided indisputable evidence to support the case of the Kew observers.

This was Lockyer's most important piece of research to date, and he clearly realised it, for it was the first that he published as a paper in the *Proceedings of the Royal Society.* As he was not a Fellow of the Society, he had it communicated by Dr Sharpey, who was one of the Secretaries; the two had probably first met when Lockyer was reporting Royal Society meetings. But the report was immediately disputed: this time by Lockyer's erstwhile friend, Huggins. Huggins reported to the Royal Astronomical Society in 1867 that, contrary to Lockyer's statements, he could detect no certain difference between the spectrum of spots and that of the ordinary solar surface. However, much to Lockyer's delight, Huggins found it necessary to modify his statement in the following year, after making further observations of spot spectra.

Perhaps the most significant part of Lockyer's paper was, however, a brief comment at the end. He finished by pointing out some of the topics that might be investigated by examining the spectra of surface details instead of light from the Sun as a whole. He queried particularly, '. . . may not the spectroscope afford us evidence of the existence of the "red flames" which total eclipses have revealed to us in the sun's atmosphere; although they escape all other methods of observations at other times?"[7]

In order to understand the significance of this remark, and also to understand why in subsequent years Lockyer came to be so involved in observing solar eclipses, we must now consider what was known of the Sun's atmosphere in the mid-1860s. When we look at the Sun from Earth any atmosphere round the limb of the Sun is generally invisible owing to the fact that the much brighter light from the solar surface is scattered in the terrestrial atmosphere and masks the atmospheric light of the Sun. At a total eclipse of the Sun, however, the photospheric light is blocked out by the Moon, and it is possible for an observer on Earth to see a faint atmosphere surrounding the Sun. Although this had been observed prior to the nineteenth century, the earlier observations did not arouse any very great curiosity. But about the middle of the nineteenth century, the importance of solar eclipses as a method of obtaining information on the physical nature of the solar atmosphere became widely recognised amongst astronomers. Total eclipses are not easy to observe: they occur at intervals of a few years, are only visible over a restricted part of the Earth, and even there can be seen only for a few minutes. Nevertheless, they soon became so important in solar studies, that expeditions to solar eclipse sites became a standard part of astronomy in the latter half of the nineteenth century. Fortunately, the development of more rapid and safer methods of transport during the nineteenth century made such expeditions feasible, if not necessarily comfortable.

By the mid-sixties, three components of the solar atmosphere had been noted from eclipse observations. There was first of all the extended corona. (This was the region that Kirchhoff suggested—erroneously, as it was soon shown—to be the place of formation of the Fraunhofer lines.) Projecting into the corona were bright clouds of material variously referred to as flames, prominences, or protuberances. Finally, and much less certain, there might be a layer of small flames not projecting very far above the photosphere. The Astronomer Royal, G. B. Airy, had noted such a layer, which he called a 'sierra', during his observations at the total eclipse of 1851.

None of these three phenomena could be seen out of eclipse. What Lockyer was speculating about in his 1866 paper was whether prominences at the limb of the Sun might not be made visible out of eclipse by an appropriate use of the spectroscope. His argument depended on the assumption, not universally accepted at the time, that prominences were clouds of hot gas. If this were true, then they ought to show a spectrum consisting of bright lines only. The scattered light in the Earth's atmosphere, which normally prevented the prominences from being seen, would have a continuous spectrum. If light from just outside the limb of the Sun was passed through a spectroscope, this continuous scattered light would be spread out into a coloured band, and, hence, weakened. On the other hand, a spectrum consisting of bright lines only would hardly be affected by passage through a prism. The more the continuous spectrum was spread out (i.e. the more powerful the spectroscope) the greater would be the weakening of the scattered light as compared with the bright lines. Given a powerful enough spectroscope, so Lockyer argued, it should be possible to weaken the background so much that prominence lines could be seen even out of eclipse. If this were so, then sweeping the spectroscope slit round the solar limb would reveal the presence of any prominences by the appearance of bright lines at the corresponding places. These could be examined from day to day to see what changes occurred. This would be an important new advance, for the time afforded by a solar eclipse was too short for alterations in the shape of prominences to be noted.

The spectroscope Lockyer had used for his sunspot investigation was not sufficiently powerful for this new project. The next step was evidently to have a larger and more powerful one built. Lockyer, though financially better off than in the early sixties, could not afford the money required for such instrumentation, for spectroscopes usable on telescopes were still novelties which had to be individually designed and built. Fortunately, his work had by this time gained him a reputation as one of the rising generation of scientists, and he could hope to obtain money from one of the very few sources of funds then available for scientific research. This was the Government Grant, given annually to the Royal Society.

The Royal Society, after a good deal of internal strife in the first part of the nineteenth century, had by 1850 reached something more nearly resembling that position as the influential voice of British science which it enjoys today. The Government recognised it in this capacity, and, from the mid-century onwards, provided it with a small annuity (£1,000 initially) to aid scientific investigations. Two of the main aims of the fund were to help research by private individuals, and to help purchase new instruments for the observational sciences, such as astronomy and meteorology. Lockyer, of course, qualified under both these headings, and so obtained a grant without much difficulty. He initially commissioned his old friend Cooke to make the spectroscope, but Cooke, worn out by business worries, died at about this time. Lockyer therefore turned to another instrument maker, John Browning of London, who had made the spectroscope that Lockyer used for his sunspot observations. Browning undertook the work, but proceeded at a rather slow pace. Moreover, when the instrument was nearing completion in the summer of 1868, Lockyer was suffering from ill-health, and went abroad to recuperate. As a result, he did not obtain the spectroscope finally, and even then still not entirely complete, until 16 October 1868.

Lockyer's attempts to see bright lines with the instrument failed because, apparently, the instrument was out of adjustment until 20 October. On that day, after an hour's vain searching,

> I saw a bright line flash into the field.
>
> My eye was so fatigued at the time that I at first doubted its evidence, although, unconsciously, I exclaimed 'At last'! The line, however, remained—an exquisitely coloured line absolutely coincident with the line C of the solar spectrum [a Fraunhofer line due to hydrogen in the red], and, as I saw it, a prolongation of that line. Leaving the telescope to be driven by the clock, I quitted the observatory to fetch my wife to endorse my observation.[8]

Lockyer immediately dispatched a brief note to his friend Dr Sharpey at the Royal Society, which appeared very shortly afterwards in the *Proceedings.*

> I beg to anticipate a more detailed communication by informing you that, after a number of failures, which made the attempt seem hopeless, I have this morning perfectly succeeded in obtaining and observing part of the spectrum of a solar prominence.
>
> As a result I have established the existence of three bright lines in the following positions:—
>
> I. Absolutely coincident with C.
> II. Nearly coincident with F.
> III. Near D.
>
> The third line (the one near D) is more refrangible than the more refrangible of the two darkest lines by eight or nine degrees of Kirchhoff's scale. I cannot speak with exactness, as this part of the spectrum requires re-mapping.
>
> I have evidence that the prominence was a very fine one.
>
> The instrument employed is the solar spectroscope, the funds for the construction of which were supplied by the Government-Grant Committee. It is to be regretted that its construction has been so long delayed.

In view of the outcome described below, there is a possibly significant difference between the original draft of Lockyer's letter to the Royal Society and the printed version. He writes in the former that, '. . . this part of the spectrum requires re-mapping, Janssen shows some lines where Kirchhoff shows none and there are lines recorded by neither.' It will be noted that Janssen's name has been dropped from the final draft, and it may be that Lockyer did not wish to have the Frenchman's name appear in the communication of this new discovery.

He dispatched a similar account to his friend Warren De la Rue, for him to read at the Académie des Sciences in Paris. (De la Rue had been educated in Paris, and was a frequent visitor there.) Balfour Stewart who had called in on Lockyer the day after his discovery, and had been shown the bright lines, also wrote enthusiastically to De la Rue.

Lockyer was not to celebrate this major discovery without disturbance. On 18 August 1868, a total solar eclipse, visible in parts of India and Malaya, had occurred. Various European societies—including the Royal Society and the Royal Astronomical Society—dispatched parties to observe the event. One of the main objectives was to look for prominences, and, if possible, to determine the nature of their spectra. The weather for once was good, and the various parties all came to the same conclusion that prominences showed bright lines. There was less agreement on the number and positions of these lines; from three to nine lines were seen by different observers, and most of the main Fraunhofer lines were claimed to have been seen in emission by somebody. The various groups telegraphed back to Europe a précis of their results, which thus became known before Lockyer succeeded in seeing the prominence lines out of eclipse. In other words, Lockyer already knew that there were bright lines to be seen. This may have been important, for Lockyer had remarked in an article published in *Macmillan's Magazine* in July 1868, that the lack of bright lines visible in his original, less powerful spectroscope suggested that prominences were not gaseous. He also had some not very reliable guide as to which lines might be visible. In fact, Lockyer later asserted that he knew of three sets of observations from the eclipse, but that they disagreed between themselves so much that they were of little use to him in his own search for bright lines.

One set of observations of which Lockyer seems not to have been aware were those made by the French astrophysicist, Pierre Janssen. Janssen had examined two large prominences visible during the eclipse, and had noted that the bright lines he saw indicated that they were essentially clouds of hydrogen gas. He was particularly struck by the intense brightness of the lines and immediately began to wonder whether it would be possible to see them again out of eclipse. Unfortunately, the sky clouded over after the eclipse, so he could do nothing further that day, but the following day was bright. He scanned the limb of the Sun again with his spectroscope, and saw the lines at once. For the next 17 days he examined the limb of the Sun, and from the positions and lengths of the emission lines managed to construct a map of the solar prominences present, noting how their appearance changed with time. Finally, he sent off a letter describing his results to the Académie des Sciences. The letter took a month—the usual transit time from India—to reach Paris, and was placed in the hands of the President of the Académie a few minutes before De la Rue announced the same discovery by Lockyer.

Disputes over questions of priority in the making of discoveries have been a concomitant of modern science ever since its inception, but the violence of such disputes in the nineteenth century tended to be particularly great. There are a variety of reasons which could be offered for this. For instance, Victorian scientists often extended their belief in the sanctity of property to the realm of ideas. In the extreme form this corresponded to a belief that a discovery could only be exploited by its discoverer and by those people to whom he gave permission. More importantly for the present situation, Victorians tended to regard simultaneous discovery of the same piece of information as an oddity. When it occurred, therefore, suspicions of foul play by one of the protagonists were often aroused. When the two scientists concerned were from different countries, nationalistic feelings also came into play—as in the simultaneous prediction of Neptune's existence by Leverrier and Adams. It is therefore very much to the credit of both Janssen and Lockyer that their simultaneous observation of prominences out of eclipse did not for once lead to a priority dispute. Instead, Lockyer struck up a friendship with Janssen which lasted till the latter's death. The two were hailed as joint discoverers, and the French Government ultimately struck a medal, bearing the portraits of both Janssen and Lockyer, to commemorate the event.

Browning, the manufacturer of Lockyer's spectroscope, felt that Lockyer would certainly have gained absolute priority in the discovery had the instrument been finished earlier, and said so. Lockyer, however, was prepared to overlook the delay. When, subsequently, a dispute over Lockyer's prominence observations did arise, Browning became one of his most fervent supporters. For if Lockyer had managed to avoid controversy with Janssen, he was not so fortunate within his own country.

In February 1868, Huggins in reporting the progress of work at his observatory had stated that one of the uses he had been making of his spectroscope had been to examine the Sun's limb in an attempt to observe bright lines of prominences out of eclipse. (Oddly enough, the instrument he used then was theoretically more powerful than the one with which Lockyer eventually succeeded in seeing the lines.) After Lockyer and Janssen's results had been announced, Huggins wrote a further article in which he implied that Lockyer had obtained the idea of looking for the lines from the experiences of the observers at the 1868 eclipse. Lockyer was immediately up in arms, and a quarrel between the former friends developed which smouldered on for some time. Huggins ultimately alleged that there was nothing in Lockyer's paper of 1866 to show that he had had any idea of how to observe prominences out of eclipse. In the early 1870s, R. A. Proctor, who by this time was a bitter enemy, went a stage further, and stated that Lockyer had got the idea for his work from Huggins. Lockyer called on the services of his friend Balfour Stewart to explain publicly that the two of them had had several conversations on the subject before the 1866 paper appeared. Stewart pointed out that he had suggested the query on prominences to Lockyer, and concluded: 'I think that anyone well acquainted with spectra, on reading the question put, could not fail to see what was meant; and, if he were previously ignorant of the principle, he could not fail to perceive it. I therefore feel rather astonished that anyone should claim the statement made by Mr Huggins two years afterwards as being the commencement of a new principle.'[9]

The priority dispute between Lockyer and Huggins over the basic idea of observing prominence lines out of eclipse was compounded and confused by a concurrent controversy as to which of them had originated the method for observing entire prominences. The bright lines seen through the spectroscope represented the light from that part of the prominence which had been projected onto the slit of the spectroscope. In order to map out the size and shape of the entire prominence, it was therefore necessary to move the slit gradually across the prominence, recording how the appearance of the bright lines changed. This was a laborious procedure, and both Janssen and Lockyer very quickly began to look for some method of seeing the prominence as a whole. Lockyer introduced an oscillating slit, relying on persistence of vision to see the prominence. If the slit scanned sufficiently rapidly backwards and forwards across the prominence, the eye retained an impression of the whole. This device, although it worked, was not satisfactory, because the oscillations of the slit caused the whole telescope to vibrate. At this point (February 1869), Huggins published a suggestion that instead of using a narrow oscillating slit, a stationary wide slit, which would allow the light from the entire prominence to enter the spectroscope, could be used. Since the wider the slit, the more the scattered background light which would enter the apparatus, he also proposed that the prominence should be examined in the red line of hydrogen, and that a ruby-coloured glass filter should be used to cut off most of the background light of other colours. Lockyer noted immediately that with a reasonably powerful spectroscope the background light should be weak, so that it would be unnecessary to use a filter—a wide slit alone would do. He immediately published this emendation, and it was in this form, without the filter, that the method came into general use.

As remarked, Huggins and Lockyer immediately disputed which of them deserved the credit for this advance. The argument this time was fairly brief, for it turned out that they had both been anticipated. One of the leading German astronomers, J. C. F. Zöllner, had suggested the entire basis of the method a short time before Huggins' original paper appeared.

Some time later, in 1873, Lockyer and his friend Seabroke, who observed primarily at the Temple Observatory of Rugby School, introduced a circular slit extending all round the solar limb. This allowed the entire chromosphere to be scanned at once. The idea, though sound in principle, never turned out as satisfactorily as had been hoped, and the widened-slit method remained the most popular way of looking at the chromosphere.

Lockyer's discovery of the bright lines in prominences had been made before his new spectroscope was entirely finished or adjusted. In the next two weeks it was finally tuned up, and, almost at once, Lockyer made a new discovery. Early in November 1868, he recorded, 'I have this morning obtained evidence that the solar prominences are merely the expansion, in certain regions, of an envelope which surrounds the sun on all sides.'[10] This gaseous envelope, which Lockyer next day estimated to extend for about 5,000 miles above the solar surface, was christened—on the suggestion of his friend Dr Sharpey—the chromosphere. The name was derived from the intensity of the main hydrogen line, which, being in the red part of the spectrum, gave a reddish colour to the entire envelope. This new discovery very quickly involved Lockyer in another dispute, this time with one of his best friends, Warren De la Rue. Lockyer's extended account of his observations was submitted to the Royal Society, from whence it was sent by Professor Stokes, as a Secretary of the Society, to De la Rue for refereeing. De la Rue was, for the most part, warmly in favour of publishing the paper; he commented to Stokes, however,

> It does not appear to me that he [Lockyer] has brought forward any observations which establish the existence of an envelope different in any way from that of the luminous prominences, *that is, of an envelope between the photosphere and the luminous prominences* . . . I am quite certain that any luminous layer different from the luminous prominences could neither have escaped me, while observing with the telescope [at the total eclipse of 1860], nor photographic depiction in the Kew heliograph.[11]

Before writing this report, De la Rue had called in at the War Office to try and induce Lockyer to withdraw the announcement of his discovery, saying that he thought it quite impossible that there should be a continuous layer of the sort described round the Sun. Lockyer offered to show De la Rue the chromosphere at any point on the solar limb he specified, but De la Rue remained unconvinced. After De la Rue had left, Lockyer, with the possibility of another scientific dispute in prospect, obtained a signed statement of what had passed between them from his colleague in the Army Regulations Department, who had been present during the interview. He then went back and restudied De la Rue's 1860 photographs, and claimed that it was indeed possible to distinguish the chromosphere even on them. The dispute actually fizzled out quite rapidly, for it became clear that Lockyer's envelope had been observed before during some eclipses. It proved to be simply the spectroscopic version of the 'sierra' that Airy had first noted with the naked eye at the 1851 eclipse. Secchi, too, had seen the envelope before, and he soon confirmed Lockyer's spectroscopic discovery. De la Rue and Lockyer remained good friends, despite this incident.

One of the first tasks that Lockyer set himself, after the discovery of the bright-line spectrum of prominences, was the accurate measurement of the position of the lines in the spectrum, so that the elements present might be identified. This was particularly important since the measurements of the bright lines at the Indian eclipse of 1868 had varied so much between different observers. It became evident that the lines were mainly due to hydrogen, with the one obvious exception of a bright line in the yellow, not far from the sodium D lines (it had indeed been mistaken for them by some of the eclipse observers). Since the two sodium lines were known as D_1 and D_2, the new line came to be known as D_3.

In making these measurements, Lockyer also noticed that the thickness of the F line of hydrogen varied. Near the solar surface it appeared much wider than up in the chromosphere. He decided that the only way of finding out more concerning these spectroscopic differences was by laboratory investigation. More likely than not, he also felt favourably inclined towards the laboratory approach to the problem because his work at the War Office prevented him from observing the Sun during much of the day, whereas the laboratory work could be done at night. Nevertheless, it was by no means a matter of expediency only; he always regarded astronomical and laboratory spectroscopy as interdependent. Lockyer, himself, had no experience of laboratory work in spectroscopy and realised the need to cooperate with someone who had. Spectroscopic work of the sort that Lockyer was interested in was regarded as the domain of chemists at this time. Huggins, earlier in the sixties, when he had felt the need for a collaborator had turned to a leading chemical spectroscopist, W. A. Miller. Under similar circumstances, Lockyer turned to Edward Frankland.

Frankland was a considerable figure in the scientific world of the period. Older than Lockyer, he had been appointed first as Professor of Chemistry at the Royal Institution, and then, a few years later, had transferred to the Royal College of Chemistry in Oxford Street. Lockyer had probably encountered Frankland first *via* his literary activities, for Frankland was a member of the X Club. From Lockyer's viewpoint, scientific cooperation with Frankland had two great advantages: Frankland was very interested in astronomy, having a private observatory of his own, and he was already experimenting on the hydrogen spectrum.

Starting in the winter of 1868, Frankland lent Lockyer a room at the Royal College of Chemistry where he could work, and gave him part-time use of his research assistants.

Lockyer already had some idea as to why the hydrogen F line became thicker towards the photosphere. In 1865, two German scientists, Plücker and Hittorf, had published an account of some observed variations in the spectra of gases and vapours. In particular, they had shown that the width of the hydrogen lines depended on the temperature and the pressure to which the hydrogen was subjected. Frankland, meanwhile, had been working on the nature of flames. Not long before Lockyer joined forces with him, he had shown that when a jet of hydrogen gas was burnt in oxygen under increasing pressure, the hydrogen lines became broadened until eventually the light from the flame was spread out into a continuous spectrum. The apparatus which he had used for this work consisted basically of an iron container with a thick plate-glass window through which the light of the flame could be observed. The gas pressure within could be changed to values either above or below the atmospheric. Frankland and Lockyer pressed this apparatus into service for their new experiments, and within a few months had reached the important conclusion that the thickening of the F line was due to increase in pressure rather than to temperature changes. They were also able to demonstrate that the pressure in the chromosphere was much less than that of the Earth's atmosphere.

They remained puzzled by the D_3 lines, however, for whereas the bright hydrogen lines in the chromosphere appeared as dark Fraunhofer lines in the normal solar spectrum, this line did not. They wondered whether it might be caused by the very great depth of hydrogen in the chromosphere. Frankland commented, 'There was nothing about that splendid mountain of glowing hydrogen you showed me last Sunday, that impressed me so deeply as the brilliancy of the yellow line and I think that we ought not so easily to give up all efforts to get it from terrestrial hydrogen.'[12] Lockyer was initially confused by the fact that Plücker and Hittorf had said that old tubes of hydrogen, which had been left for some time, occasionally showed instances of new bright lines, especially near the sodium D line. But gradually he came to the quite different conclusion that the D_3 line was emitted by an element hitherto undiscovered on Earth. This conviction came partly from the lack of success in laboratory identification, but even more from his solar observations. For he noticed that when, as often happened, the bright lines were distorted or displaced, the known hydrogen lines all changed in the same way, whereas the D_3 line did not.

The possibility of discovering new elements by spectroscopic analysis was a very live issue in the 1860s. At the beginning of the decade, Kirchhoff and Bunsen had detected the two metals, rubidium and caesium, in the laboratory by this method. Similarly, in 1861, William Crookes in England and Lamy in France had detected the existence of the element thallium from the observation in the laboratory of a single bright line in the green region of the spectrum. Despite these precedents Frankland was not prepared to agree with Lockyer that they had between them discovered a new element on the Sun. As the letter quoted above shows, he felt that there was still a chance that the yellow line came from hydrogen. (Frankland was, perhaps, made more cautious by the experience of Henry Sorby, an eminent Yorkshire scientist, at just this time. Sorby announced, in 1869, the discovery of a new element—Jargonium—by spectroscopic analysis, but was forced in the following year to withdraw his claim as it was found that the interpretation of the spectrum was more complicated than he had supposed.) Lockyer, characteristically convinced of the correctness of his own reasoning, stuck to his opinion and coined the name helium, from its solar origin, for the hypothetical element. But as a result of Frankland's caution, it was a long time before he used this name in public. Indeed, the first public mention of 'helium' seems to have been by Sir William Thomson in his Presidential address to the British Association at Edinburgh in 1871. Thomson actually said that both Frankland and Lockyer had proposed the existence of the new element. Frankland seems to have let this pass, but in the following year the next President of the British Association, W. B. Carpenter, a London physiologist, referred to the hypothesis again, less favourably this time, and Frankland immediately wrote to Lockyer to disclaim responsibility for the idea. 'Surely Dr Carpenter is wrong in coupling my name with yours in connection with Helium as I remember always protesting in our conversations about the yellow line, against making this assumption, until we had exhausted every effort to get the line out of hydrogen.'[13]

From Lockyer's point of view, the problem was that time passed and the new element obstinately failed to produce its bright yellow line in any terrestrial laboratory spectrum. The only advance concerned its appearance on the Sun. The leading solar observer in the United States, C. A. Young, pointed out that some fainter chromospheric lines appeared under the same circumstances as the D_3 line, and were therefore presumably associated with it. Despite this, most scientists in the seventies preferred to reserve judgment on Lockyer's proposed new element.

Frankland and Lockyer's laboratory experiments led to a further brush with Huggins over a question of priority. They had found that, given appropriate conditions of temperature and pressure, the spectrum of certain gases, such as hydrogen and nitrogen, could simplify to a single line. During the 1860s Huggins had begun to investigate the spectra of nebulae. 'Nebulae' was a portmanteau word at this time, but it included, amongst other things, clouds of luminous gas in interstellar space. Huggins' preliminary observations suggested that this particular type of nebula had a spectrum of a single bright line only, which he believed to be connected with nitrogen. He proposed that the spectrum was simplified as compared with ordinary nitrogen either because nitrogen in the nebulae took on a more elementary form than on Earth, or because some of the nebular light had been absorbed in interstellar space. Frankland and Lockyer now pointed out that these assumptions were unnecessary, the simplification of the spectrum simply indicated that the nebulae were cooler and more tenuous than the solar chromosphere. Huggins claimed that Frankland and Lockyer had added nothing to the work he had already published. Again both sides refused to alter their statements. As Lockyer briefly observed, 'In this case, as in the former one, Mr Huggins has neither substantiated his statement nor withdrawn it. Instead, of this, he has reiterated it, taking no notice of our reply.'[14]

It was obviously of prime importance to discover whether (apart from the D_3 line) prominences and the chromosphere consisted of hydrogen only. By the first few months of 1869, Lockyer had obtained evidence that they did not. In March of that year he found that when the F line of hydrogen was very bright, indicating increased solar activity, lines of magnesium, sodium and barium also appeared bright in the chromosphere, as well as many other lines whose source he could not identify. He reasoned that this indicated that new materials were being injected from the photosphere into the chromosphere, the height which they reached in the latter being indicated by the maximum extension of their bright lines. These new lines were subsequently seen by other observers; just over a year later, C. A. Young, in the United States, working in a much clearer atmosphere than Lockyer enjoyed in London, listed over a hundred bright lines he had observed in the chromosphere. These apparent uprushes of material, together with the known rapid changes in the forms of prominences, intrigued Lockyer and turned his attention to the question of motions in the Sun's atmosphere.

During the 1860s, one of the talking points in the new science of astrophysics was whether the speeds at which astronomical objects were moving could be determined spectroscopically. Christian Doppler, an otherwise rather obscure Professor at the University of Prague, had proposed in the 1840s that motion towards or away from an observer could affect the light coming from an object (just as motion of a source emitting sound, a railway locomotive, for example, can change the pitch of the sound). The concept had subsequently been refined by the Frenchman,

Fizeau, who pointed out that the Fraunhofer lines provided a means of detecting a small shift in the wavelength of light. From laboratory measurements, the positions of the lines at zero relative velocity could be determined; any shift in the positions of the corresponding lines in an astronomical object would then be a measure of its speed in the line of sight. In April 1868, Huggins announced to the Royal Society that he had managed to measure velocities in the line of sight ('radial velocities' —as they came to be called) for some of the brighter stars. His measures were necessarily inaccurate, but the demonstration that such measurements were possible was undoubtedly important in spurring a new interest in the subject. Lockyer subsequently claimed that he was unaware of this work by Huggins, which may well be true, for the idea of looking for Doppler shifts was in the air towards the end of the sixties. Lockyer's interests, moreover, were directed towards the Sun, not the stars.

From his earliest observations of the chromospheric spectrum, Lockyer had been puzzled by peculiarities in the shape of the hydrogen F line: sometimes it bulged outwards on both sides, sometimes it was shifted towards the red or the blue ends of the spectrum, sometimes it was even split in two. These appearances could not be explained in terms of Frankland and Lockyer's laboratory experiments, for these seemed to suggest that temperature and pressure affected the line symmetrically. Lockyer consulted with his friends over possible causes of asymmetrical distortion of the line, but they, too, were puzzled. Finally it dawned on Lockyer that he might be dealing with Doppler effects. Lockyer's measurements of chromospheric lines were, of course, being made at the limb of the Sun. Here, motions up and down in the atmosphere would appear as motions across the line of sight, and hence would not result in shifts in wavelength of the spectral lines. Lockyer therefore could not hope to see the immediate injection of material upwards into the chromosphere. Nevertheless, it was firmly believed in the 1860s (and later) that the events in the solar atmosphere were in many ways directly comparable with happenings in the terrestrial atmosphere, i.e. that solar 'meteorology' was similar to terrestrial meteorology. Since it was known by this time that up and down currents in the Earth's atmosphere generally also had a circular motion as well, it was expected that this would be true of the Sun. Hence, there would be winds in the solar atmosphere, parallel to the surface, just as winds on Earth were parallel to the surface; it was, apparently, these winds that were being revealed by the changes in the F line. But the wind speed on the Sun was typically a hundred kilometres per second, as compared with perhaps the same speed per hour on Earth.

Hardly surprisingly, this first application of the Doppler effect to the Sun met initially with scepticism. In Italy, Respighi suggested that the distortions were due to disturbances in the Earth's atmosphere or in the telescope; Secchi, on the other hand, agreed that the changes were a Doppler effect, but claimed that they were caused by the rotation of the Sun. Lockyer countered the former criticism by pointing out that, if Respighi were right, all the lines visible should be affected, contrary to observation. Secchi's criticism did not require refutation for it was found that he had miscalculated the rate at which the Sun rotated, making it 200 times faster than it really was. Meanwhile, Young, in America, confirmed Lockyer's observations, and it came to be generally accepted that this interpretation of the line distortions was the correct one.

Lockyer carried out an extensive survey of the shape of prominences once the widened slit method had been introduced. He was fascinated by their infinitely varying appearances. 'Here one is reminded, by the fleecy, infinitely delicate cloud-films, of an English hedge-row with luxuriant elms: here of a densely intertwined tropical forest, the intimately interwoven branches threading in all directions.'[15] But he soon decided that, despite their variety, only two basic types need be distinguished—the eruptive and the nebulous. This finding was rapidly confirmed by Zöllner and, in modified form, by Respighi. To Lockyer's annoyance, however, a year after he had published his classification, Secchi in Italy published a very similar system without referring to Lockyer's work at all. Lockyer immediately leapt in to establish the correct priorities. The tone is indicated by a comment from Stokes when he returned one of Lockyer's anti-Secchi papers. 'We [the Royal Society] are not in the habit of printing controversial papers, and yours is of that character.'[16] Stokes was, in fact, rather harassed by Lockyer's frequent contributions to the *Proceedings of the Royal Society* at this time; for, apart from being occasionally contentious, Lockyer also constantly attempted to add to, and to change, the contents of his papers right up to their final printing.

This particular paper was published, on Stokes' advice, in the *Philosophical Magazine.* Apart from the reply to Secchi, it contains an interesting sidelight on Lockyer's approach to scientific research—one worth noting as representative of the un-Baconian approach of many British astronomers to their science in the nineteenth century.

> M. Secchi objects to my mixing up theory and observations. I plead guilty. I confess a remark made some time ago by M. Faye is always present with me when I am observing. The remark is, 'A good theory is as necessary as a good telescope'. Without a working hypothesis, I should certainly have cross-questioned the Sun much less than I have done; and it should be a truism that in a research, such as the one we are now conducting, it will not do to observe blindly or haphazard.[17]

Lockyer subsequently clashed with Secchi again, reproaching him publicly for his treatment of a fellow solar astronomer in Italy, Respighi. Secchi was stung to a passionate response, and this led to an exchange of correspondence. Lockyer asserted firmly,

> How have you treated him [Respighi]? As an impartial witness I am bound to say that I think no treatment could have been worse.
>
> It is because I believe with you that the shameful things to which you refer are intolerable among men of science, that I have attempted to prevent them in future.[18]

It is evident from Lockyer's correspondence with Respighi, with whom he was on very cordial terms, that relations between Italian astrophysicists were noticeably strained in the 1870s.

> You are doubtless aware that a society of spectroscopists has been established in Italy for the purpose of making observations of the Sun. I have hitherto hesitated to take part in it, although I have been already nominated as a member because the programme of the Society drawn up by Tacchini and Secchi has been published without being first communicated to the members; second, because this programme appears to me somewhat defective; and thirdly because with regard to some of the members I cannot entertain that harmony of feeling and mutual confidence which are indispensable in a society of persons formed for the pursuit of a common object.[19]

As a result of the laboratory work with Frankland, and his own solar observations, Lockyer was led to formulate a fairly detailed picture of the Sun's physical nature by the end of the sixties. This differed considerably from the original model based on solar observation which Kirchhoff had proposed at the beginning of the decade, but it tied in closely with the new consensus of opinion amongst solar physicists. Whereas Kirchhoff had supposed the photosphere to be the surface of a molten Sun, the new view regarded it as a cloud layer, on the lines suggested by Secchi and Sir John Herschel, with the addition that the Fraunhofer lines were now thought to be produced just above the cloud level, and not in the corona as Kirchhoff had supposed. The sunspots were cavities in the clouds, regions of higher pressure and greater absorption. Above the photosphere and the spots was the chromosphere, which consisted mainly of hydrogen, except when disturbances occurred and other elements were injected into it from the photosphere.

The most important contention of this model was the suggested place of formation of the Fraunhofer lines. This received widespread support: even Kirchhoff was prepared to countenance it. Henry Roscoe, one of the leading British chemists and a long-term friend of Lockyer's, wrote to him of a visit to Heidelberg.

> I had several long and interesting talks with Kirchoff about the Sun. I told him how much you regretted having missed him in the autumn. He seems willing to accept as possible your theory of the absorption taking place below the chromosphere but in general expresses himself with caution as to the physical condition.[20]

A question that remained was the significance and, indeed, the existence of the solar corona. Lockyer was sure that there was something there, for it had been too commonly observed to doubt, but he was loth to count it as an outer appendage of the Sun. It seemed to him that the upper chromosphere was so tenuous that the existence of another extensive envelope of gas beyond it was quite unacceptable. This led him in the early seventies to take a special interest in attempts to determine the nature of the corona. Unlike the chromosphere, however, there seemed no way of studying the faint light of the corona except during eclipses. For this reason, Lockyer now began to contemplate participating personally in eclipse observations. Fortunately for him, three eclipses occurred in quick succession, in 1869, 1870 and 1871. The first of these could be seen conveniently from the United States. Lockyer was too busy to have any hope of attending it personally, for the first issue of *Nature* was about to be launched. But he had already acquired several friends amongst the American astronomers, and they were happy to keep him informed of the observational results. (As we shall see later, two attempts were made to attract him to astronomical positions in the United States during the seventies. Although he accepted neither, Lockyer seems frequently to have been on rather better terms with the astronomical community in America than with that in Britain.)

The results of this 1869 eclipse provided more, if rather confusing, information about the corona. For the first time, bright coronal lines were observed spectroscopically. So far as the very rough measures of position allowed, it seemed that the lines were similar in wavelength to some of the lines which had previously been detected in the terrestrial aurorae. This suggested that the corona was not a normal extension of the Sun's atmosphere, but a vast solar aurora, with the implication that it was an electrical phenomenon of some sort according to contemporary ideas. The brightest of the coronal lines was measured as having a wavelength of 1474 on Kirchhoff's old scale. (Although Ångström's scale of measurement came in during the next few years, this line continued to be referred to as '1474K' in the literature.) The same line had been observed in the upper chromosphere by Lockyer only a few weeks before the eclipse, and he had equated it with a Fraunhofer line of iron at apparently the same wavelength in the photosphere. He therefore felt that the coronal line should be similarly assigned.*

* One of the fascinating coincidences of astronomy is that the final identification of the 1474 line—by Edlén in the early 1940s—showed it to originate from iron atoms. They were not, however, iron atoms in any sense that would have been understood in the 1870s. Edlén demonstrated that the atoms concerned had been stripped of thirteen of their electrons, and therefore emitted a spectrum completely different from the ordinary laboratory spectrum of iron.

The situation now appeared most confusing, as a letter to Lockyer from C. A. Young, written about this time, indicates.

> The idea that 1474 might represent some new element occurred to me at once when I found it in the Corona, but of late I own I have more inclined to the opinion that it might possibly be a true iron line, and caused by meteoric iron dust of almost infinitesimal fineness. Yet I have always felt the difficulty of supposing the complicated iron spectrum reducible to *this* one line. Perhaps you may recollect I urged this upon you at your house on one occasion. At that time, you did not seem to feel the difficulty so much as I did . . .[21]

Apart from the spectroscopic difficulty that Young mentions, the major problem in accepting the presence of iron in the corona as an ordinary gas was its high vapour density. It seemed improbable that heavy iron atoms should rise higher in the solar atmosphere than the much lighter hydrogen atoms. Lockyer, in fact, was not so greatly worried by these questions for he was becoming more and more convinced that the corona was not of solar origin, at all. Instead he believed that it was a halo in the Earth's atmosphere formed by the scattering of sunlight.

Whereas Lockyer had not been personally involved in the 1869 eclipse, he was a very active participant in the next one. This was due to take place just before Christmas, 1870, and was visible in Spain, Sicily, and a part of North Africa. The question of making arrangements to send British parties to the eclipse was discussed by the Royal Astronomical Society in March 1870, and in the following month a joint committee with the Royal Society was formed. Both societies voted sums of money towards the cost of an eclipse expedition, and the Royal Astronomical Society solicited volunteers to assist on the expedition. The response was gratifying, with over fifty applicants, including Lord Tennyson. Airy, as Astronomer Royal, was naturally elected the spokesman of the joint committee, and approached the Admiralty for help in transporting the various proposed eclipse parties to their destinations on Her Majesty's ships, as had been done before. This time, however, the request was turned down. Lockyer initially suspected that some of the members of the organising committee did not care particularly whether or not the planned expedition got under way, and so were not making much effort in its support. Later, however, he modified his opinion, deciding that the refusal of transport had occurred mainly because the wrong department of State had been approached in the first place.

It now looked as if there might be no British observers at the eclipse, but at this point Benjamin Peirce, who was in charge of the American parties observing the eclipse, visited Lockyer in London. When he heard of the British plight he obtained permission from the United States Government to invite Lockyer and other British solar physicists to join the

United States eclipse parties. Lockyer got leave from the War Office to go; but meanwhile Peirce and the American ambassador had been visiting the Prime Minister, Mr Gladstone. As a result Lowe, the Chancellor of the Exchequer, called Lockyer to see him and enquired about the transportation difficulties, saying that it was the first time that he had heard of the matter. Lockyer was allowed to state the British requirements for the eclipse, which he put at £2000 and two ships, and just five weeks before the eclipse an organising committee was set up; this time, hardly surprisingly, with Lockyer as the Secretary, in place of Stokes, and A. C. Ranyard, a young amateur astronomer, as his assistant. Four parties, to Spain, Gibraltar, Algeria and Sicily, were organised, with Lockyer in charge of the Sicilian party.

His group included, amongst others, his wife, Henry Roscoe and W. K. Clifford, one of the outstanding young British mathematicians of the period. (Up to his early death at the end of the seventies, Clifford was a close friend of Lockyer's. This may well indicate how far Lockyer had drifted away from his early religious convictions, for Clifford was widely regarded as the most materialist thinker amongst British scientists by Christian intellectuals.) The other main party was the one to Algeria under the direction of Huggins, and included Tyndall and Crookes. Lockyer's party went overland to Naples, but the other three parties took their way southwards from Portsmouth by HMS *Urgent*. They had a fairly hectic journey, being in danger of sinking at one point. In the end, however, they were more fortunate than Lockyer's party which embarked on HMS *Psyche* at Naples. Within a few hours their ship had struck a submerged rock and was immovable. Still, no one was lost, and all the passengers with their equipment were rowed ashore. As Clifford later wrote '. . . well, if ever a shipwreck was nicely and comfortably managed, without any fuss—but I can't speak calmly about it because I am so angry at the idiots who failed to save the dear ship.'[22]

Despite this misadventure, Lockyer's party was in plenty of time to set up camp at Catania, but their luck did not hold for the eclipse itself. The sky over their station turned cloudy, and the eclipse was only visible for a few seconds. Ironically, spectators on the boats which were trying to salvage the wrecked *Psyche* had a perfect view of the whole eclipse. The Algerian party also failed to see anything of the eclipse. They had been joined by Lockyer's friend, Janssen, who had had an adventurous journey out of Paris, then besieged by the Germans, in a balloon. Lockyer was for a while annoyed by Janssen's flight, for he had gone to some trouble to obtain a permit from the Prussian Government to allow Janssen's free passage through the enemy lines without need for adventure. Many years later it transpired that Janssen was carrying secret messages to the French Government outside Paris, and therefore felt that he could not use the pass.

Despite the disappointing results from Lockyer's and Huggins' groups, the other two British parties, and the American observers especially, managed to make worthwhile observations. In the first place, some reasonably good photographs of the solar corona were obtained for the first time. Previously the only method of recording had been to employ amateur or professional artists to sketch what they saw as rapidly as possible. An examination of the photographs and drawings only served to convince Lockyer that he was right in considering the outer corona to be an effect of the Earth's atmosphere, though he seems to have been prepared to allow a narrow inner corona as a real solar appendage. More importantly, Young for the first time definitely observed the 'reversing layer'. It had long been realised that if a certain depth of the solar atmosphere normally produced the dark Fraunhofer lines, the same layer seen tangentially at the limb of the Sun during eclipse, when the bright sunlight had been cut off, should show a similar spectrum, but with the lines bright. This 'reversal' of the lines had not previously been seen, but Young now observed it, though only for two seconds. The shortness of this time confirmed the belief that the layer concerned in the production of the Fraunhofer lines was a thin shell near the photosphere, as Lockyer and others had thought.

The next eclipse was due to take place in December 1871, and would be visible from South India, Ceylon, Java, and Australia. Lockyer, who was now on the Council of the Royal Astronomical Society, expected that that body would take some steps towards preparing for an eclipse, though, as the youngest member, he did not think that he should be the first to bring the matter up at a meeting. When no mention had been made of it by the last meeting of the Council before the summer recess, however, he proposed that the Governments at home and in India should be asked to help a few volunteer observers to go to India and Ceylon, free of charge, to observe the eclipse. A committee appointed to consider the proposal decided that Huggins and Lockyer should both be invited to go, but that only a restricted range of coronal observations, specifically spectroscopic ones, should be carried out. Huggins, in the event, could not go; and Lockyer refused to go if the type of coronal investigation was to be limited. Meanwhile, he had approached the Royal Society, and he now told the committee that the Royal Society were prepared to approach the Treasury jointly with the Royal Astronomical Society. Airy was very pessimistic of success. He wrote to Lockyer, 'I doubt much whether the Government will undertake for any special expedition. They seem to be disgusted with Astronomical enterprises.'[23] The committee was evidently irked both by Lockyer's refusal to follow their suggestions as regards observations, and for going over their heads to the Royal Society. They now refused to alter their conditions; nor, they said, could they authorize a joint approach to the Treasury. As a result, the preparations, both of the Royal Astronomical Society and of the Royal Society collapsed, and again, as for the 1870 eclipse, it looked as if no official British expedition would be sent.

Lockyer became extremely bitter, especially towards the Astronomer Royal, but at this point Sir William Thomson intervened. As President of the British Association for the year, he emphasized in his address to the Association how important eclipse expeditions were. The British Association therefore decided to try their own hand at organising an expedition, and accordingly they approached the Government for a grant. The money was swiftly granted, Lockyer was put in charge of the organisation, and the P. & O. line offered much reduced passages to India and back. Invitations were sent out to a large number of potential observers at home and abroad, but it was too late for most of them to accept, and the party that Lockyer took out contained few well-known names. C. A. Young, with whom Lockyer had struck up a close acquaintance at the 1870 eclipse, wrote sadly from New Hampshire. 'Your letter . . . stirred up almost fiercely the desires which I thought I had quieted. To be one of your happy dozen and in such company for three months at least this *would* be life.'[24] Lockyer himself led the main group which was to observe from an old fort on the coast at Bekul, in the north of what is now Kerala.

The party's arrival caused a considerable flurry. The ship to which Lockyer's party had transferred in India wished to have some firing practice after the party's equipment had been unloaded.

> This tremendous proceeding on the part of the ship, and the wonderful similitude of the telescopes to the native idea of a big gun, soon wrought a wonderful change in the ideas of the dwellers along the coast. The Eclipse was a pretence. There was war! If otherwise, why the firing with shot? Why occupy the fort? Why erect big guns [i.e. the eclipse telescopes] in the most commanding place in its, to them, vast extent? Why these soldiers from Cannanore? Instant action! All high-caste women and all gold into the interior.[25]

Lockyer's party gradually managed to calm these fears, and preparations for the eclipse went forward peacefully. The day's timetable was to work from about 5.00 to 8.00 in the morning, and then again from 16.00 to 20.00 in the evening. It is evident from Lockyer's account that he was revelling in the new atmosphere, and in the challenge of commanding an expedition. After a continuous spell of clear weather, the clouds gathered on the evening before the eclipse, and Lockyer feared that he might once again be cheated of his observations. To add to his problems, he had contracted a fever, and was dosing himself with opium. However, the next morning was fine and the observations of the eclipse went off like clockwork. The only unusual incident was provided once more by the local population.

> The natives see in the eclipse their favourite god devoured by the monster Rahoo, and they, like the observers, are not there for nothing. Yells, moans, and hideous lamentations rent the air as the monster seems to them to get the upper hand; the excitement increases, and evidently something is afoot. Mr McIvor's sharp eye detected an intended sacrifice of fire, the intended fuel being the long, parched grass covering the landscape exactly between the fort and the eclipsed sun. In a moment he pointed this out to Captain Christie; in a few more a posse of police was stamping out the flames, and the smoke-bank, which threatened to upset all the work, gradually died away; the moans, however, still continued, and Rahoo worked its wicked will.[26]

The most important observation at the 1871 eclipse was nevertheless not made by any member of the British expedition, but by Janssen. He detected for the first time that presence of dark Fraunhofer lines in the coronal spectrum, besides the bright lines which had been seen before. This new observation suggested that besides gas in the corona, there must also be solid material which reflected the photospheric light. But if Lockyer did not make the most dramatic observation, he, and also Respighi, introduced a new piece of instrumentation at this eclipse which was to be extensively used at later ones. This was the slitless spectroscope. As the name implies, instead of feeding the light from the corona through a slit, and so viewing the spectrum as a series of bright lines, no slit was used, so that the corona round the Sun was seen instead as a series of coloured rings. With this system not only could the whole corona be viewed at once, but differences of height of the corona in different spectral lines could be observed very easily. Apart from indicating the presence of hydrogen out to greater distances than had hitherto been measured, the new instrument did not provide a great deal of new information on this, its first appearance.

The results of the eclipse intensified one problem. The corona observed photographically seemed to differ from that observed spectroscopically. It was a matter of speculation which gave the 'true' representation. Lockyer came to be convinced that the difference between the photographic and spectroscopic corona was related to a peculiarity in the apparent chemical composition of the Sun. As had been noted from Kirchhoff's first observation of solar chemistry, the elements detectable in the Sun were almost entirely metals, the obvious exception, hydrogen, having many of the properties of a metal. Lockyer decided that the reason was that the metalloids, i.e. the non-metals, were all congregated above the chromosphere. There they were not heated sufficiently to produce spectra, but they could act as reflectors of sunlight. Hence this part of the corona would show up photographically but not spectroscopically, and this explained the differences produced by the two methods at eclipses.

It will be noted that Lockyer's viewpoint had now changed in one respect. The eclipse observations convinced him that there was a genuine solar corona, larger than he had previously thought possible, though he still believed that terrestrial atmospheric effects were also at work. Moreover, his thoughts on the composition of the corona were modified by this eclipse. Two Indian army officers, Herschel and Tennant, both astronomers, the former being a third-generation member of the famous Herschel family, had observed that the 1474 line could be seen all over the corona, being apparently independent of the underlying coronal structure. This strengthened a general suspicion amongst astronomers that the line had nothing to do with iron, as had originally been supposed. During the following twelve months Lockyer convinced himself from laboratory experiments that this was indeed the case, and that his former identification must be modified.

The 1871 eclipse added to a growing coolness between Lockyer and the Royal Astronomical Society. Since it was the British Association which had persuaded the Government to finance the expedition, Lockyer felt himself under no obligation to speak of his work at the Royal Astronomical Society, although the Society had loaned him some of the instruments used. His main lecture on the eclipse results was, indeed, given at the Royal Institution—much to the annoyance of some Fellows of the Society.

Lockyer's success in scientific research was marked by a corresponding rise in his importance in the scientific world. He joined the Royal Astronomical Society in March 1862, and in 1866 was elected a member of the Council of the Society, a position he occupied for the next seven years. Charles Pritchard, with whom he came to be on close terms of friendship in the next few years, was elected President of the Society in the year Lockyer joined the Council. Writing just before their election, Pritchard advised Lockyer on the current problems in the Royal Astronomical Society. He ended ruefully, 'Practically "Routine Business" and Assistant Secretary and Carrington stop the way.'[27] (The Assistant Secretary, Williams, was growing too old for the job, but there was no statutory retirement age, and he refused to resign. Carrington, the leading solar astronomer of the fifties and early sixties, was in the middle of a dispute within the Society: a forerunner of the violent quarrels of the 1870s.) Lockyer, typically, was eager for reform, and Pritchard, a much older man, cautioned him, '. . . my friend—remember—"festina lente".'[28]

Much more importantly, he was elected a Fellow of the Royal Society in the summer of 1869. Lockyer had obviously been hoping for this; he had been sending his papers to the Royal Society for publication rather than to the Royal Astronomical Society (a point which was taken up against him in the 1870s when he was under attack in the latter Society). His successful work in solar spectroscopy obviously provided a very solid reason for his election, and it was one of his co-workers in solar physics, Balfour Stewart, who took the initiative in proposing him. Stewart wrote to Airy in January 1869, saying, 'In conjunction with yourself and others I have been very much interested in Mr. Lockyer's researches and I think it only right that an attempt should be made to get him into the Royal Society.'[29]

Airy agreed to sign the certificate in favour of Lockyer, but insisted that the reason given for proposing Lockyer should be changed. Balfour Stewart had placed Lockyer's discovery of the chromosphere as his main claim to fame, but Airy insisted that this was not an original discovery. The existence of such an envelope had been known to himself ever since the eclipse of 1851. Instead, he said,

> . . . the thing which Lockyer did and which you have not specified is this. He made a very large and laborious series of observations expressly directed to the discovery, if possible, of prominences. He did not stumble on them as Janssen did, from observing their nature in an eclipse, but with rare merit attacked them while yet obscure, and successfully.[30]

An immediate result of Lockyer's leap into scientific eminence was that he became inundated with requests to give scientific lectures at various levels of popularity. One result of the lack of formal scientific training at all levels of English education at this time was that the public lecture formed a very important channel for arousing public interest in science, and for communicating new scientific results to the public at large. Indeed, popular lectures were used for announcing results which were new even to the scientific world. In a similar way, it was then still possible to write books which were semi-popular, and which yet contained original scientific results. Lecturing could also be quite profitable and this was, of course, an aspect of especial importance still to Lockyer. Lockyer combined his lecturing with his literary work. Having prepared with great care a set of lectures, he might reproduce these as a series of articles in some journal, and then publish them as a book. Indeed, since he often quoted from one of his books in another, the same material might appear in book form more than once. This is a partial explanation of Lockyer's remarkable literary productivity. He similarly combined the art of illustration. Having made a set of lantern slides for his lectures, being one of the early exponents of such visual aids in scientific lecturing, he would use the same figures for his books.

As has been remarked, these public lectures provided an important outlet through which the voice of science could make itself heard, and for the lecturer to press the importance of scientific matters. Thus Lockyer was invited to Cambridge to give the Rede Lecture in 1871, and Frankland wrote to him saying, '[I] Hope that you will give a vigorous push to experimental science at Cambridge.'[31] Sure enough, Lockyer ended his lecture with a strong plea to his audience to support the building of an astrophysical observatory at Cambridge, as distinct from the observatory already there which specialized in classical astronomy.

Lockyer's research ability, his energy and enthusiasm, brought him considerable recognition and admiration amongst his fellow-scientists by the early 1870s. At the same time, his skill in controversy and his pugnacious determination to have the priority and significance of his ideas thoroughly recognised, was noticed and recorded even by his friends. De la Rue, writing to Lockyer concerning an article on solar physics, remarked gently,

> I venture to make the suggestion that it might read better if 'the worker' were substituted for the personal pronoun in some parts to avoid too frequent repetition.[32]

A surprising number of mathematicians and scientists of the nineteenth century were spare-time versifiers. Some, such as J. J. Sylvester, were serious in intent, but more were avowedly humorous. Of this latter group, the best known today is the Rev. C. L. Dodgson, but in the scientific community of Lockyer's time the most eminent was certainly James Clerk Maxwell, the first head of the Cavendish Laboratory at Cambridge. Maxwell had a good eye for the salient features of his scientific contemporaries, and some time in the early seventies he summed up Lockyer in the following lines, which were to dog the latter for the remainder of his days.

> And Lockyer, and Lockyer,
> Gets cockier, and cockier
> For he thinks he's the owner
> Of the solar corona.[33]

Given the rather tense state of British astronomy as the new science of astrophysics struggled into existence, spasmodic altercations in the small community of British astronomers were almost unavoidable. Given Lockyer's temperament, it was almost equally inevitable that he would be at the centre of the disputes. However, the basic scientific controversy of the seventies, in which Lockyer was greatly involved, actually concerned a much broader question than just the development of his own speciality.

IV

THE DEVONSHIRE COMMISSION

The late 1860s were a period of great heart searching for those interested in the development of British science and technology. At the beginning of the fifties it had seemed to the casual gaze that Britain had developed a magnificent technological lead over other countries. The Great Exhibition of 1851 seemed to confirm this picture, for the British exhibits gained numerous awards. This superiority of British manufactures was generally supposed to show how right successive British governments were in interfering as little as possible in the workings of the State. But many scientists felt less happy. If the British Government was letting science and technology (or 'arts' as it was then called) evolve as private initiative dictated, some other governments were not.

Scientific research was thought then in England to be mainly an activity for those who could afford it. It might be a reasonable hobby for a gentleman, but to treat it with extreme seriousness—as the Prince Consort did—was to invite ridicule. (It was recognised that, as a German, he was liable to such aberrations.) Scientists who were actually paid by the State in Britain—the Astronomer Royal, for example—were extremely few. As was emphasized several times during the 1850s, this amateur approach to science was typically English, and was happily accepted by the average Victorian as a part of the English tradition. But many British scientists in the sixties felt that a continuance of this approach would soon lead to a marked decline in the relative industrial importance of Britain. The manufacturers as a body took little notice of these jeremiads. Britain was still the workshop of the world and would continue to be so. Besides, there were always good reasons for not encouraging scientific and technical education. They could argue, for example, that a scientifically trained artisan would be a bad thing. He would be able to understand the principles of the process on which he was working, and could then go to a competitor and sell these industrial secrets. Similarly, the State should not provide financial backing for any science of commercial interest. This would be favouring one industry at the expense of another, would therefore interfere with the free interplay of the market place, and hence disrupt the British economy.

Despite this opposition, some advances in the provision of scientific and technical education were made in the 1850s, several of them stemming directly or indirectly from the 1851 Exhibition. In November 1851, Prince Albert formally opened the Government School of Mines and of Science Applied to the Arts in Jermyn Street, just to the south of Piccadilly. It had been argued in favour of its foundation that, since much of the wealth of Britain depended ultimately on mining, it was scandalous that there should be no centre for the diffusion of knowledge about the subject. Although State-supported, the new institution charged fees in the same way as the existing constituent colleges of the University of London, University College and King's College. But the number of full-time students it recruited in its early years was distinctly disappointing. On the other hand, the evening lectures on science for working men were tremendously successful. Not far away, in Oxford Street, was the Royal College of Chemistry which had been in existence since 1845. The Professor there, Hofmann, had been brought over from Germany by the Prince Consort specifically to get the College going. He, too, had only a relatively small number of students, though many of them, including De la Rue and Crookes, were to become leading figures in the world of British science during the next two decades.

After the Great Exhibition the feeling grew that London should have a technical university, as distinct from the all-purpose university already in existence. As a first step in this direction, it was decided to merge the Metropolitan School of Mines, as the Government School of Mines had been renamed, and the Royal College of Chemistry, which had run into financial difficulties. In 1853, despite the opposition of De la Beche, the director of the School of Mines, and of members of his staff, this union was carried through, and the joint institution was put under the control of the newly formed Department of Science and Art. The latter department, which was subsequently to figure largely in Lockyer's life, had been formed under the Prince Consort's influence as an expanded version of the already existing Department of Practical Art. The new department, like the old, came under the aegis of the Board of Trade, whose political chief at that time was Edward Cardwell. Cardwell, who later, when he transferred to the War Office, became Lockyer's superior, disliked the new department. But the two main movers there were Henry Cole and Lyon Playfair, who had been the Prince Consort's two main executives in the organization of the 1851 Exhibition, and they were quite capable of upholding the interests of their department against Cardwell. Their main desire—to develop scientific and technical education in Britain—did not, however, flourish in the fifties. Apart from the creation of a few schools devoted to science teaching, the Department of Science and Art did not become a major influence in promoting science education till the sixties.

At a higher educational level, Oxford and Cambridge began to take a gingerly interest in science in the 1850s. Oxford introduced an Honours School of Natural Science in 1850 and Cambridge in 1851. But these had little immediate impact. The degree which retained the most respect was the old Mathematics Tripos at Cambridge, and many of the leading British scientists continued to receive their university training in this highly specialized manner. A large number of Lockyer's contemporaries in mathematics and physics had been to Cambridge, whereas most of the chemists had been trained elsewhere in Britain, or on the Continent. This difference in background may help to explain why Lockyer seems to have been more at home with the chemists. Partly he did not belong to the Cambridge network, partly he was completely lacking in a mathematical background. The presence of a number of influential Cambridge mathematicians in the astronomical community was sometimes a source of annoyance to astronomers who were not so connected, and who suspected that they were thereby being disadvantaged.

In London, both University College and King's College had taught science from the beginning. But the most important advance in the metropolis in the 1850s was made by the University of London, which was an administrative entity with rather little to do with the London colleges. In 1858, the University threw open its examinations to all comers, not merely those taught at the London colleges, and at the same time, instituted a new system of science degrees. For struggling provincial institutions, such as Owen's College in Manchester, the possibility of taking these London science degrees proved particularly helpful. For example, it helped Roscoe, who replaced Frankland at Manchester when the latter moved to London, to build up one of the major schools of chemistry in Britain by the time Lockyer became acquainted with him in the 1860s.

Finally, the presence of science teaching at Scottish universities should be mentioned. Although no longer so uniquely attractive to English students as they had been earlier in the century, they were still much livelier as scientific centres than Oxford or even Cambridge.

Despite the existence of these various institutions offering higher education in science, the number of people receiving any such form of training remained very small. Indeed, the problem was that although advanced education in science in England was scarce, it more than satisfied the demand. On the one hand, there was so little elementary teaching of science that few could go on to an advanced course; on the other, scientifically qualified personnel were hardly required at all, so there was little impetus to read science at university. Huxley once remarked on the difficulties in the way of a man who wished to be a scientist (he had

Charles Darwin particularly in mind, but his words fit Lockyer just as well),

> But in the majority of cases, these men are what they are in virtue of their native intellectual force, and of a strength of character which will not recognise impediments. They are not trained in the courts of the Temple of Science, but storm the walls of that edifice in all sorts of irregular ways, and with much loss of time and power, in order to obtain their legitimate positions.
>
> Our universities . . . do not encourage such men; do not offer them positions, in which it should be their highest duty to do, thoroughly, that which they are most capable of doing . . .[1]

An analysis of the members of the learned societies at this period certainly suggests that a high proportion of their membership was composed of moderately interested amateurs rather than dedicated professionals. This was undoubtedly true of the first society Lockyer joined, the Royal Astronomical Society.

There was one other development in the 1850s, stemming solely from the 1851 Exhibition, which was greatly to affect not only the future of scientific education, but also Lockyer's personal future. The Exhibition had made a very substantial profit, and this money, together with another large sum voted by the Government, was set aside to be used by the Commissioners of the 1851 Exhibition for the advancement of science and technology. They soon invested a good part of the money in purchasing a large area of land at South Kensington, for they had in mind the possibility of creating there a technical university, together with facilities for the display of various types of collections, and housing for learned societies. (The idea of a technical university frequently occupied the thoughts of British educators during the latter half of the nineteenth century. This was, in effect, a recurring sideways glance at the German success in this type of endeavour.)

The Commissioners' hope that the South Kensington area would form a complete scientific and cultural complex hardly materialized in the fifties. The Science and Art Department duly established itself at South Kensington, but the only other important addition there was the establishment of a South Kensington Museum in 1857. This contained various exhibits left from the Exhibition of 1851, plus some additions, in a rather ill-assorted display. These various objects were housed in allegedly temporary corrugated iron buildings which soon became a local landmark, popularly known as the 'Brompton Boilers'.

Many who were involved with scientific and technical matters, especially those who had close scientific contacts with the Continent, felt that the rate of progress in the fifties was quite inadequate. In 1855, the British Association meeting at Glasgow considered a detailed report on the need to improve the position of science in Britain. Two years later the

Royal Society presented a memorandum to the Prime Minister on the same question, and throughout the fifties the Society of Arts was agitating for the improvement of technical education.

This, then, was the situation when Lockyer first became interested in science. Almost from the start of his career, he was concerned with the need for increased support of science teaching and research in Britain. It was one of the subjects that he took up in the *Reader*, adopting there a viewpoint that was to characterise him throughout his life, that the State must step in and support science. A comparison with other countries such as Germany or France, where there was State support, showed that Britain was lagging behind. However, during the heyday of the *Reader*, pressure for State interference on behalf of science seems actually to have declined a little as compared with the fifties. Perhaps this was because the International Exhibition in London in 1862 showed, so it was believed, that Britain still led the world in technology. (Lockyer, incidentally, was involved in this exhibition. He used his influence to have some of Thomas Cooke's optical instruments included in the British section.) On the other hand, it may have appeared that some progress in science teaching was at last being made.

Lord Salisbury, one of the few politicians of the day genuinely interested in science, had found on his appointment as Lord President of the Council, that the Science and Arts Department now came under his control. The Education Department of the Privy Council had been set up early in 1856, and the Science and Art Department was transferred to it from the Board of Trade. Murchison, who had replaced De la Beche as Director-General of the Geological Survey in 1855, was greatly opposed to this transfer. He was annoyed that he could no longer deal directly with the Minister of State but had to go through Henry Cole. From the point of view of science teaching, however, the transfer was valuable. Cole records a visit by Salisbury to South Kensington in the Spring of 1860, 'Lord Salisbury became impatient the Science instruction had not advanced like Art, and said, If we could not find out how to teach his carpenters at Hatfield some Science useful to them, he would abolish the name of "Science" from the title of the Department.'[2] Although Lord Salisbury's concept of science was distinctly limited, his pressure on the department to expand its science side was effective.

In 1859, the department had introduced a Science Teachers' Certificate. A person who gained this certificate could set up as a teacher of science, and present students for the elementary scientific and technical examinations which the department also set. According to the students' success in these examinations, so the teacher was reimbursed. The system came therefore to be called 'payment by results'. There were, from the first, doubts as to the efficiency of teaching which revolved so closely round examinations. But there could be no doubts about the rapid increase in the number of people, of a wide range of ages, who took the department's elementary examinations as a result of the increased number of science teachers soon available. In 1860, 500 students were being prepared for the Science and Art Department's examinations, in 1865 the number was nearly 5500.

In the latter part of the 1860s pressure for State assistance to scientific research and science teaching once again increased rapidly. The next International Exhibition was in Paris in 1867, and this time the British exhibits fared relatively badly. The prophets who had warned of a coming lack of competitiveness on the part of British industry, because of the neglect of technical education, suddenly began to be heard. Perhaps the most weighty of these prophets was Lyon Playfair, one of the jurors at the Exhibition. He got in touch with Lord Taunton, the President of the Schools Inquiry Commission which was then sitting, and urged the need for a further study of the advance of scientific and technical education on the Continent as compared with Britain. The difference was indeed becoming evident to any intelligent inquirer, whether he was a scientist or not. Matthew Arnold, touring the Continent in the sixties to obtain comparative information on education, noted, 'In nothing do England and the Continent at the present moment more strikingly differ than in the prominence which is now given to the idea of science there, and the neglect in which this idea still lies here.'[3]

A fellow-juror with Playfair at the 1867 Exhibition was Alexander Strange, who shortly became closely associated with Lockyer. Strange had spent his early life as a member of the Indian Army, where he was noted as one of the leading surveyors. He retired finally to England in the early 1860s with the rank of Lieutenant-Colonel. Back in London, he took charge of a new section of the India Office for the design and construction of surveying instruments, and was soon recognised as a leading expert in this field. He became a Fellow both of the Royal Society and of the Royal Astronomical Society. (Surveying, like navigation, had long been regarded as an adjunct of astronomy.) Lockyer and Strange served together on the Council of the latter society in the 1860s. They had numerous mutual friends, for example, Cooke acted as adviser to both on questions of instrument construction. Strange, like Lockyer, was a great believer in the need for State aid for both education and research in science, and, after the 1867 Exhibition, began to agitate actively for such support.

Almost immediately the Society of Arts returned to the fray, and announced a major conference on scientific and technical education to be held in January 1868. All the main participants in the struggle for a greater emphasis on such education were there—Strange, Playfair, Huxley, Henry Cole, as head of the Science and Art Department, and with him Captain Donnelly, a Royal Engineer officer, who had become second-in-command to Cole at the department after Playfair had resigned to become Professor of Chemistry at Edinburgh. The topics debated at this conference accurately reflect the major educational problems which scientists, and those technologists who were interested, saw confronting them. The main complaints, as we have seen, concerned the lack of science throughout nearly all schools, from State elementaries to public schools, the low status of science at universities (especially Oxford), and the general absence of higher technical education. The conference concluded that all these deficiencies could be overcome if only the Government could be persuaded to take the initiative and provide the necessary aid.

Two months later the Government did make a hesitant step forward, a Select Committee on Scientific Instruction was set up under the chairmanship of Bernhard Samuelson, a leading industrialist and Member of Parliament. This committee examined a number of expert witnesses, and published a report in the summer. The major problems it pinpointed proved to be very much the same as those already specified by the Society of Arts.

Whilst this agitation for increased scientific and technical teaching got under way, the question of State support for scientific research simultaneously became the subject of a renewed discussion. The moving spirit in this was Strange. The British Association met in Norwich in 1868, and Strange presented a paper there 'On the Necessity for State Intervention to secure the Progress of Physical Science'. It had become clear, he claimed, that the development of scientific research by private enterprise alone was quite insufficient. What was needed was finance by the State. In particular, it was important that research institutions should be formed. The British Association was receptive to Strange's arguments, and appointed a committee to inquire into the contemporary state of research in physical science and what improved provisions might be needed in the future. The committee contained a sparkling display of scientific policy-makers. Strange was there, of course, so were Frankland, Huggins, Huxley, Balfour Stewart, Stokes, Sir William Thomson, Tyndall, and half a dozen others. Subsequently, a few other members, Lyon Playfair, Tennyson and Lockyer, were coopted.

This was an important event for Lockyer. He was now publicly associated with the group of scientists who were trying to force Government support of science. Moreover, he was brought into close contact with Strange, who perceived that Lockyer's knowledge of the literary world could be put to good use in the effort to obtain general recognition of the need for such support. Strange plied Lockyer with relevant material and suggestions.

> Here are, I am sure, much more than enough to set your ready pen going on a theme which I know is so congenial to you.
>
> I find there is a prevalent notion that 'extension of existing Institutions'—and 'increased help to individuals' will do all that is needed—this might with advantage be combated—nothing can be more futile. Another point I would suggest for your consideration is the small comparative amount necessary to produce an enormous effect—a mere song, compared with the resources of a nation like England, would suffice to erect and endow Experimental Institutions in every large town.[4]

The committee reported back in 1869, saying that, on the one hand, scientists believed that an increase in the amount of research would be of great value to the community, but that, on the other, they were equally sure that current facilities for such research were quite inadequate. The committee concluded that specific recommendations on the changes desirable to encourage research would require a much fuller investigation than they could manage. They therefore proposed that the British Association should approach the Government, and request that a Royal Commission be set up to make a full inquiry. The Association did not follow up this suggestion immediately, for in the meantime, other members had raised the further question of State aid to higher education in science. Another sub-committee was established to consider this, and only after it too had reported back did the Council of the British Association approach the Government. This meeting, in February 1870, together with continued pressure from other sources produced the desired effect. The Government—Gladstone's first ministry—agreed to set up a Royal Commission on Scientific Instruction and the Advancement of Science under the seventh Duke of Devonshire.

The choice of the Duke was an obvious one for Gladstone to make. There was a scientific tradition in the family, and the Duke was, himself, a distinguished product of the Mathematical Tripos at Cambridge. Since his undergraduate days he had devoted considerable attention to the application of science to agriculture and industry. Only the year before he had been elected the first president of the Iron and Steel Institute. He had also just agreed to provide backing for the establishment of the Cavendish laboratory at Cambridge. More important, perhaps, from Gladstone's point of view was the political support that the Duke had given not long before in the House of Lords to the Liberal Bill to disestablish the Irish Church. His support there had been valuable in defusing a potentially dangerous clash between the Lords and the Commons.

The Duke was ultimately assisted by eight other commission members, who were both acceptable to the Liberal party and also interested in scientific education, scientific research or science-based industry. The list included the Treasurer of the Royal Society, W. A. Miller (who died not long after the Commission started its meetings), and both of the Secretaries, Stokes and Sharpey, as well as the inevitable Huxley.

It was required as usual to appoint a Secretary for the Commission from the permanent Civil Service. What was needed was a man experienced in committee work and also in scientific research. If there is any surprise that Lockyer should have been summoned from the War Office to take up the post, it can only be because Government appointments do not always go to the obvious choice. However, with his two close friends Huxley and Sharpey as Commissioners, Lockyer would have been singularly unfortunate not to have been chosen, and chosen he was. He was utterly delighted. After his demotion at the War Office and the contrasting success of his scientific research, he had been looking seriously for a full-time scientific position for himself. But it was, of course, one of the major complaints of the scientific community that such positions were far too few.

The Savilian Chair of Astronomy at Oxford had fallen vacant some months before, and one of Lockyer's friends at Oxford, Appleton, had suggested that Lockyer should apply.

> The electors are a curious body of heterogeneous persons, all however resembling one another in their special knowledge of Astronomy:— Archbishop of Canterbury, Lord Chancellor, Chancellor of Oxford (Lord Salisbury), Bishop of London, Home Secretary, two chief justices, Chief Baron of Exchequer, Dean of Arches, Warden of New College . . . If you could induce these people to elect you, you would I am sure stir us all up in a subject which is practically dead among us. At all events it is worth thinking about. In case I am promoted to any of the above offices in the meanwhile you may calculate on my vote.[5]

But the competition was strong. Charles Pritchard, writing to Lockyer about Royal Astronomical Society business, noted: 'Todhunter went off [the Council] at *his own request.* A pretty matter for a man who aspires to the Savilian Professorship of *Astronomy* which he does, and may obtain.'[6] Todhunter is best remembered today as the editor of Euclid, whose work was used by every budding mathematician in the latter part of the nineteenth century. He was also interested, however, in the theory of the shape of the Earth and published a massive review of this subject. Pritchard was hardly unbiased in his comments for he, too, was a candidate for the Chair, and was in the end appointed. This was an odd appointment in a way for Pritchard was in his early sixties and had been a schoolmaster for most of his life. (The school he founded—Clapham Grammar School—catered for the sons of many of his friends amongst the scientists and liberal theologians. Some of these sons themselves became well-known scientists; one might instance two of Lockyer's friends —George Darwin and Alexander Herschel.) In fact, Pritchard survived to a ripe old age and managed to get astronomy moving once again at Oxford. Lockyer's ability to compete for academic positions was certainly hindered by his lack of a degree, and this was another reason why the appointment to the Devonshire Commission was a great relief to him.

The Commission examined its first witness in June 1870 (the sessions were held at Westminster), and for the next five years Lockyer was employed full-time on the work of the Commission. ('Full-time' was, of course, a relative term in those relaxed days. Apart from his research work, editorial work for Macmillan and various public lectures, Lockyer also participated in two eclipse expeditions during this period.)

The various reports of the Devonshire Commission are of major importance for our understanding of the way in which science developed in Britain during the vital middle years of the nineteenth century. Because of the long public debate beforehand, many of the leading scientists had had their thoughts turned towards what we would nowadays call science policy. As a result, there had been time both for individuals to draw up their own blueprints for the future of science, and for a general consensus on the main points at issue to be established. However, this is not the place to examine the investigations of the Commission in the detail they deserve. Instead, the overall pattern will be outlined, but detailed reference will be made only to the two parts of the Commission's work which had a direct and important bearing on Lockyer's future : the development of the South Kensington site and the idea of State aid for science.

The Commissioners first inquired into the Government-controlled institutions concerned with scientific and technical education in London, the Science and Art Department, the Royal College of Chemistry, the Royal School of Mines, etc. The Science and Art Department was continuing to expand its examination system. By 1870 the number of students under instruction had risen to nearly 30 000, although the subjects they were taking were sometimes motivated as much by the likelihood of passing the examination as by their possible future utility. The most popular were physical geography, machine construction and drawing, practical plane and solid geometry, and animal physiology, in that order. Although the increasing numbers were in one sense satisfactory, in another they were worrying, for the Department was becoming seriously perturbed by the amount of money which was being disbursed to successful teachers. To be more accurate, the Department was disturbed by the cries of anguish coming from the Treasury. A teacher working full-time (day and evening) might earn £250 per annum. Moreover, it had become easier to qualify as an approved teacher. The Science Teachers' Certificate had been abolished in 1867, and teachers now simply had to pass the ordinary examination with sufficiently high grades. As a result, people could, and did, pass the examination one year, and set up immediately as teachers for the next. To a certain extent this development had been offset by the provision of grants to enable a certain number of teachers to visit London during the long vacation each year for courses of lectures and practical instruction given by members of the Royal College of Chemistry and the Royal School of Mines, including Frankland and Huxley. Formerly, when there had been only a small number of science teachers, they had been brought up to the Science and Art Department itself for instruction; but, as Henry Cole explained to the Commission '. . . we found that although the numbers increased our premises did not increase, and that the stinks they made were such a horrible nuisance to the whole museum that we were obliged, and most unwillingly, to give it up.'[7]

The Science and Art Department had to rely largely on Army officers, especially Royal Engineers, to keep the examination system going. Some 60 to 70 were employed each year to tour the schools and check on the examination arrangements. The size of this band reflects the fact that the Department was mainly concerned with elementary education, an area in which Cole felt that their progress was satisfactory. He was less happy with the state of the higher education establishments which came into his orbit. The Royal School of Mines and the Royal College of Chemistry were turning away students through lack of space, and a lease connected with the former's building in Oxford Street was due to expire very shortly. Cole had for some time been trying to unite the two institutions on the South Kensington site together with the Royal School of Naval Architecture, the three to form a new Metropolitan College of Science. The Royal School of Naval Architecture was, in fact, already installed in South Kensington, though in temporary sheds which, according to the Superintendent of the School, 'were against the law in point of security, and . . . nothing but a Government department would dare to have things of that kind up in use.'[8] According to Cole, one advantage of combining teaching resources in this way would be that the specialists under training in mining and naval architecture would receive a general scientific instruction which at present they lacked. Cole also felt that a joint institution at South Kensington would fill another gap : it could act as a Training College providing fairly advanced tuition in science for intending school teachers.

Cole emphasized what was one of the constant beliefs of Victorian educationalists, that the essential task of scientific education was not to teach applications, which could be picked up later on the job, but to provide a background of general science, which would train the mind in the right kind of thought process. This was, indeed, a noticeable theme of most of the witnesses before the Devonshire Commission. Equally typical of a good many of the witnesses was Cole's assessment of the extent to which State aid should be given to Science, 'I think that the Government should not do what it finds its subjects doing quite as well without its aid, but I have yet to learn that the teaching, even of the three R's, might be accomplished without the aid of the State . . .'[9] Education was an important area where the traditional Victorian policy of *laissez faire* had evidently failed.

Cole's evidence, a good deal of which was probably prepared by Donnelly, has been quoted in some detail because many of the other witnesses were in general agreement with him. Huxley, for example, who gave evidence, as well as being a Commissioner, agreed that a Normal School of Science should be set up for training teachers, and that science teaching should not be slanted towards practical ends. But, unlike Cole, he by no means believed that elementary science teaching was in a reasonable condition.

> Unless I am greatly misinformed, the scientific instruction which is now current in the kingdom was the result, so to speak, of a battle between two official departments. I do not know whether I am rightly or wrongly informed, that it was in the teeth of the Educational Department that this scientific instruction was introduced; but, if such be the case, the fact accounts, I think, for the nocturnal and somewhat surreptitious, position which science at present occupies.[10]

One slightly dissident voice was that of A. W. Williamson, who did not think that there ought to be direct Government support of institutions for teaching science. Williamson was, however, a rather biased witness. He was Professor of Chemistry at University College London and, as he subsequently revealed, was losing students to the Royal College of Chemistry and the Royal School of Mines. The students believed that a Government position was more easily obtained by a candidate trained at a Government establishment. Williamson also commented extensively on the question of university salaries, a sore point for several of the witnesses. He pointed out that many university scientists in England had to make up their income by fees from consultancy work, whereas Professors of science in France and Germany could live like gentlemen on their teaching salaries.

Besides keeping the minutes at the meetings of the Commission, Lockyer had the additional task of collecting any extra evidence that was felt to be relevant to the points under consideration. For the first report, his major task under this heading was the circulation of a letter to the heads of colleges at Oxford and Cambridge enquiring what parliamentary grants for science they were receiving. There appears to have been no Government information available on this. The replies revealed that virtually no money was going from the State to either university for the support of science. As had been expected, it was also clearly demonstrated that Oxford was much behind Cambridge in the cultivation of science.

As Editor of *Nature*, Lockyer occupied an odd position in the Commission's investigations. On the one hand, it was his duty to gather information impartially for the Commission, acting simply as its instrument. On the other, he controlled the comment on matters under consideration by the Commission in what had rapidly become the most influential journal in the world of British science. In fact, Lockyer, himself, seems to have avoided, where possible, writing on topics currently under review by the Commission, but it was widely believed at the time that some of the anonymous editorials in *Nature*, which did comment on these topics, came from his pen. Moreover, the correspondence columns of *Nature* were the site from time to time of controversies concerning points under investigation by the Commission. For example, the correspondence in *Nature* over the future of the herbaria at the British Museum and Kew, which engendered the controversy between Hooker and Owen described previously, eventually formed the basis for one of the appendices to the Commission's first report.

The first report appeared in March 1871, and its recommendations followed very closely the lines which had been suggested by a majority of the witnesses. In particular, it was proposed that the Royal College of Chemistry and the Royal School of Mines should be transferred to South Kensington and provided with the appropriate laboratories and technical assistance. This suggestion was strongly opposed by Murchison at the Royal School of Mines together with four members of his staff who specialized in the more practical areas of mining. It was their contention that a merger of the sort proposed would erode the special function of the School, and turn it into an institution for general elementary science teaching. To some extent, this was precisely what the pure scientists at the School wanted. The tensions in the Royal School of Mines which had been there throughout the sixties erupted. At this crucial moment Murchison died, and the School was left without a Director. The faction opposed to the move now rallied round Percy, the Professor of Metallurgy, whilst those in favour of the move rallied round Huxley. The battle between these two and their followers extended throughout the seventies, but the crucial point was decided in 1872, when it was agreed that chemistry, physics and biology teaching should move to South Kensington, whilst the mining sciences should remain at Jermyn Street.

The second report of the Devonshire Commission came out in March 1872, not long after Lockyer's return from his second eclipse expedition, but it contained little of relevance to Lockyer's future career, being concerned with scientific and technical instruction in elementary schools and training colleges. One point of interest which emerged was that the introduction of a revised code of education in 1861, which had been intended to ensure that the teaching of the three R's was improved, had resulted in a rapid lowering of the standard of elementary science teaching.

The third report of the Commission, which appeared in August 1873, investigated the state of science at Oxford and Cambridge, but contained some material of more general interest. For example, it indicated what were then considered to be the possible careers open to trained scientists. There were only three : a man could either go into teaching, or he could specialize in medicine, or in engineering. The lack of employment was, indeed, acting as a bottleneck to the professionalization of science in Britain. To the contemporary view this appeared particularly clearly when comparison was made with the state of affairs in Germany. Yet the way ahead was just beginning to appear at this time. The 1870 Education Act had set up local school boards empowered to levy rates in aid of education. Most of the boards naturally wished to keep the rates as low as possible to avoid annoying the electors. They therefore looked with great interest at the money which could be obtained from the Science and Art Department for science teaching. As a result, there was from the seventies onwards a growing demand for science teachers, which stimulated, in turn, an increased, though still small, demand for higher education in science.

Another point covered in this report was the place which scientific research should take at a university. It was argued that the traditional picture of a university professor as basically a teacher with research as an optional extra should be abandoned. Instead, research should be seen as a primary duty of a university, as it was in Germany. Frankland explained to the Commission,

> In my opinion the cause of this slow progress of original research depends, in the first place, upon the want of suitable buildings for conducting the necessary experiments connected with research; secondly upon the want of funds to defray the expenses of those enquiries, these expenses sometimes being very considerable; but thirdly and chiefly, I believe that the cause lies in the entire non-recognition of original research by any of our Universities.[11]

The fourth report, in January 1874, dealt with science in museums. It was an admirable characteristic of the Victorians that they had a very high concept of the place of museums in education and research. Whereas we nowadays tend to equate such scientific research as occurs in museums with biology or geology, the Victorian concept was that a museum could be of value to research in any field of science and technology. This fourth report approved the idea that the natural history collections should be removed from the main British Museum and re-located in South Kensington; in other words, it supported the establishment of a British Museum (Natural History). It also recommended that a collection of physical and mechanical instruments should be established which should be joined with the exhibits already in the South Kensington Museum and in the Patent Museum to form a new comprehensive science museum.

Lockyer's expertise was called on to a major extent in this report, for he was given the task of assessing the current state of science in France for comparison with that in England. Lockyer's firsthand knowledge of France, and acquaintance with many of the leading French scientists, served him well here.

The fifth report, in August 1874, and the seventh report, in June 1875, contain little material of relevance to Lockyer. They were concerned with universities in Britain and Ireland other than Oxford and Cambridge. One point of interest from the fifth report deserves mention, however; it appears that members of the teaching staff at University College and King's College, London, felt that the Royal School of Mines and the Royal College of Chemistry had a higher prestige and so could attract higher quality staff.

The sixth report, also dated June 1875, was the one which involved the most work on Lockyer's part. It was concerned with the teaching of science in public and endowed schools, and as is noted at the beginning of the report, 'Mr Lockyer was appointed Assistant Commissioner with reference to this branch of the Inquiry, and by means of personal visits, and by the circulation of various forms of questions agreed on between him and the Commissioners, much information has been obtained.'[12] Lockyer travelled over much of England in pursuit of his enquiries, visiting almost all the major public schools. He made detailed studies at each place both of the science teaching available and of the laboratory space, if any. Thus, he spent some time at his home town, Rugby, talking with J. M. Wilson, about his work as science master at Rugby School where there was an advanced interest in science for an old public school. Lockyer and Wilson already knew each other, and had both been involved in the British Association investigation prior to the setting up of the Devonshire Commission. Moreover, Wilson, who was widely regarded as one of the leading experts on science teaching in schools, and is extensively quoted in this sixth report, was another of Macmillan's advisers. It was Wilson who introduced Lockyer to G. M. Seabroke, a Rugby solicitor who, as we have seen, collaborated with Lockyer in his solar work. For, besides the laboratories, Rugby also possessed the Temple Observatory (called after Frederick Temple who had been headmaster of Rugby shortly before) which Seabroke used for solar observations.

To those schools that he was unable to visit, Lockyer sent a questionnaire which he had drawn up. His major conclusion was that the time and space devoted to science in these schools was still extremely limited although two previous Royal Commissions had recommended that this side of their teaching should be extended.

'Among the 128 Endowed Schools from which we received returns, Science is taught in only 63, and of these only 13 have a laboratory, and only 18 Apparatus, often very scanty.'[13] Of the 128 schools which had been approached, 87 had provided Lockyer with information on their science curriculum. 'Of these 30 allot no regular time whatever to scientific study; 7 only one hour a week; 16 only two hours; while out of the whole number only 18 devote as much as four hours to it.'[14]

It was the eighth, and final, report of the Commission, dated June 1875, as were the sixth and seventh reports, which was to have the most influence on Lockyer's future. Whereas the previous reports had been mainly concerned with scientific education, this last report concerned itself solely with the relationship between the State and the advancement of scientific research.

The first question investigated was the extent and nature of the scientific work which was already being carried out by Government establishments. There proved to be five main groups : the various Surveys (Geological, Topographical, Hydrographical), the astronomical observatories (Greenwich, the Cape of Good Hope and Edinburgh), the botanic gardens (Kew, Edinburgh and Dublin), the meteorological observatories (mainly connected with the astronomical observatories, but including the Meteorological Office, which was financed like a public department though run privately), and finally, physical and chemical sections in such departments as the Board of Trade. Despite this varied scientific activity, the almost unanimous opinion of the witnesses was that the extent to which scientific and technical knowledge was available within the Civil Service was inadequate, and that the knowledge which was available was being improperly used. One expert witness from the Office of Works told the Commission, 'Our statesmen do not appreciate properly the value of scientific advice or scientific inquiry . . . they are very much fonder of experiments made upon a large scale with no defined system, than they are of experiments which have been brought out as the result of a carefully studied previous inquiry.'[15] The contemporary method of obtaining expert scientific or technical advice on occasions when it was evidently needed was to convene an *ad hoc* committee to consider that specific problem only. Several witnesses pointed out the deficiencies of this approach, for example that, 'The members of such committees must be selected more or less to fulfil certain political conditions . . .'.[16]

As an example of the inadequacy of Civil Service science, the experience of a previous Chief Constructor of the Navy may be quoted.

> In this country the earlier ironclads were made of a form involving very large and fine lines, in fact a form analogous to that of mercantile steamships, and the consequence was that although in the 'Minotaur' type of ship armour and backing equivalent only to that used in the first instance in the 'Warrior' was employed, yet we got a ship 400 feet long, costing nearly half a million sterling. The impropriety of that course impressed itself upon my mind, and I believe it was more for that reason than for any other that I ventured to propose to the Admiralty a great change in that respect and placed before them the design for the 'Bellerophon' as an example of a vessel which should be as fast as those long ships, and more effectually armoured, and much more handy, carrying at least as efficient an armament, and yet should cost about £100,000 less. That policy was sanctioned . . . and it was stated . . . that by adopting that modification of form at least a million sterling had been saved to the country in the course of a very short time. But I wish to impress upon the Commission . . . that that economy resulted from a mere tentative and limited application of a scientific principle, which has never been developed, and which the organisation of the Admiralty furnishes no means for developing.[17]

Apart from Government establishments, state aid to science was on an infinitesimal scale. The grant to the Royal Society had remained unchanged at £1000 ever since the middle of the century, and there was general agreement amongst the witnesses, in which the Commission concurred, that this amount needed to be substantially increased. It seems, however, that the sum could have grown had the Royal Society so wished, but some of the senior Fellows were against an increase. Indeed, the Society did not always distribute completely the £1000 already being received. Special Government grants had also been given at intervals for scientific expeditions, such as Lockyer's eclipse parties, but as the Commission pointed out (in a passage almost certainly drafted by Lockyer), 'These contributions are of great value, but they do not appear to be granted or refused on any sufficiently well defined principle.'[18]

One of the questions the Commission considered at length was whether some type of central organisation might be set up to help the Government decide on questions concerning science. Most witnesses agreed that the only satisfactory solution would be the appointment of a Minister of Science. As one of them observed, 'I conceive that the recommendation by Bentham in the last century of such a minister can hardly fail to be practically adopted before the close of the present century.'[19] A majority of the scientists examined also felt that the proposed minister should be advised by a Council of Science, selected from their own ranks. Strange was a strong advocate of such an arrangement. He instanced the confusions over the organisation of the 1870 eclipse expedition, which had so goaded Lockyer, as an example of the need for a Department of Science providing a means of communication between scientists and the Government. It is most noticeable, however, that the three politicians who were examined by the Commission—Lord Salisbury, Lord Derby and Sir Stafford Northcote—were all opposed to this proposal. Despite this political opposition, the Commission recommended in its final report that a ministry should be established to deal jointly with science and education, and that a science advisory council should be formed.

The Commission finally turned to the most important question of all: what assistance, if any, ought the State to give to science? In view of the political attitudes still prevalent in the 1870s, it was first necessary to establish that there was a case for State aid at all. Strange, when examined by the Commission, quoted from a speech that Gladstone had made a year or two before, to illustrate the sort of political attitudes which had to be overcome. Gladstone had warned his listeners that, 'It was in the growth of individual and local energy and in freedom from all artificial and extraneous interference that the secret of the greatness of this country was to be found. That danger of centralization which had been a formidable and fatal difficulty in other lands had not yet obtained serious dimensions among ourselves, but it had lifted its head.'[20] This may be compared with Lord Salisbury's answer to the Commission when he was asked whether he thought that the State might legitimately interfere in the giving of aid for the advancement of science. He replied: 'I certainly do. It is a very orthodox doctrine to hold, and one which could be supported if necessary by quotations out of Adam Smith, the essence of the doctrine being, that the State is perfectly justified in stimulating that kind of industry which will not find its reward from the preference of individuals, but which is useful to the community at large.'[21]

Although many of the leading scientists in the latter part of the nineteenth century were Liberals (of the handful of scientists who became Members of Parliament nearly all were Liberals), the leading Liberal during this period, Gladstone, was regarded with distaste by many of them. Conservative ministries proved, if anything, marginally more sympathetic to science than the Liberals. Not, of course, that either party was very positively inclined towards the advancement of science, although the scientists examined by the Commission strongly emphasized the political importance of scientific development, claiming that if Britain fell too far behind in scientific research it would have serious consequences for the economy of the country.

When the Commission turned to detailed suggestions for the distribution of state aid, it found a majority of the witnesses in favour of the State creating and supporting new research institutions, which should preferably be sited in, or near, London. Not everyone agreed on this point, however. Some felt that the money should rather be spent on the improvement of existing institutions. Strange, who naturally appeared as the leading witness on scientific institutions, was strongly opposed to this latter viewpoint. He suggested, instead, that there were nine new research institutions which were urgently needed, '(1) an observatory for physics of astronomy; (2) an observatory for terrestrial physics, namely, meteorology, magnetism, etc.; (3) a physical laboratory; (4) an extension of the Standards Office; (5) a metallurgical laboratory; (6) a chemical laboratory; (7) an extension of collections of natural history, and an able staff of naturalists; (8) a physiological laboratory; (9) a museum of machines, scientific instruments, etc.'[22]

Although there were differences in detail, Strange's main suggestions were supported by several other witnesses. Sir William Thomson, for example, when similarly asked what scientific institutions should be founded, replied: 'There should be five. One at present exists, namely the Royal Observatory at Greenwich. Another in my opinion is very much wanted, an observatory for astronomical physics, then again a physical laboratory, and a laboratory for chemical research, and a physiological laboratory are needed.'[23] It was generally conceived by their proponents that the laboratories should not only house full-time staff supported by the State, but that they should also provide facilities for private investigators who would otherwise be hampered by their lack of means. As Frankland pointed out,

> Men of this class [i.e. private investigators with no teaching commitments—he mentions several, including Lockyer, De la Rue and Huggins] are really peculiar to England, for I have never known any such instance in Germany or in France, of men altogether disconnected with teaching taking up research in the way it is done in England. I think that for such men the establishment of national institutions such as those which are recommended by Colonel Strange would be peculiarly useful. In fact, I have heard several of these gentlemen express strong opinions as to the great advantage it would be to them if they could go to some institution of that kind to conduct research.

The data collected by the Commission indicated clearly that natural history and geology received more attention from the State than the physical sciences. Professor H. J. S. Smith, who became a member of the Commission when Miller died, had spoken of this with some heat at the British Association meeting in 1873, pointing out particularly how little financial support physics received. It may be noted that Commission members in those days felt little restraint in expressing strong opinions on matters officially under consideration. Indeed, during the lifetime of the Commission Lockyer had himself joined a short-lived Association for the Organisation of Academical Study which was started in 1872 by his friend Appleton at Oxford. This was an avowed pressure group, one of whose aims was to have men supported full-time by the State to do research.

One of the requirements for furthering the physical sciences noted by many witnesses was the need for an observatory devoted to astronomical physics. This figured on the lists of both Strange and Thomson. Many witnesses, indeed, believed that more than one such observatory was needed; the others being scattered throughout the British Empire.

Strange had circulated a questionnaire concerning the establishment of an observatory of this type in Britain to four leading scientists—Balfour Stewart and Sir William Thomson in this country, Faye (who had become President of the French Académie des Sciences) and Hillgard (Secretary of the American National Academy of Sciences). The first question on his questionnaire points directly to one of the major reasons why astronomical physics, in the specific guise of solar physics, was thought to be so important at the time. He asked, 'Is the systematic study of the solar constitution likely to throw light on subjects of Terrestrial Physics, such as Meteorology and Magnetism?'[24] All four of his correspondents answered affirmatively. The economic importance of accurate weather predictions is too obvious to need much elaboration, except to note that the need then was less for the British Isles than for the Empire, both to cover the sea routes and to help agriculture, especially in India. The existing Meteorological Office had not progressed very far towards prediction in the 1870s. It was still mainly concerned with the accumulation of data, rather than with a detailed theoretical attempt to understand basic causes. One member of the Meteorological Committee, which controlled the Meteorological Office, appeared as a witness before the Devonshire Commission, and was asked whether any results of scientific interest had yet been obtained. He answered, 'I should say none at all.'[25]

Although there was general agreement over the need for an observatory devoted to physical astronomy, this did not extend to the question of who was to control it. Some witnesses, De la Rue and Sir William Thomson, for example, were quite clear that the new observatory should be separate from any establishment devoted to classical astronomy: meaning, essentially, that it should not be linked to Greenwich. Others were undecided, and one key witness, the Astronomer Royal, was utterly determined that any new national observatory should come under his own control. Airy tended to be pessimistic about innovations. He was noticeably cooler about State assistance to science than most of the other scientists interviewed. It seems reasonable to suppose that some witnesses stressed the need to separate the new observatory from Greenwich because it was generally doubted whether Airy would back the development with sufficient enthusiasm. The final recommendation of the Devonshire Commission was that an observatory for astronomical physics should be established which should be completely separate from all existing institutions, and preferably not situated in the London area. Similar observatories should be set up elsewhere, especially in India. One can fairly certainly see Lockyer's hand in these proposals, and also in the amount of space given to astronomical physics in the final report.

The final recommendations of the Devonshire Commission represented an overwhelming endorsement of the views that Strange and, therefore, Lockyer had been advocating. But the general trend of the Commission's thinking had been known long before its sittings ended. By the time the final report appeared, the concept of State aid for science had already led Lockyer into one of the liveliest battles of his career.

By the early 1870s, Lockyer had become highly unpopular with some Fellows of the Royal Astronomical Society, who felt he had far too high an opinion of himself and of his own work. It was only reasonable in view of the importance of Lockyer's researches, however, that his name should soon come up for consideration for the Society's Gold Medal, awarded each year for specified meritorious investigations into astronomy. So, at the end of 1870, Dunkin and De la Rue proposed that Lockyer should receive the medal for his work in solar physics. Various other names were suggested until, finally, Browning and Pritchard proposed that Lockyer and Frankland should be awarded the medal jointly for the investigations they had done together. The regulations governing the award of the Gold Medal were, and are, rather complex. Candidates were proposed at the November meeting of the Council, and one of the names was selected by a simple majority vote at the December meeting. Finally, the Council voted on the individual merits of the selected candidate at the January meeting, where he had to receive three-quarters of the votes cast in order to qualify for the medal. In this case, Lockyer's name was selected to go forward at the December meeting, but at the January meeting he failed to get the required three-quarters majority. As a result, no Gold Medal at all was awarded in 1871.

Although this was not a unique event in the annals of the Royal Astronomical Society, it was sufficiently unusual to cause some comment. The President of the Society, William Lassell, a well-known amateur astronomer, tried to smooth over the situation, by saying that Lockyer's name had been coupled with that of Frankland, and the bye-laws of the Society did not permit the medal to be awarded jointly to two investigators at once. However, as Lockyer must have been only too well aware, despite this bye-law, the medal had already been awarded jointly once, to Huggins and Miller in 1867 for their astrophysical researches. To ensure that the same argument would not be used again, a new bye-law was passed during the next summer whereby the Society was permitted to award the medal to two or more people at once so long as they had been working on the same piece of research. At the subsequent November meeting of the Council of the Royal Astronomical Society, De la Rue and Browning again proposed that a joint award should be made to Lockyer and Frankland. It appears that Frankland's name was coupled with Lockyer's because as a senior and respected scientist it was thought by the proposers that Lockyer's opponents might hesitate before voting against the combination, whereas they would have no hesitation about voting against Lockyer alone. On this occasion, the name of a leading Italian astronomer, Schiaparelli, was chosen to go forward, and in due course he received the medal.

Early in 1872 Huggins, who had been one of the Secretaries of the Society for some years, retired, and R. A. Proctor was chosen to succeed him. Proctor, who was much the same age as Lockyer, had read Mathematics at Cambridge, and had then taken up astronomy. Initially, he was a typical fairly well-to-do amateur, but in the mid-sixties he lost a good deal of money in a bank failure, and subsequently turned to writing to make his living. Besides turning out a number of popular science books, Proctor was a prolific contributor of papers to the Royal Astronomical Society journals in the early seventies. Conflict between Proctor and Lockyer was, perhaps, inevitable. Apart from the similarity of their circumstances and their mutual inclination towards pugnacity, both were seeking to compensate for an uncertain financial background by achieving public recognition of their scientific eminence. Their first collision occurred at the beginning of the 1870s over the question of the nature of the solar corona. Lockyer thought that it was at least partly caused by the scattering of light in the Earth's atmosphere; Proctor believed that it was a genuine solar appendage. (It is an interesting fact that the two main scientific points debated by Lockyer and Proctor during the seventies—the nature of the corona, and the possibility of detecting a relationship between solar activity and terrestrial meteorology—show Proctor in what we today would regard as the more tenable position, although, as a scientist, he was undoubtedly much inferior to Lockyer.)

In April 1872, Strange turned the attention of the Society to the question of State assistance for the physics of astronomy by reading a paper before it on, 'The Insufficiency of Existing National Observatories'. Strange argued that he was following up a suggestion by Airy, who had proposed a few months before that a special observatory should be set up for the observation of Jupiter's satellites. It seemed evident to his contemporaries that this was simply a politeness to Airy, since Strange's essential purpose was divined to be the establishment of an astrophysical observatory with, so it was generally expected, Lockyer as its head. At the next meeting of the Royal Astronomical Society Council in May, Strange proposed that the President of the Society, now Arthur Cayley (mainly remembered today as a pure mathematician, but he also worked on some of the problems of lunar and planetary theory), should be authorised to bring to the attention of the Devonshire Commission the need for more research in astronomical physics. This was debated at the May meeting, again at the June meeting, then later in the month there was another special meeting on the topic. By this time the Council had become polarized into two main groups: those in favour of a solar physics observatory in England separate from Greenwich (and, implicitly, to be headed by Lockyer), and those who favoured a more general approach to astronomical physics (meaning essentially the mixture of solar and stellar work that Huggins was doing) to be under the general control of Greenwich. As a result, the special meeting of the Council at the end of June had the following motions before it.

Proposed by Dr. Huggins, Seconded by Mr. Burr:—
1. That the President be authorised, on behalf of the Council and Fellows of the Royal Astronomical Society, to bring before the Royal Commissioners on Scientific Instruction and on the Advancement of Science now sitting, the importance of further aid being afforded to the cultivation of the Physics of Astronomy.
2. They think such aid would be most effectually given by increased assistance, where needed, to existing Public Observatories, in the direction recommended by the heads of those Observatories, especially that at the Cape of Good Hope, and by the establishment of a new Observatory on the Highlands of India, or in some other part of the British dominions where the climate is favourable for the use of large Instruments.
3. The Council do not recommend the establishment of an independent Government Observatory for the cultivation of astronomical Physics in England, especially as they have been informed that the Board of Visitors of the Royal Observatory at Greenwich, at their recent meeting, recommended the taking of Photographic and Spectroscopic records of the Sun at that Observatory.

Proposed in amendment by Colonel Strange, Seconded by Mr De la Rue, that the following be substituted for the second clause:—
The following seems to the Council to be the provision now requisite:—
1. An Observatory, with a Laboratory and Workshop of moderate extent attached to it, to be established in England for the above researches.
2. A certain number of Branch Observatories, to be established in carefully-selected positions, in British territory, in communication with the Central Observatory in England, for the purpose of, First, giving to Photographic Solar Registry that certainty which experience has already proved to be necessary; and, Secondly, to investigate the effect of the Earth's Atmosphere on Physico-Astronomical Researches in different geographical regions, and at different altitudes. For these purposes India and the Colonies offer peculiar advantages.[26]

During the complex discussions which followed on these proposals, Strange conceded that the proposed main observatory in England might with advantage investigate other fields of astronomical physics besides the Sun. Despite this measure of agreement, however, his amendment was defeated, and this must be regarded as an adverse vote of the Council against Lockyer. A very clear idea of the factors involved in the debate can be had from a letter which Proctor wrote to Airy just a few days before the special meeting.

> It is pretty generally understood that whatever Colonel Strange's own views may be, the real object of the resolution is to set up a certain junior member of our council as a sort of 'Astronomer Royal for the Physics of Astronomy'. This is touched on (rather more boldly than was expected) in the words 'that no individual has as yet distinguished himself equally in these researches and in the more exact department of astronomy'. But the gentleman himself whose interests are in question openly admits that the resolution is to have the effect mentioned; and as openly complains that hitherto the Royal Astronomical Society, the Royal Society, the British Association, and the Astronomer Royal have shown a want of zeal in the cause of the physics of astronomy.
>
> You will find that as understood by the supporters of the resolution, the 'physics of astronomy' signifies, in the main, any branch of astronomy associated with what Dr. De la Rue *has* done and Mr. Lockyer *might* do. The study of stars and nebulae, for instance, is only set as a subsidiary matter. Solar Physics and Solar photography become the leading features,—and *they* are associated with the preposterous idea (worthy of Zadkiel and the editor of Moore's Almanac) that the phenomena of weather may one day be predicted from the records of solar spots, faculae, and prominences!
>
> On the other hand, Dr. Huggins' idea is that the physics of astronomy includes in an equal degree every branch of observation by which our knowledge of the nature of the celestial bodies, individually and collectively may be increased. . . .
>
> Should this matter come before the commission in the sense proposed by Colonel Strange, our action will be interpreted (i) before the commission (ii) before the scientific public by Mr. Lockyer, Secretary of the Commission though not on it, and editor of Nature. In what sense it will be interpreted is pretty generally known; for he is not silent about these matters. . . .
>
> Mr Lockyer never appears at our Council meetings unless such matters [concerning the financing of science] are under discussion. You may remember how regularly he attended when eclipse matters were in question. Hence one can interpret his apparition last Friday.[27]

If the Royal Astronomical Society had shown itself conservative on the question of solar physics, the British Association, which Lockyer had always preferred, proved a slightly more receptive forum during this same period. The British Association had formed a committee to consider the provision of science lectures and education in science. The members included Roscoe, Huxley, Sir William Thomson and Balfour Stewart, as well as Lockyer himself. The topics discussed were actually rather wider in scope than this brief might suggest, and one of the sub-committees set up concerned itself with the organization of meteorology in Britain. Balfour Stewart and Lockyer both strongly urged on it the case for supporting solar physics. Lockyer ended his memorandum with the words, 'I see no way of having the work done by private effort. I have tried hard to continue the work; and in the fact that it was begun in this country by myself I had the strongest inducement to carry it on; but nothing short of one's whole time will suffice for such inquiries.'[28]

The dissensions amongst the Council members of the Royal Astronomical Society had not hitherto spread throughout the general membership of the Society, but the stresses were now too large to be easily glossed over. In November 1872, matters came to a head when De la Rue, Strange and Lockyer all resigned from the Council. Airy immediately wrote to Lockyer asking him to reconsider.[29] Lockyer, in his reply, proceeded to outline the reasons for his resignation.

> Permit me then to tell you at once in confidence that my primary reason for quitting the Council is the offensive manner in which Mr. Proctor is conducting himself towards me. Week after week in more or less obscure journals which as Editor of Nature I must see I find myself attacked by one who takes good care to advertise himself as 'Honorary Secretary of the Royal Astronomical Society'. Now to Mr. Proctor as an individual I do not care to reply. To deal with him as the Honorary Secretary of an honourable Society would cause a scandal and on these grounds I have determined not to reply to him.
>
> But to one whose every spare minute, as mine is, is given to research it is almost maddening to think of these things. . . . I certainly should never have left the Council because my scientific work is unappreciated I might add underrated and misrepresented continually by Mr. Proctor or because I differed with the majority on such a question for instance as that of the physical Observatory in which question my name was mentioned in a most discreditable way to those who so used it.[30]

Just at the time that Lockyer was withdrawing from the Council, his name came before it again as he was proposed once more for the Society's Gold Medal, this time in conjunction with Janssen and Respighi. The proposer was his old friend Pritchard. More interestingly, the seconder was Huggins. Round about this time there was evidently a certain change in the balance of interests on the Council. Huggins, in particular, though perhaps he did not regard Lockyer with any greater joy as a person, had come to be distinctly suspicious of Proctor's intrigues. Lockyer received a letter from Balfour Stewart saying, 'I have had a note from Huggins who has evidently had enough of Proctor.'[31] During 1872, Huggins and Lockyer had, in fact, come into closer contact again, as they had both been appointed to a British Association committee to investigate problems relating to tables of the wavelengths of spectral lines.

Pritchard's nomination was not the only one before the Council. Another name put forward, by Denison, one of Lockyer's opponents on the Council, was that of Proctor. It was subsequently alleged by Strange that Proctor had solicited the nomination. Whether or not this was true, it is evident that Proctor was very keen to get the medal, and was worried that he would fail to obtain the final majority required. He feared that he would be selected on the first vote and then rejected on the second, as Lockyer had been before. He therefore wrote to Airy suggesting that some form of transferable vote should be introduced which would ease the voting requirements for the medal.[32]

Now Proctor had antagonized the Astronomer Royal a few years before by attacking him on the subject of the transits of Venus. This had been smoothed over at the time, but early in 1873 Proctor returned to the question again. Venus transits the Sun—that is crosses its face as seen from the Earth—approximately twice a century. Early in the eighteenth century, the English astronomer, Halley, had suggested that such transits could be used to determine the distance of the Earth from the Sun. The necessary observations had to be made from points far apart on the Earth's surface and, with the still primitive facilities available for travel in the mid-eighteenth century, this required a major effort of scientific organisation. Unfortunately, the results, when they were finally gathered, proved to be rather disappointing. The first transit of Venus during the nineteenth century was due to take place in 1874, and in the hope of improving measurements of the Sun–Earth distance, a vital parameter in classical astronomy, Airy began to prepare for the dispatch of expeditions to different parts of the globe. Proctor believed that Airy had not chosen the best stations for making the observations, and early in 1873 he published an article in the *Spectator* saying so. This was followed soon after by a letter to *The Times* from his friend Denison saying the same thing. The quarrel between Proctor and Airy over the transit expeditions went on with increasing vigour throughout 1873. Thus Airy, though by no means favouring Lockyer's ideas on the founding of an astrophysical observatory. was also by no means on friendly terms with Proctor. It is hardly surprising that Council, in the event, made no change to the voting procedure.

Strange now took the battle into the enemies' camp. At the beginning of February, shortly before the Annual General Meeting of the Royal Astronomical Society, he sent round a circular to all Fellows, drawing their attention to the resignations of Lockyer, De la Rue and himself and saying that this had been a result of the present composition of the Council. He suggested, therefore, that the present Council should not be voted back into power, but that other names—he suggested a large number—should be substituted. Strange sent a copy of this circular to Airy, who commented on its favourably, but suggested some possible amendments. As a result, Strange sent round a second circular proposing a possible alternative balloting list to that suggested by the Council.

There was just time before the meeting for the President and the two Secretaries of the Society to reply to Strange's first circular. They pointed out that some of the people named there were not available to serve on the Council, and added that, because of the resignations, further major changes in the composition of the Council would be inadvisable.

Besides the war of circulars now developing, the battle raged in the periodicals. If Lockyer controlled *Nature*, and was on friendly terms with the editors of some of the higher-class periodicals, Proctor was a regular contributor to several of the more popular journals which included scientific features. He and an anonymous 'F.R.A.S.' (who was rumoured to be another vociferous member of the anti-Lockyer group at the Royal Astronomical Society, Captain Noble) wrote especially frequently for the *English Mechanic.* This was a 'technical' journal, as compared with *Nature* which was essentially a 'scientific' journal. Nevertheless, there was evidently some competitive feeling between the two, and this may have helped exacerbate the battle of words between Proctor and Lockyer. During 1873, several anti-Lockyer letters appeared in the *English Mechanic*: one (signed 'A Little Bird') appeared just before the Annual General Meeting.

> News for the Astronomical Society! Mr Lockyer is renewing his interest in it. He has procured a list of the fellows, in order (as his messenger stated) to send a circular to a large number of them. Mr Lockyer has not so bestirred himself since, by similar scheming, he helped to oust Hodgson from an honourable though laborious office. Mr Lockyer has not spoken at any meeting since 1869, when he spoke in his own laudation. He has only contributed one page to the Society's proceedings during the last five years, and that was a page of self-glorification. He has undertaken even to 'write the Society down'; and for three years its very name has been banished from the journal he edits. Yet now he has been moved to action—in the interests of the Society, of course.[33]

The meeting, itself, was one of the stormiest in the history of the Society. 'Colonel Strange was repeatedly challenged to substantiate his statements of the incompetence of the persons objected to, and their combination for party purposes, but contented himself by stating that it was merely his own personal opinion, and that he had every respect for the individuals in question. A resolution expressing regret for Colonel Strange's action having been carried, the ballot took place and occupied nearly two hours before the result was ascertained.'[34]

Only one of Strange's alternative list was elected, but this in no way silenced the pamphlet war. In March, Proctor and his friends on the Council, Browning, Burr, Denison, and Noble, sent round a new leaflet in which they alleged that Strange had been abetted in his attack on them by Lockyer. They went on to add that Strange would neither withdraw nor substantiate his charges that they were incompetent for office, and that he had suppressed the fact that the sole difference between the three seceders and the Council had been over the question of setting up a new Government observatory for cultivating the physics of astronomy. They regretted the resignation of De la Rue, but felt that the loss of Lockyer was no great burden for the Society.

Towards the end of March, Strange replied. He admitted, in effect, that Lockyer had helped him draw up his first circular, but asserted that the attack had not been planned a long time beforehand. As one of the scrutineers at the ballot, he proceeded to reveal precisely how the voting had gone. Only 109 Fellows had voted, and 9 of these handed in invalid papers. Although only one of his opposition list was elected, there was, he claimed, a majority of 2 to 1 against the Council list as a whole. Strange went on to deride Proctor personally for his lack of success in gaining the Gold Medal, and alleged it was because some of the most distinguished living astronomers, who were members of the Council, had not considered it a proper choice.

All of the group except Proctor, who drafted a separate answer, replied immediately, saying, amongst other things, that 'the most distinguished living astronomers' referred to in Strange's circular simply meant Airy. They went on,

> '. . . as you have all doubtless read or will read the controversy on the Transit of Venus between him [Airy] and Mr. Proctor, in *The Times, Spectator* and the *Monthly Notices* [of the Royal Astronomical Society], and remember that Colonel Strange avowed that he had consulted with Sir G. Airy with reference to his second circular, we leave the whole of this matter to your own reflections.[35]

It appears, in fact, that Airy was by no means the only person on the Council who opposed the award of the medal to Proctor. J. C. Adams also objected violently, as did Pritchard; Huggins, too, had voted against Proctor.

Proctor issued an answer to Strange on his own account. In this he stated bluntly, 'I venture to express my conviction that whereas it is earnestly to be desired in the interests of the Society and of science that Mr De la Rue should resume his seat at our Council board, it would be a circumstance to be regretted if either Mr Lockyer or Colonel Strange returned.'[36] He then asserted that his nomination for the Gold Medal had been completely unsolicited, whereas, 'His [Lockyer's] nomination in 1870 is said to have been solicited, and Colonel Strange's suspicions about myself confirm the rumour.'

The exchange of circulars now died down, but simultaneously Proctor's battle with the Astronomer Royal became more intense. The climax came when Proctor published a paper in a supplementary number of the *Monthly Notices*, of which he was then the editor, poking fun at the

Admiralty in connection with the forthcoming transits of Venus. There was a storm over this apparent abuse of editorial power, and Proctor was forced to resign as Secretary of the Royal Astronomical Society before the end of 1873. Pritchard wrote to Lockyer immediately, 'I hope you have had a sight of the Resolution inserted on the minutes of the R.A.S. regarding the abominable conduct of our late friend the Editor M.N. [*Monthly Notices*] . . . He must be at least a monomaniac—but his rabidness in future will be less mischievous.'[37]

Although Lockyer and Proctor were now both separated from the workings of the Council of the Royal Astronomical Society their dispute continued to simmer. It was discovered in 1875 that certain Royal Astronomical Society instruments, which had been loaned to Lockyer for the eclipse expedition of 1871, had been given by him to an astronomer in Australia without the Society's permission. Although it also subsequently transpired that an equivalent amount of money had been made available in exchange, Proctor's group seized the opportunity to mount a new attack on Lockyer.

Lockyer, meanwhile, was in serious difficulties with his career. When the Devonshire Commission issued its final report in 1875, Lockyer's secondment from the War Office should have come to an end. Because of the gross over-staffing at the War Office, the Treasury early in 1871 had decided to offer clerks there the opportunity of resigning on favourable terms. They would be given a pension based on their number of years of service plus seven. Lockyer had immediately handed in his resignation. He had, however, been calculating the amount of pension he would receive on the basis of the salary he had been receiving as Editor of Army Regulations, but, as we have seen, this post was taken from him. The Treasury therefore insisted that he could only be paid on his basic salary as a clerk. Lockyer decided that the resulting pension was too small, even when taken in conjunction with his literary work, to support himself and his family, and he therefore withdrew his resignation.

He searched again with increasing desperation for some post in astronomical research. As the evidence collected by the Devonshire Commission emphasized, such positions in the early 1870s were few and far between in this country, especially for a man with no academic qualifications. In the United States, on the other hand, astronomy, after a slow start in the first half of the century, was beginning to build up more rapidly, and posts were appearing there with slightly greater frequency. On two occasions in the early 1870s Lockyer might have become head of a new American observatory. Joseph Henry who, as Secretary of the Smithsonian Institution, held an influential position in American science, was introduced to Lockyer in London in the early 1870s, and seems to have tried hard to persuade him to settle in the States. The first time was when a proposal arose for the creation of the Leander McCormick Observatory at the University of Virginia.

Mr McCormick wants to get a first class astronomer, to take charge of his first class telescope and is anxious we understand to secure the services of such a man *at once* in order to superintend the erection and the equipment of the concern. Some of us have managed to get the name of Norman Lockyer suggested to him, learning from Professor Henry of the Smithsonian Institution at Washington that perhaps Mr Lockyer might be willing to come. I think from a letter of Mr McCormick's I have just seen, that he will through Professor Henry probably soon make some sort of proposition on the subject to Mr Lockyer.

Like most men of plenty of money and no scientific knowledge, Mr McCormick has a good deal of reverence for a great scientific name, especially from the other side of the ocean: hence he would probably listen with much attention to any suggestion of Lockyer.[38]

This suggestion hung fire for a long time, then Henry wrote to Lockyer in 1874 to tell him it had finally fallen through, but to say, also, that there might be the possibility of a post at a new observatory to be founded in California.

In 1871 I visited San Francisco, and was introduced to Mr Lick, who was the owner of the public house at which I lodged. I had considerable conversation with him on the subject of the advance of science, and, perhaps, had some influence on him in the way of the disposition of his money. I have since at the request of the President of the Academy written to express my high appreciation of what he has done, and since the receipt of your letter have written to urge him to establish an Astro-physical observatory; not to wait till he can have the largest telescope in the world constructed, but to appoint a director or superintendent, a man of original science, and liberally supply him with the means of procuring a full set of instruments necessary for the observation, not only of solar phenomena, but also those of magnetism, electricity, meteorology, earthquakes, etc.; not to attempt to put up a large building, but merely to erect temporary ones of little expense under the direction of this superintendent.

I have also taken the liberty of suggesting that you would be the proper person as superintendent.[39]

But this position, too, went astray (although, of course, the building of the Lick Observatory went ahead, and by the end of the century it had established itself among the leading observatories in the world). Lockyer was left desperately appealing to his friends for advice. The response from Roscoe, at least, was radical.

I have been thinking a good deal about what you told me yesterday. It seems to me monstrous that you should still have to hang on to the War Office and I cannot help thinking that you had better resign and secure your pension. If people only knew that you have cut yourself loose from the drudgery of the War Office they will exert themselves to insure you a living somehow—and you cannot be worse off than attached to a place where all your time and talents are given up to routine work of the most uninteresting kind.[40]

Lockyer felt that with his family responsibilities he could not cut adrift in this way. Moreover, the Devonshire Commission had recommended the establishment of a separate astrophysical observatory, and, despite the bitter attacks Lockyer was undergoing, he was still the obvious person to head such a project. His hopes in this direction were certainly buttressed by one of the final resolutions passed by the Commission. This proposed that, on the termination of the Commission, some arrangement should be made to secure his services for the promotion of science in Britain. The Duke of Devonshire brought this resolution to the attention of the Prime Minister, Disraeli, and of the Lord President of the Council. Lockyer therefore wanted to mark time until the final decision concerning a new observatory had been made.

By the mid-seventies, he had many friends in the various departments established at South Kensington. The Royal College of Chemistry had migrated there in 1873, and with Frankland's approval, which he subsequently regretted when space became more difficult to find, the Science and Art Department offered Lockyer a room for laboratory work in the same building. The department also provided him with some space in the gardens, which were then under the day-to-day control of the Royal Horticultural Society, for a temporary observatory. When Lockyer's appointment to the Devonshire Commission lapsed, his friends in the Science and Art Department intervened again on his behalf to prevent his return to the War Office. Although the main recommendations of the Devonshire Commission were still under consideration, it had been decided to hold an exhibition of important laboratory and teaching apparatus in the South Kensington Museum as soon as possible. The Science and Art Department therefore wrote to the Treasury requesting that Lockyer should be temporarily attached to the department to help organize the loan exhibition (so called because nearly all the apparatus was to be borrowed) and also to assist in the inspection of certain schools. As a part of the desired reorganisation of the War Office, the Treasury had for some time been prepared to allow the transfer of clerks from the War Office to other departments. So, although the Treasury generally disapproved of the Science and Art Department's interest in museums, they acceded to this particular request. In the following year, after further negotiation, Disraeli made the transfer permanent.

The Science and Art Department stressed in their correspondence with the Treasury that they attached considerable importance to Lockyer's scientific researches, and wished them to continue. As a result, Lockyer had now for the first time a job which officially regarded him as a scientific investigator. To begin with, however, his work on the loan collection kept him too busy to take advantage of this. It had been intended that the collection should be exhibited to the public in 1875. But as its scope extended, and it was made more comprehensive, the opening date was gradually pushed back to the middle of 1876. The decision to make the collection international in extent sent Lockyer off to the Continent in pursuit of further exhibits. He visited France, Holland, Belgium, Germany, Austria, Switzerland and Italy, but his main task was to select and borrow British instruments.

The exhibition was a great success, and it was generally agreed that much of this was due to Lockyer's exertions. When Queen Victoria opened the exhibition, she especially requested that Lockyer be introduced to her. Lockyer's work can, indeed, be seen as one of the factors which led on to the development of the present Science Museum at South Kensington. The month after the loan collection was opened, a massive memorandum, which Lockyer supported, was presented to the Lord President of the Council calling for the establishment of a Museum of Pure and Applied Science to be based upon such parts of the loan collection as would remain after the exhibition. Ultimately, these suggestions were acted upon in part, and for a time, the idea of teaching museums in applied science became all the rage. Thus James Stuart, the Professor of Engineering* at Cambridge, decided in 1876 to centre his teaching there round a collection of models.

Lockyer's success was won at a price, for he was brought by the overwork and worries involved to one of his worst break-downs. By Spring 1877, he was so down-hearted that he began to talk of giving up science altogether, but his friends rallied round and helped. Lauder Brunton, one of the best physicians in London, as well as a close personal friend, advised him to take an extended holiday in the South of France. The Science and Art Department arranged for him to have a long leave, and Macmillan supplied him with some ready cash. Lockyer rapidly recovered from his depression, and by the end of the year was not only back in London, but had thrown himself again into the scientific fray.

The recommendations of the Devonshire Commission were still hanging fire. Indeed, when Lockyer looked back on the Commission's work towards the end of the century, one of his predominating impressions was that few of its recommendations had been implemented. The key question for Lockyer remained the possibility of the State founding a new observatory. Although his present rather vaguely defined position gave him time for his own research, he felt that the importance of astronomical physics undoubtedly warranted the creation of a permanent, State-aided establishment. In 1876, a large deputation from the British Association had

* His proper title was Professor of Mechanism and Applied Mechanics.

urged the Committe of Council on Education to carry out the relevant recommendations of the Devonshire Commission, including the creation of a physical observatory. The President of the Committee in his comments to the deputation had said that Lockyer's transfer to South Kensington, and the permission he had been given to make observations as part of his paid duties, represented a step towards such an observatory. This seemed a hopeful sign both for Lockyer personally and for the advancement of science in general, but it was recognised that the pressure on the Government had to be maintained. In the following year therefore several leading scientists (including J. C. Adams, Joule, Maxwell, Roscoe, Stewart and Sir William Thomson) submitted a memorandum to the Committee of Council in which the importance to meteorology of a solar physics observatory was stressed, and the need for an early decision pleaded. The Council passed this memorandum on to Stokes, who was not a signatory, and Balfour Stewart, and asked them to decide whether South Kensington might reasonably become a centre for solar physical research. General Strachey was also consulted to find out the attitude of the Royal Engineers.

The extensive use of Royal Engineers officers at the Science and Art Department has been mentioned. There was, too, a detachment of other ranks from the same corps stationed at South Kensington. Their primary duty was fire prevention, but since this left them with a fair amount of time on their hands, they were also frequently impressed into other jobs. Lockyer began to use them soon after his arrival at South Kensington as technical assistants, for example, in astronomical photography, a position for which the skills imparted by their Army training made them well fitted.

Lockyer had already expressed himself as strongly in favour of having regular observations of the Sun made from somewhere in India. Stokes and Balfour Stewart's report clearly supported him in this. They suggested that astronomical physics should be divided into two areas. The first involved experimental research and the perfecting of observational methods; the second consisted rather of the continuous recording of solar phenomena. The former required an observatory within easy reach of instrument makers and good laboratory facilities; the latter demanded mainly a clear atmosphere. As Stokes and Stewart observed, 'It is needless to say that the climate of the United Kingdom is not one which a person having all the fairly civilized parts of the Earth to choose from would naturally select for clearness of atmosphere.'[41] They therefore emphasized that continuous solar observations were best made from somewhere other than Britain, and decided that the most suitable site would probably be in North India. They noted that already, 'arrangements have been made for sending out to India a highly intelligent sapper of the name of Meins, who has been trained by Mr Lockyer, and will be employed in taking photographs of the Sun.'[42] Lockyer had obtained this concession from Lord Salisbury, who was Secretary for India in Disraeli's second administration. Salisbury, as we have seen, was sympathetic to the claims of science, and had accepted the case for solar photography on a continuing basis in India.

If India was to be chosen for continuous solar observation, Stokes and Stewart were quite happy to consider South Kensington as a centre for experimental work in solar physics.

> From enquiries we have made we have reason to believe that a great deal of additional valuable work could be done at South Kensington, and the requirements of the memorialists met to a considerable extent, at a very moderate outlay. The chief wants appear to be an additional temporary assistant to Mr Lockyer at a salary say of £150, an additional sapper whose services might be available when not on special duty, an allowance for chemicals and for photographic reductions, making together £250, and some additional instruments.[43]

Despite the relatively modest nature of these proposals, the months passed by and no further move was made except for the dispatch of Sapper Meins to India, where, under the direction of the Surveyor-General, he began to take daily photographs of the Sun. These were subsequently dispatched back to South Kensington for examination, but no financial assistance was given to Lockyer to aid in their reduction.

Meanwhile, opposition to the foundation of a new observatory, and in particular, to Lockyer being put in charge, continued to be expressed at the Royal Astronomical Society. Lockyer had recently lost one of his strongest advocates, for Strange had died unexpectedly in 1876. However, in the same year Huggins was elected President of the Society, and proceeded to support Lockyer's nomination yet again for the Gold Medal. Once more Lockyer was selected at the December meeting of Council, and once more he failed to secure the necessary majority at the January meeting. The opposition to Lockyer in the Royal Astronomical Society was, in fact, so long lasting that although he lived on until the First World War, he never received the medal. He was virtually the only outstanding British astronomer of his generation not to be so honoured.

After Lockyer joined the Science and Art Department, some of the dislike he had engendered rubbed off on to his new employer. One of the Royal Engineers employed at South Kensington was Captain Abney. He, too, was a well-known astronomer, and a pioneer of infra-red photography. When, in 1877, the two Secretaries of the Royal Astronomical Society, Dunkin and Ranyard, who had replaced Proctor, were due to retire, Abney was proposed to fill one of the vacancies. The anti-Lockyer group in the Royal Astronomical Society immediately put up two opposition candidates, one of them being Ranyard again. Since assisting Lockyer in the eclipse organisation at the beginning of the seventies, Ranyard had joined the anti-Lockyer camp. In the event, Ranyard was elected Secretary for another term instead of Abney. Moreover, by an anomaly in the voting system, Abney even failed to gain a seat as an ordinary member of the Council, being defeated by another anti-Lockyer candidate, Captain Noble, even though he had gained more votes overall. As a result, there was another long, inconclusive and contentious debate, lasting into 1878, on the need to amend the rules governing the voting for Council.

During 1878 an important move forward was made. The Duke of Devonshire had been regarding with some displeasure the lack of action on the recommendations of his Commission. He now appealed again to the Committee of Council on Education, asking them specifically to provide funds for the development of astronomical physics. A response was evoked at last. A Solar Physics Committee was established and charged with the duties of trying out new methods of observation, of determining what solar researches were under way in other countries, and of collecting and reducing observations of the Sun, especially those being made by Meins in India. The Committee consisted of the three advisers of the Committee of Council on Education, Stokes, Balfour Stewart, and Strachey, also three representatives of the Science and Art Department, one of whom was Lockyer, and one representative from Greenwich.

One of the difficulties in the way of State support to solar physics was the interest in the subject at Greenwich. Since the early 1870s, the Royal Observatory had taken over the task of daily photography of the Sun's surface from the Kew Observatory. De la Rue who had been in charge of this work at Kew had given up active observing in 1873, when he presented his instruments to Pritchard at Oxford. Airy was therefore hardly happy at the thought of another State-aided institution doing similar work.

But, initially, the amount of overlap seemed to be fairly small. The Solar Physics Committee was established on a temporary basis, and was provided with an annual grant of only £500, of which £300 was for paying the fees and travelling expenses of Committee members. Moreover, although Lockyer was recognised as in charge of solar physics work at South Kensington, he was simply *primus inter pares*. Solar physics research carried out anywhere by any member of the Committee was sent to the Royal Society for publication as Committee work.

Because of its voluntary, co-operative structure, the establishment of the Solar Physics Committee did not lead to any great immediate change in Lockyer's circumstances. It did, however, lead to a new outburst of campaigning at the Royal Astronomical Society. In 1879, Huggins, strongly supported by Airy, was chosen to be the next Gold Medallist, but, like Lockyer before him, was voted down at the final selection meeting. This need not be entirely explained as an indirect thrust at Lockyer, but it certainly indicates the continued existence of a clique antagonistic to him in the Society. Huggins' rejection as medallist proved to be, indeed, simply a curtain-raiser to a grand debate on the question of State aid for an astrophysical observatory, which rocked the Royal Astronomical Society during the session 1880–1. This was sparked off by a preliminary report which the Solar Physics Committee published in 1880 on the initial phases of its work. It was explained that the members of the Committee saw the current state of affairs as purely an interim stage before the introduction of something bigger and better. Subsequently, at the Annual General Meeting of the Royal Astronomical Society, early in 1881, the anti-Lockyer group in the Society put forward a motion asking for a special meeting to discuss State endowment of research. After an acrimonious discussion on the legality of such a meeting, since it was not specifically concerned with astronomy, the motion was passed, and a meeting was called for April 1881.

Four resolutions were put down for discussion at the meeting by the anti-Lockyer group. Proctor was no longer involved, having retired into the background after his resignation as Secretary.

> 1. That, in the opinion of this Society, the granting of public money for research in cases where it does not appear that results useful to the public will be obtained, or where the researches proposed are likely to be undertaken by private individuals or public bodies, does not tend to the real advancement of scence.
> 2. That this Meeting considers it inexpedient that a Physical Observatory should be founded at the national expenses.
> 3. That this Meeting is of the opinion that the Government grant to the Committee on Solar Physics at South Kensington should be discontinued.
> 4. That in the opinion of this Meeting, full accounts should be published of all money expended by the Government for scientific purposes, and that in all cases the nature of the work to be undertaken should be as clearly defined as possible.[44]

Just before the meeting was due, a circular opposing these resolutions was sent round to Fellows of the Society by A. A. Common. Common was an amateur astronomer, an expert on photography and the manufacture of large reflecting telescopes, who came to be a close friend of Lockyer during the 1880s. He pointed out how ridiculous it would be if a Society, which, according to its Charter, had been instituted for the encouragement and promotion of astronomy, was to pass a resolution which would effectively place a limitation on the amount of astronomical research in Britain.

Let us think how we stand with regard to other nations; pass them through your mind; we are all behind, doing nothing, and in a state of stagnation, and while America, France, Austria, and others are founding Observatories and promoting the Science in a large and liberal manner, we are asked to assist in this state of stagnation by leaving all in the hands of a few private individuals who cannot, if they would, undertake the work that requires doing, and now, because certain Fellows of the Society have an idea that the *real* advancement of Science can only take place in a certain way, we must lag behind and cease to lead as we did.[45]

Indeed, one of the bitterest pills that Lockyer had to swallow during the 1870s was the sight of new astrophysical observatories going up on the Continent whilst nothing was being done in this country. His friend Janssen, for example, became director of a fine new astrophysical observatory at Meudon in 1875.

When the special meeting of the Royal Astronomical Society took place, the proposers of the resolution appeared to have gained a notable adherent, for Airy's signature was found to be added to theirs. There can be no doubt that Airy was unhappy about any further extension of State aid to astronomy. He wrote at this time in the *English Mechanic*, the traditionally anti-Lockyer magazine,

I think that successful researches have in nearly every instance originated with private persons, or with persons whose positions were so nearly private that the investigators acted under private influence, without incurring the danger attending connection with the State. Certainly I do not consider a Government is justified in endeavouring to force, at public expense, investigations of undefined character, and, at best, of doubtful validity: and I think it probable that any such attempt will lead to consequences disreputable to science. The very utmost, in my opinion, to which the State should be expected to contribute, is exhibited in the large grant entrusted to the Royal Society.[46]

It transpired, however, that Airy was only partly on the opposing side. According to his letter to the meeting, at which he was unable to be present, he simply objected to the Solar Physics Committee receiving any money apart from members' expenses. He did not object to the existence of the Committee as such, and he was prepared to co-operate with it.

The first two speeches in support of the resolutions at the meeting were made by old opponents of Lockyer, Denison (now Sir Edmund Beckett) and Captain Noble. The first speech against was by Professor H. J. S. Smith, Lockyer's erstwhile colleague on the Devonshire Commission. Smith was known for occasional sardonic remarks concerning Lockyer. On one occasion he suggested that Lockyer did not always distinguish clearly between the editor of *Nature* and the Author of Nature.[47] More importantly, he proposed that the debate on the endowment of research simply cloaked research for endowment.[48] But these comments seem to have arisen as much from a delight in epigrams as from any particular opposition to Lockyer, and here, at the meeting, he moved an amendment, in opposition to the resolutions, that the Society should not voice any corporate opinion concerning State aid to science. After a long debate, the resolutions were turned down, and Smith's amendment carried by a large majority.

This clear-cut decision more or less marked the end of Lockyer's decade of controversy with the Royal Astronomical Society. The Society now turned to other matters, and, if Lockyer can hardly be said to have become reconciled to it, at least he no longer found himself in constant collision with some of its members. Indeed, Lockyer tacitly acknowledged this change for the better. A month after the special meeting he attended an ordinary meeting of the Society and read a paper on his recent solar work at South Kensington—a most unusual event.

V

SOUTH KENSINGTON AND METEOROLOGY

Most British universities and colleges have had variegated histories, but the teaching institutions at South Kensington would probably take any prize for complexity of development within a short span of time. As one observer of the London scientific scene in the seventies remarked concerning them, 'I would fain have discoursed in separate essays on the rise, progress and ultimate destiny of these several institutions, but this Gordian knot is all too toughly interlaced to yield to my blunted penknife'.[1] We have seen that the Devonshire Commission endorsed the transfer of both the Royal School of Mines and the Royal College of Chemistry to South Kensington, but that, mainly because of internal tensions, the compromise was reached that only the basic sciences should remove there. As a result, by the end of 1872, Huxley, Frankland, and Guthrie (who replaced Tyndall as the physicist) had taken possession of a new building, later to be known as the Huxley building, in Exhibition Road. The basic organ of government of the new combined institution was a council of the professors, but Donnelly, who was appointed Director of Science at South Kensington in 1874, had ultimate responsibility.

John Donnelly was the most influential of the Royal Engineers officers to be attached to the Science and Art Department. Having served meritoriously in the Crimean War, he was posted on his return to the London military district, and put in charge of the preparations for the new buildings then being planned at South Kensington. When Playfair left for Edinburgh, Cole offered an inspectorship to Donnelly. As Cole was particularly interested in the work of the department on the Art side, Donnelly, in conjunction with Captain Fowke, concentrated on the Science side. After Fowke's death in 1865, Donnelly came to be in charge of science interests in the department, and was accepted as Cole's second-in-command. During the latter part of the 1860s, he began to press both for the establishment of a central science college at South Kensington, and also for a fully fledged science museum, where the exhibits could be studied by teachers and loaned out to schools for practical demonstrations. It was natural that, when Cole retired in 1873, Donnelly should take charge of South Kensington. Cole's real successor was Sir Francis Sandford, but he was already fully occupied as Secretary of the Education Department, and left Donnelly very much to his own devices. From this point on, till the end of the century, Donnelly played a key role in the developments at South Kensington. Although he never again saw active service, he remained an Army officer, and ended his days as a major-general; it is possible that Gilbert had him in mind when writing *The Pirates of Penzance.*

> 'I'm very good at integral and differential calculus,
> I know the scientific names of beings animalculous,
> In short, in matters vegetable, animal and mineral,
> I am the very model of a modern Major-General.'

By the latter part of the 1870s, the South Kensington site was proliferating with institutions. According to the Sixth Report of the Commissioners for the Exhibition of 1851, it was used in 1878 for the following purposes.

1. The South Kensington Museum, the School of Science, and the Department of Science and Art.
2. The Natural History Museum.
3. The India Museum.
4. The Queensland Museum, and objects from other colonies, forming the nucleus of the proposed general Colonial Museum.
5. The Patent Museum.
6. The Royal Horticultural Society.
7. The Royal Albert Hall.
8. The National Portrait Gallery.
9. The National Training School for Music.
10. The School of Art Needlework.
11. The School of Cookery.
12. The Museum of Fish-culture.
13. Various portions of the Commissioners' buildings have also been taken advantage of for temporary Exhibitions, for Army and Civil Service Examinations, and for the Cambridge Local Examinations, in addition to those conducted by the Science and Art Department.

But the development of advanced science teaching was still hindered by the conservative attitude of the majority of the professors at the Royal School of Mines. This was particularly galling for Huxley, who was continuing to press for the formation of a Normal School of Science, the primary objective of which would be the provision of a thorough training for future science teachers. It was not until 1881, however, that the basic science departments finally broke away from the influence of the mining curriculum and set up the sort of School that Huxley had envisaged.

Although something of the old connection was retained—the entire institution was referred to officially as 'The Normal School of Science and the Royal School of Mines' and there were joint courses in the first two years—the Normal School was now enabled to award its own qualifications based entirely on its own requirements. The qualification awarded was not a degree, but a diploma, the associateship of the School. Huxley derived the name 'Normal School' from the École Normale in Paris, but it proved to be an unpopular title in Britain, and subsequently, in 1890, the School was renamed the Royal College of Science.

It was apparent that the new School needed additional staff, for the existing members could hardly provide the comprehensive tuition that a school science teacher might be expected to require. Additional lectureships were therefore established in mathematics, botany, agriculture and astronomy. Lockyer, almost inevitably, was appointed to the astronomy post. At the age of forty-five, he had finally attained an official appointment in astronomy. The main credit for these developments must certainly go to Huxley, who had been planning for this expansion of teaching at South Kensington ever since the Devonshire Commission. But he was well backed by Donnelly, who proved to be a very good friend to Lockyer.

The course for the new associateship of the School was planned to last for three years, of which the first two would be taken by all students, and would cover a wide range of scientific subjects. The general scheme for these first two years was:

First Year

Both terms	*First term only*	*Second term only*
Mathematics Practical Geometry Mechanical Drawing	Chemistry	Mechanics

Second Year

Both terms	*First term only*	*Second term only*
Mathematics Mechanical Drawing	Physics Astronomy	Geology Mineralogy

In the third year of the course, the students could select one field of science for specialized study.

Since astronomy appeared in the second year of the new course, Lockyer's first year after appointment was virtually free from teaching and he was required to provide only a few sessions at the telescope. Even when teaching did begin, however, he was not overburdened with lecturing and since he was also doing less public lecturing at this time than in the early seventies, he found it possible to continue acting as an occasional inspector and examiner in science for the Science and Art Department. Some of the practical work for the astronomy course was carried out in the small observatory that Lockyer had established at South Kensington on the initiative of the Science and Art Department. (The site is now covered by part of the Science Museum.)

The course that Lockyer gave was by no means limited to astronomy. According to the prospectus of the School for 1882–3 'One of the chief objects of the course . . . is to enable teachers to become practically familiar with the instruments, methods and results dealt with in the teaching of Physiography. It is also intended to give the general student an idea of the conditions of extra terrestrial matter. The course will include 12 lectures and there will be demonstrations both in the Physico-Chemical Laboratory and in the various observatories.'

Physiography was a name that Huxley had invented for a subject which covered much of what would now be described as physical and mathematical geography. Lockyer and his colleague, Judd, were the chief Science and Art Department examiners for this subject from the end of the seventies onwards. (A. R. Wallace was for many years an assistant examiner under them.) It was the one scientific subject which had become popular for teaching in elementary schools by the 1880s. Considerable weight therefore needed to be given to it in a course designed for school teachers, and, as a result, astronomy formed a significant part of the total scheme at the School. Lockyer had between 30 and 40 students during the first few years. This increased to over a hundred in 1885, and the average up to 1890 was eighty students per year. He was also required to give a three-week summer course in alternate years, subsidised by the Science and Art Department, for visiting teachers.

Obviously Lockyer could not handle all these students alone, and the School rapidly provided him with demonstrators to help with the teaching. Lockyer had had personal assistants to help with his researches for over a decade by this time. The new demonstrators were naturally also employed in this capacity, and came to be closely involved in his research activities as well as his teaching commitments. Captain Abney, who with Donnelly and Lockyer represented South Kensington on the Solar Physics Committee, was appointed occasional lecturer in photography at the School from 1883 onwards, and also helped with some of the astronomy teaching.

Mention of the Committee points to the curious ambiguity of Lockyer's position at South Kensington. On the one hand, he was lecturer in the Normal School; on the other, he was the guiding spirit of the Solar

Physics Observatory, which was a completely independent entity. He, himself, was quite clear that his research career could develop satisfactorily only if both the School and the Observatory flourished together. Hence, he received a jolt in the mid-eighties when the future of the Obsevatory came seriously into question. The Treasury always regarded Lockyer's financial activities with suspicion, and this was compounded by an even deeper suspicion of the Science and Art Department. In January 1887, this antagonism reached such a pitch that the Treasury wrote to the Solar Physics Committee saying that the grant towards its work would be discontinued. This move had been expected at South Kensington, and Donnelly had been busy in the previous month trying to find additional funds for Lockyer, whose salary as lecturer was £450 per year, though examining fees raised this to £750. The members of the Committee reacted swiftly. They drew up a memorandum which stressed not only the importance of their researches, but also the necessity for continuity of observation. The circulation of this memorandum, together with political pressure from members of the Committee and their friends, finally caused the Treasury to retreat, and it was agreed that the grant should continue. In return, the working of the Solar Physics Committee would be reorganised.

From Lockyer's viewpoint, this crisis, which had seemed so serious a threat, turned out to be a major advance, for the reorganisation that was ultimately agreed on proved to be highly advantageous for him personally. Whereas, before, the Solar Physics Committee had been the central entity and Lockyer simply a leading member, it was now agreed that the Committee should become essentially an advisory and supervisory board for the Solar Physics Observatory, fulfilling a regulatory function similar to that of the Board of Visitors at the Royal Observatory, Greenwich. Lockyer's position became correspondingly similar in its scope to that of the Astronomer Royal. He was given complete control of the day-to-day running of the Observatory, and authority to decide which publications should be issued as officially emanating from the Committee. Equally importantly, members of the Solar Physics Committee ceased to be paid for their services, and the money thus made available was transferred to the running costs of the Observatory instead. At about this same time, Lockyer was promoted to Professor of Astronomical Physics in the Normal School. This led to an increased basic salary of £800 per year, plus fees.

Lockyer was now the independent head of an officially recognised British astrophysical observatory; the position he had wanted for so long. It could not be said, however, that his observatory presented a particularly impressive aspect to visitors. The instruments had originally been set up in the gardens, but during the 1880s they were all grouped together in a motley assemblage of wood and canvas huts behind a temporary post office which had been set up in Exhibition Road. These huts, later nicknamed the 'Sun-spotteries', built as a temporary expedient until such time as money became available for more permanent buildings, remained in use throughout the whole of Lockyer's career at South Kensington. (It should be added that the 'temporary' post office building lasted for almost as long.)

The instrumentation, like the huts, formed an oddly assorted selection. Much of it either belonged to Lockyer, or had been loaned to him personally by his friends. Despite this, some of it was of very high quality. Rowland, the maker of the best diffraction gratings in the nineteenth century, wrote to Lockyer in 1885,

> I have at last got you a 6in. grating equal to mine in definition but not so bright. I have set it up and have been photographing with it. It is worth its weight in gold as there are now only two others equal to it in the world. I have spoken to the University authorities about it and they have suggested that it be presented to the Royal Society—with a request to let you use it as long as you wish. They thought this would be best as it is too valuable to be sold . . .[2]

Lockyer's close contacts with the United States paid off strongly in the matter of diffraction gratings, the manufacture of which was a definitely American speciality. A few years before, Lockyer had obtained a grating, against strong competition, from another leading maker, Rutherfurd of New York, as the following letter from Langley reveals.

> On reaching New York I called at Mr. Rutherfurds to see about your grating. He is away, but Chapman told me he (Chapman) had sent you one, two weeks ago which I suppose you have now got. I seized upon two large ones I found there and bore them off, that being the only way of getting them, as he has apparently promised so many people that he has no longer any very exact idea of the order of priority of claimants for the few he makes.[3]

Thus Lockyer had at his disposal some pieces of instrumentation that even his friend Janssen, in his much better equipped observatory at Meudon, might envy. As one of Lockyer's research assistants observed many years later, 'After all the superiority of modern appliances is largely in the matter of convenience and time-saving, and even now there are but few spectroscopic problems which could not have been successfully undertaken in Lockyer's laboratory as soon as good photographic dry plates become available.'[4] One major problem was the most efficient use of this instrumentation. Lockyer, himself, seems to have been a rather clumsy experimenter. His gift lay more in seeing how an experiment should be done than in actual instrumental dexterity. His assistants helped in the latter respect, but their training did not normally include much optical work. It was here that the members of the Solar Physics Committee proved especially helpful to Lockyer. Stokes, in particular, advised on the adjustment and development of Lockyer's optical instruments, and during the seventies and eighties was often at South Kensington overseeing the work.

By the end of the 1880s, accommodation at South Kensington was becoming very cramped, and it was no longer possible to house all the students who wished to attend courses. The need may be judged from a memorandum submitted by Judd, the Professor of Geology, 'In spite of the greatest care in ventilation, the effects of the insanitary and over-crowded condition of these laboratories are beginning to make themselves painfully felt. Of the four female students two have broken down, and some of the least healthy of the men are beginning to be affected.'[5] Not long after Lockyer had moved to South Kensington, the Commissioners for the 1851 Exhibition had offered to provide a site and buildings there for teaching purposes at their own expense, if the Government would then take them over and pay running costs. This the Treasury had refused to do. Now, in 1890, Parliament voted a sum of £70 000 to provide new buildings for the Science Museum and for laboratories on the same site. It was decided that chemistry, physics and astronomy should be taught in a new building, whilst the other departments expanded into the space thus vacated. Rücker, who had become Professor of Physics at South Kensington on the death of Guthrie in 1886, and Lockyer therefore began to draw up joint plans for a set of new physical laboratories. Whilst they were engaged on this, however, it was announced that an anonymous donor (who later proved to be Mr Tate) had offered the Government £80 000 for the erection of an art gallery on part of the site, and that the Government had decided to accept the offer.

Lockyer was immediately up in arms and wrote furiously against the new proposal in *Nature*, for he had always held that this particular stretch of land should be reserved for science alone. He helped organise a memorandum, signed by most of the leading British scientists, against the proposed art gallery, and this was presented to the Prime Minister, Lord Salisbury, in April 1891. Donnelly, who was necessarily in the middle of the battle, was unhappy with Lockyer's attitude. He felt that the scientists were being, as usual, excessively intransigent. He wrote to Lockyer

> It is only possible to get things done by opportunism and *Nature* won't help the getting things done by letting Science have a jealous scream at Art. Getting a building for Art does not diminish the chances of getting one for Science but increases them greatly.
>
> If it had not been for Art, Science would not have had a pied à terre at S.K. and if it had not been for the insane jealousy of scientific people, Jermyn Street etc., the provision for Science at S.K. would have been much greater long ago. Who stopped the Science Museum Scheme after the Sc.[ience]. Loan Collection? Not the artists surely but the weak kneed Science men dominated by Percy.[6]

But Lockyer and the other scientists continued to press their point, and, with Salisbury still sympathetic to science, it was decided after several months of discussion that the Tate Gallery should not go ahead on the proposed site. Instead, a survey of other sites was made, and eventually the gallery was built at Millbank.

Despite this satisfactory agreement over the site, the Government made no move to proceed with the erection of the laboratories. Year after year the Treasury turned down eloquent appeals from the Science and Art Department for building to commence. Finally, in April 1898, a Select Committee of the House of Commons reported on the building plans for South Kensington, and gave their approval to the proposals of Lockyer and his fellow-professors. It seemed that at last things were beginning to move, but when later in the year the Government proposals finally appeared it was found that laboratories were to be erected in an area which had previously been earmarked for an extension to the Victoria and Albert Museum. If Lockyer held as a basic tenet that the land to the west of Exhibition Road must be reserved for science, he also believed that the land to the east of the road should be devoted to buildings for the arts. He therefore once again lifted up his voice in protest at the Government proposals. A Parliamentary Committee was established to consider the question, and Lockyer appeared before it to explain the reasoning behind his protests.

A major factor was his belief, shared by the other professors at the Royal College of Science, that the laboratories must not be separated from the Science Museum. Insistence on this had been one of the causes of the delay in commencing work on the new laboratories, but as we have seen, Lockyer was convinced of the importance of museums for science teaching. This had been a commonplace assumption a couple of decades before, but it was now coming under suspicion. Donnelly had been pressing for more museum space at South Kensington as well as for more laboratory space, and this too was considered by a House of Commons select committee. Almost immediately, however, Donnelly found himself under attack for his handling of affairs at South Kensington. One of the points stressed by his opponents was the worthlessness of museums for scientific education. It must be admitted that this charge was to some extent just a pretext. Dislike of the Science and Art Department for its freedom from Government control, and especially for its financial independence, had finally come to a head in the late nineties, and took this opportunity to boil over. Donnelly, who was just about due to retire, suddenly found himself under a political cloud. Lockyer sprang to Donnelly's defence, as did many of the other scientists at South Kensington, but Donnelly's departure remained an unhappy one. As he left, the Government performed a *volte-face* over the laboratories, and agreed that they should be erected in the position that Lockyer and his colleagues desired. In due course, this came to pass, but by the time they were ready for use, Lockyer was no longer involved. In 1901 he reached the statutory retirement age for teaching staff, and ceased to be a professor.

Squabbles over buildings were not the only matters connected with the South Kensington site that occupied Lockyer's time and energy in the eighties and nineties. In 1889, Rücker learnt that plans were to be submitted to Parliament for approval of the construction of an underground railway running below Exhibition Road. The professors approached Donnelly to point out to him the danger that vibration and, if the proposed railway were to be electrified, electrical currents would interfere with measurements in the physical laboratories, and in the observatory. Lockyer got in touch with the veteran Lyon Playfair, since he was the senior member of the 1851 Commissioners, and they had the ultimate say on how the site should be used. Playfair was sympathetic. He pointed out that when a similar proposal had been made by the Metropolitan Railway in 1884 he had stipulated that the line should not run so close to the laboratories that any significant vibration might result. The application had, in fact, been ultimately quashed altogether. 'Railways', Playfair wrote, 'are horrid, material, unaesthetic institutions'; and added that he thought the scheme was probably being promoted by the Royal Albert Hall to gain more customers.[7] Donnelly was somewhat less sympathetic. He recommended that the scientists should not be too extreme in their demands. The building of the railway should not be absolutely opposed, rather some agreement should be reached with the company (the Clapham Junction and Paddington Railway) to limit the effects. The professors at the Normal School refused to accept this advice, however, and they were reinforced in their opinion by the staff at the adjacent Central Institution of the City and Guilds.

This latter institution, one of Huxley's enthusiasms, had been established at South Kensington in 1884 as a complementary teaching body to the Normal School. Whereas the latter trained scientists and science teachers, the Central Institution was intended to train engineers and technical teachers. But some of the teaching inevitably overlapped, and there were close informal links between the two bodies. Thus H. E. Armstrong and W. E. Ayrton, who were employed by the City and Guilds to teach chemistry and physics respectively, were both on excellent terms with Lockyer. Indeed, Armstrong and Lockyer were old acquaintances, as Armstrong had been working with Frankland in the late sixties, at the same time as Lockyer, though on a different topic.

During the next three years, the staff of the Central Institution (which became the Central Technical College in 1893) and the Normal School (which became the Royal College of Science) cooperated in a study of the probable disturbances to their work which would result from the construction of a nearby railway. Lockyer, himself, approached the Royal Observatory for advice, for they had been involved for some time in the attempt to divert the flood of London rail traffic away from Greenwich. H. H. Turner told him,

> We have had several tough fights about railways and the correspondence and papers are very voluminous. In 1865 Sir G. Airy in the teeth of a strong public agitation, drove the S.E. railway to a minimum distance of 1720 feet; they wanted to go through the Park.
>
> Lately we made some careful experiments on the subject . . . As a result of these experiments a railway which was to pass at a minimum distance of 840 yards from the Observatory was successfully opposed.[8]

In March 1893, Rücker and Lockyer were called as witnesses before a House of Commons select committee, and their evidence, backed by that of other scientists, proved sufficient to rebuff the railway company. But the respite was only temporary. Early in 1897, Rücker reported again to the Council of Professors, saying that the construction of three electric railways in the near proximity of the College was known to be under consideration. A new committee was convened, this time under the auspices of the Board of Trade, and all the old arguments were repeated. Lockyer was less involved on this occasion, perhaps because electrical disturbances were less important to him than mechanical vibration.

As we have seen, a major reason for the formation of a Solar Physics Committee, centred on South Kensington, and for its support by the State was the hope that its researches might provide a greater understanding of terrestrial meteorology. Astronomy and meteorology were linked together by tradition. The word 'meteor' originally meant any unusual phenomenon in the sky. By the time Lockyer started his research career, the word had come to be restricted to shooting stars, material that entered the Earth's atmosphere from space and was heated to incandescence by the friction. But the two subjects continued to occupy adjacent niches. Astronomical observatories were generally also used for meteorological observation, and several of the leading meteorologists amongst Lockyer's contemporaries were Fellows of the Royal Astronomical Society. James Glaisher has been mentioned earlier as an example. He was a member of the Royal Observatory staff and the main exponent in this country of meteorological observations from manned balloons. Astronomers might concern themselves with meteorology for a variety of reasons—because, for example, the Earth's atmosphere affected the seeing conditions, and so had a major influence on observations—but Lockyer's interest in the subject was almost certainly aroused by his friendship with the observers at Kew.

When Lockyer first became acquainted with the Kew Observatory in the 1860s it already possessed a long, if not particularly distinguished, history. Its beginnings were propitious enough, for it had originated early in the eighteenth century as the observatory of Samuel Molyneux. He had started there with James Bradley the series of measurements which ultimately led the latter to two of the major observational discoveries of the century—the aberration of light and the nutation of the Earth. Subsequently, however, Kew became the private observatory of the reigning monarch, and little work of scientific interest was carried out there. In 1841, the Government decided to cease maintaining Kew as an observatory. The instruments were dispersed, and the building handed over to the Royal Society. The Society was then at a low ebb scientifically, and was not prepared to make any use of the building. It therefore passed on into the hands of the currently much more active British Association, who turned it into a centre for magnetic and meteorological measurements. One of the leading figures in this takeover was Sabine, and in the fifties his interests turned especially to the relationship between terrestrial magnetism and solar activity. This led to the intensive studies of solar physics which were carried out at Kew under Balfour Stewart during the 1860s. Early in the 1870s the Royal Society resumed control of Kew, and Stewart, unable to agree with the new controlling committee, left for a post in Manchester. The observatory ceased detailed solar measurements, and concentrated once again on terrestrial magnetism. Thus Lockyer's first encounter with it in the sixties occurred during a very active period, when solar physics, meteorology and terrestrial magnetism were all being investigated simultaneously. The possibility that these three different areas of investigation might be inter-related, which was then impressed on his mind, remained one of the driving forces behind his research for the rest of his life.

The 1860s were a time of great popular interest in meteorology, mainly because there was an upsurge of enthusiasm for weather forecasting by the use of weather charts. No official weather charts were produced in Britain during the sixties, though France started issuing daily charts in 1863, but Admiral Fitzroy provided unofficial weather forecasts for the newspapers during the earlier part of the decade. Fitzroy is best remembered as captain of the *Beagle*, the ship in which Darwin sailed round the world in the 1830s, but in the latter part of his life he became one of the leading meteorological advisers to the Government. His attempts at weather forecasting met with severe criticism from scientific circles however, and after his death in 1865, no one came forward to take over his role. The prevalent Victorian ethos naturally led, instead, to unsuccessful attempts to provide a weather service by private enterprise. Thus Glaisher headed a group in the 1860s which proposed the formation of a Daily Weather Map Company. The prospectus of the company shows that Glaisher's plans were wider than this name might suggest, and that he also had an eye on the same scientifically interested public Lockyer had in mind for *Nature*.

> Besides the Map, the publication will contain two pages of letterpress, presenting a carefully prepared Epitome of the News of the Day, foreign and domestic, with ORIGINAL ARTICLES upon topics relating to Commerce, Agriculture, Literature, and Popular Science. In this department the columns of the *Weather Map* will furnish the means of intercommunication between all classes of the scientific public at home and abroad, and provide a medium through which all discoveries can obtain a wide and immediate publicity. It will thus supply what Science has long wanted, and the vast increase in the number of its professors and students so amply deserves, a DAILY ORGAN.[9]

Lockyer was in no way unique or original in his willingness to countenance a relationship between the state of the Sun and terrestrial weather. William Herschel, at the beginning of the nineteenth century, had actually claimed to have found a definite connection between the number of sunspots visible each year and the price of corn, this latter being a reflection of the average annual weather. But Herschel's figures remained dubious, and the first solar–terrestrial relationship generally accepted as proved was that between sunspot numbers and terrestrial magnetism. The establishment of this relationship in the fifties, followed by the growth of interest in scientific weather forecasting in the sixties, almost inevitably led to a new search for possible connections between solar activity and terrestrial weather in the seventies. There were good reasons for this type of research being pressed most strongly in Britain. The responsibilities of a widespread Empire had resulted in a corresponding interest in climatic variations on a world-wide scale. All the colonies depended to a great extent on agriculture for their livelihood; their governments were correspondingly concerned with meteorology, and were prepared to be interested in meteorological research. This was particularly true of India where famine was a major recurrent disaster. It was also realised that the Indian sub-continent experienced a considerably simpler annual climatic cycle than, say, England, so that the chance of detecting a significant relationship with the solar cycle was much increased there.

It was in the early 1870s that Lockyer first began to consider seriously what connections between solar and terrestrial variations might exist. In 1871, shortly before he left on the eclipse expedition to India, he met the editor of a Ceylon newspaper, who was then in England. The editor mentioned in passing that the intensity of the monsoon rains in Ceylon seemed to vary in a 13-year cycle. Lockyer immediately seized on this, and enquired whether an 11-year period, like that of the Sun, might not provide an even better fit. To his delight it transpired subsequently that an 11-year period was, indeed, possible.

Just about this time several papers by different authors appeared, all asserting that significant correlations between solar activity and terrestrial weather existed. First of all, Piazzi Smyth published the results of a long series of measurements, started in 1837, of temperatures just below the Earth's surface, using thermometers buried in the rock at Edinburgh. He concluded that there was a long-period variation in these temperatures with the same period as the solar cycle. Next, J. Baxendell at the Radcliffe Observatory, Oxford, published an analysis of the meteorological records there for the previous eleven years, and claimed that there were variations of both atmospheric pressure and temperature in step with the solar activity. At about the same time Stone, at the Cape of Good Hope, examined the previous thirty years' observations there and concluded that the mean annual temperature at the Cape was related to the solar cycle. Finally, in 1872, Meldrum in Mauritius used the many years of observations there to show that cyclones in the Indian Ocean occurred most frequently when the sunspot number was largest, and that the rainfalls in Mauritius, Adelaide, and Brisbane were generally greater at sunspot maximum than at minimum.

Encouraged by this spate of publication, Lockyer now entered the field himself. For his first investigation, he followed Meldrum's approach and examined rainfall data. He used the records from the Cape and from Madras, where meteorological observations had been under way for some time, and found a variation similar to that claimed by Meldrum. The importance of this result, if it could be fully established, was very obvious, for the years of minimum rainfall in India naturally also tended to be famine years. If a cycle in the rainfall level could be established, forward planning against future famines would become possible.

One result of Lockyer's growing interest in meteorology was that from this time on he became involved as an adviser on meteorological matters. In 1875, for example, he spent some time instructing two members of an Arctic expedition in relevant meteorological techniques. In recognition of his assistance the expedition named an island in Smith's Sound after him. On the Solar Physics Committee, however, he was not at first the leading expert in the study of solar-terrestrial relations. This was Balfour Stewart's sphere. Stewart was especially interested in magnetic variations and solar effects, and his work in this area led him on, in the 1880s, to the realization that there was an ionised layer in the upper atmosphere of the Earth (now called the ionosphere). Stewart also worked in areas which excited Lockyer's more immediate attention. In 1882, he examined the annual average depth of the Nile and the Thames, and found that the maximum depths of both rivers occurred just after sunspot maximum, with another maximum just after sunspot minimum. This could be interpreted as meaning that the volume of water was a maximum at these times, presumably a result of the increased rainfall. Stewart soon extended his work to the Elbe and the Seine, and found a similar result for them. This type of analysis was not entirely new; comparable results had been obtained elsewhere in the early 1870s.

Balfour Stewart was also concerned with measurements of the amount of solar heat incident on the Earth (the so-called solar constant) in the hope of detecting variations which might lead to corresponding changes in climatic conditions. This work, known as actinometry, was equally relevant to Lockyer's meteorological interests. It also tied in with his wish to see solar physics developed in India, since one of the important requirements for this type of work was a clear atmosphere, which could be found in India but not in England. Two types of actinometer, one devised by Stewart, the other by Roscoe, were sent out to India for a series of observations of solar radiation. But the work was delayed by the death of Meins, the R.E. sapper who had been dispatched from South Kensington to India to carry out the measurements. Even when the necessary observations were finally made, during the mid-eighties, the inexperience of the Indian observers and recurrent haze from dust storms vitiated the work. No funds were forthcoming to continue the measurements further. This aspect of the Solar Physics Committee's work therefore lapsed, and the first important series of measurements of solar-heat flux were made later by Lockyer's friend, Langley, in the United States.

Lockyer was particularly interested in the possibility of the solar-heat emission varying in step with the solar cycle. Contemporary opinion accepted this as a reasonable possibility, but on one crucial detail Lockyer came to take a view diametrically opposed to that of his contemporaries. He later claimed that this difference of opinion was a major reason why he was slow to become deeply involved in meteorological investigations during the seventies and eighties.

> I hesitated for a time to take up the meteorological inquiry as a part of the Observatory work, for the reason, that in the earlier stages of my researches I had to combat the view, expressed by many influential workers, that the times of greatest sun-spot activity were those during which the earth received *least heat*, the reason given for this view being that the spots exercised a kind of screening action.[10]

Lockyer believed, on the contrary, that the Sun actually gave out more heat at maximum, because it was more active at this time in all ways. Since the spots must obviously screen out some of the Sun's radiation, he had therefore to look round for some other source of the supposed extra radiation at maximum. The prominences seemed to him an obvious possibility, and his work with Seabroke at Rugby in the early 1870s, in which they made daily tabulations of the solar prominence numbers, had the testing of this hypothesis as a main objective. However, the weather conditions in Britain prevented any statistically very useful data being accumulated, and helped enhance Lockyer's desire for a programme of continuous solar observations from India.

Lockyer's interest in meteorology and solar physics continued throughout the 1870s. His visit to the United States to observe the 1878 eclipse proceeded in good part from his desire to study the activities of American meteorologists. As we have seen, the connection between meteorology and solar physics was also a major theme in his pressure for the creation of the Solar Physics Observatory. Although this emphasis on the practical utility of the proposed observatory may well have been somewhat exaggerated in the hope of gaining greater official sympathy for the project (as Proctor and his supporters claimed), there can be no doubt that Lockyer genuinely believed in the economic importance of his work. In 1879, he combined with the Director-General of Statistics to the Indian Government and the Professor of Mathematics at Patna College to present to the Indian Famine Commission a report dealing with the relationship between the sunspot cycle and the rainfall in Southern India. This pointed out that famines in Madras since 1810 could be correlated very closely with sunspot minima. The authors emphasized that the exact timing of the famines was probably not as predictable as this correlation might suggest, but that the important feature was the cyclical nature of famine occurrence. But in 1881, H. F. Blandford, the main Indian Government meteorologist, reported to the Famine Commissioners that, although he, too, believed that solar heat varied cyclically, he had been unable to find any simple correlations of the sort that Lockyer had reported. He concluded that there appeared to be no immediate possibility of using solar observations to predict weather in India.

This negative result, together with Balfour Stewart's recognised preeminence in the study of solar-terrestrial relations, may partially explain why Lockyer's interest in solar-terrestrial relationships seems to have waned in the eighties. But these cannot be the only reasons—after all, Stewart died in 1887—a more pressing cause was probably his inability to demonstrate that the solar temperature was actually higher at sunspot maximum. The solution of this problem came later, from work on one of Lockyer's recurring interests, sunspot spectra.

As soon as the Solar Physics Committee was created in 1879, Lockyer started regular measurements of sunspot spectra, at daily intervals, if the weather permitted. This collection of data was mainly undertaken to try and clarify his understanding of the physical nature of the spots, but there was from the start the additional possibility that they might provide some indication of variations in the solar heat. It was obviously going to be a long business to distinguish significant variations, if they were related to the eleven-year solar cycle. The work had to continue methodically throughout the eighties and nineties before Lockyer could definitely claim to have found some distinctive change. He finally discovered that the width of certain spot lines altered with the solar cycle. At solar minimum, lines of various metals, especially iron, were widest, whilst at maximum certain other lines, whose origin he was unable to determine, were most broadened. According to Lockyer's dissociation hypothesis (discussed in the next chapter) these variations could be understood if the Sun was hotter at spot maximum than at minimum. This was in just the direction he had predicted years before. The observational backing for his belief seemed finally to have appeared, and he once again turned to a consideration of solar effects in terrestrial meteorology.

The approach that Lockyer now took was naturally based on his spectroscopic method, and it began with the observation of an unexpected peculiarity in the sunspot lines during the latter part of the nineties. Since one set of spot lines widened during the solar cycle whilst the other set became narrower, somewhere between maximum and minimum the two sets were necessarily the same width. Lockyer labelled this the 'crossover point'. From his previous observations he expected such a crossover towards the end of the nineties, but much to his surprise it failed to occur. It seemed that for some reason the Sun was staying hotter than it should be. To Lockyer's great excitement reports from India at the same time told of famine consequent on an irregular rainfall. This coincidence in the two sets of observations was sufficient to convince Lockyer that he should take up again the analysis of meteorological observations.

The first fruit of this renewed effort was a joint paper with his son, Jim Lockyer, in which they examined rainfall variations in the vicinity of the Indian Ocean as a function of Lockyer's spectroscopic estimates of solar temperature. The Indian Meteorological Department had been established in 1875, and by this time had accumulated a considerable quantity of good data for study. Lockyer chose India again, as for his previous studies, both for this reason, and because he believed that weather in the tropics should be simpler to understand. Lockyer explained the Indian data in terms of what he called 'heat pulses' (a positive pulse at spot maximum, and a negative pulse at minimum). To both these solar heat pulses there corresponded rain pulses on Earth. Famines would tend to occur in between these latter : that is between maximum and minimum in the spot cycle, or just before the crossover in Lockyer's spectra.

This work was warmly welcomed in India, and Lockyer, feeling now that success in prediction was just round the corner, pressed on with his analyses. His subsequent work falls into the period after his retirement from the Royal College of Science, and will be dealt with in a later chapter; but it may be noted that the Solar Physics Observatory at the beginning of the twentieth century had returned to the intensive study of solar-terrestrial relationships which had been given originally as the main reason for its establishment.

We have seen that there was always some uncertainty from year to year whether the Solar Physics Observatory would be able to continue financially. But the annual bout with the Treasury was not Lockyer's only struggle, since he also encountered occasional resistance amongst his own colleagues at South Kensington. More than one of his fellow-professors hoped on occasion that solar physics would be brought to a summary halt. Maybe Lockyer's need to use the facilities of other departments acted as an irritant. Finally, in 1897, Lockyer protested to the Solar Physics Committee that the observatory had been in an experimental stage for long enough, and it must now be put on a continuing basis with additional permanent staff. Lockyer complained that he was being forced to carry out the work of the observatory with only two assistant computers on short-term appointments. (He was exaggerating a little here. Although Fowler, the Demonstrator in Astronomical Physics, certainly had to do a good deal of teaching work in the College, nevertheless he was also a mainstay of the research at the observatory. Indeed, Fowler was just at this time leading Lockyer towards a major new break-through in spectroscopy.) However, the Solar Physics Committee was impressed by his case, and took up the matter with the Science and Art Department, who wrote in turn to the Treasury. The Treasury, for once, was generous, and increased the annual grant from £500 to £1000, so that during the next few years Lockyer was able to increase the number of members of staff at the observatory appreciably. Despite this favourable response, the observatory, although in a happier state, was still not placed on the same sort of permanent footing enjoyed, for example, by the Royal Observatory.

Some idea of the complexity of the observatory finance and control can be gathered from the draft shown on page 130, which Lockyer prepared in 1882[11] (though his problems were much eased by Donnelly's membership of the Solar Physics Committee).

In view of the great important of Lockyer's assistants to his researches, it seems fitting to end this chapter with a discussion of some of them as individuals, and of their relationships with Lockyer. It must be remembered that posts in science were few in number and generally badly paid throughout the whole of the nineteenth century. As a result, people who

Diagram showing how the work of the Solar Physics Observatory was provided for after 1882

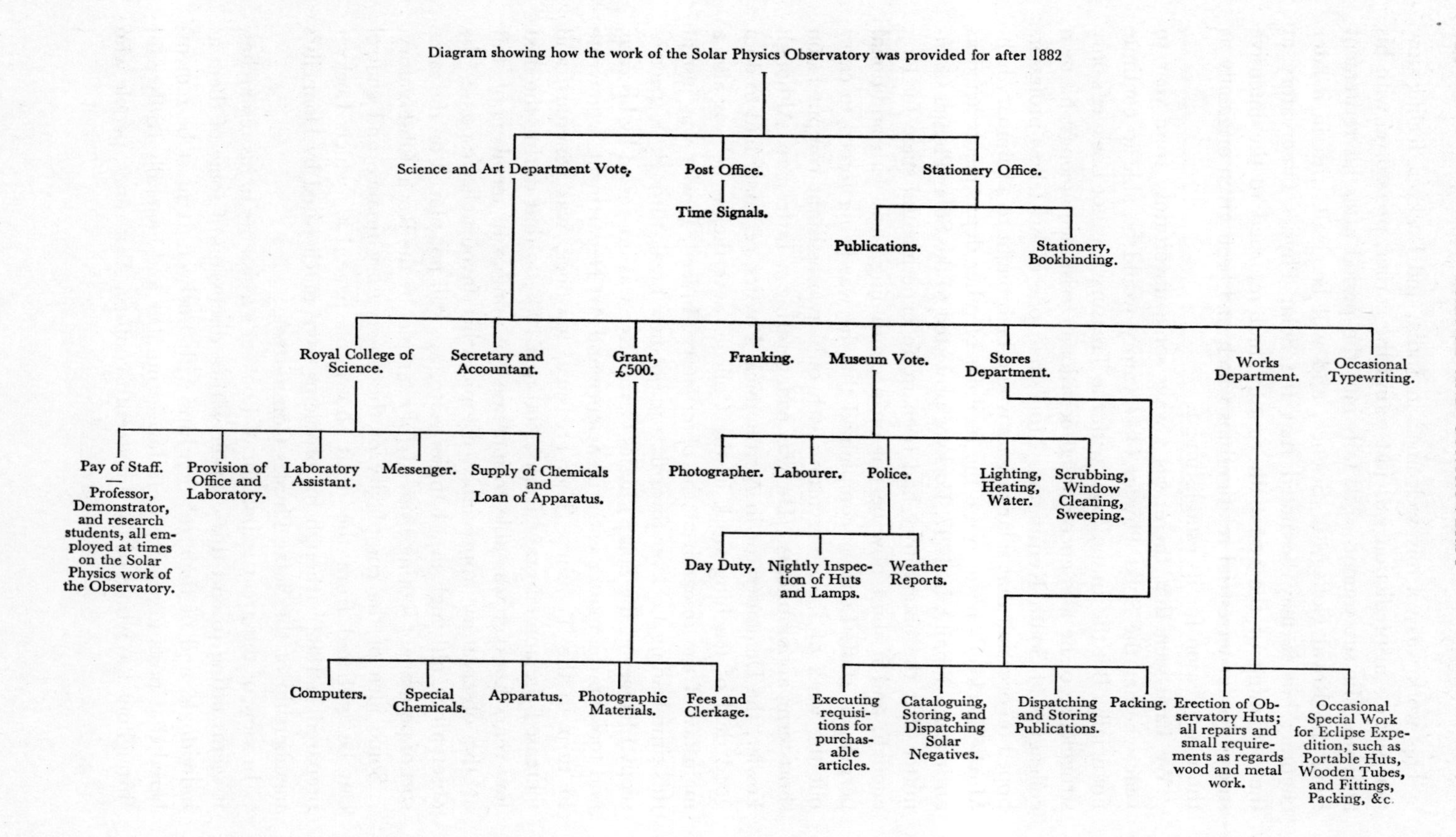

became research assistants were more often than not men highly dedicated to science, and of above average ability. H. G. Wells in one of his science-fiction stories comments on the attraction of such posts for the students at South Kensington. The scientist he is describing has just begun his studies there, and, '. . . before he had held this scholarship a year, was tempted by the possibility of a small increase in his immediate income, to abandon it in order to become one of the nine-pence-an-hour computers employed by a well-known Professor in his vicarious conduct of those extensive researches of his in solar physics—researches which are still a matter of perplexity to astronomers.'[12]

Despite their ability, assistants were commonly regarded as something akin to superior servants, a type of relationship which can be found exemplified in Davy's attitude to Faraday earlier in the century. Fowler, probably the greatest of Lockyer's research assistants, described his experiences of observing with Lockyer in the following way. 'On such occasions, while perfectly friendly, Lockyer rarely allowed me to forget that I was his assistant, and spoke of little outside the work in hand . . .'[13]

In a similar vein, the work produced by a research assistant was frequently regarded as the property of his employer, who could publish it, or not, as he wished. If he did publish, it could be under his own name alone. For example, Meldola, who was Lockyer's assistant for a time in the mid-seventies, recorded the following in his notebook.

> June 2nd, 1875. Mr. L[ockyer] . . . stated that he wished to get out paper before end of session on Li[thium] spectrum.
>
> June 14th, 1875. Then [Lockyer] asked, 'Have you got that paper on Li[thium] spec[trum] ready?' (*Note*:—The evidence of the existence of Li[thium] in the sun is in my opinion not sufficiently strong at present to warrant publication . . .)[14]

Assistants would often go far beyond routine research, and would formulate and test their own hypotheses, but any results could still be taken over and published by their employers. Thus it has been said of Fowler that,

> One forms the picture of an able and faithful servant, fitted for more than a subordinate position, but patiently accepting the conditions of service until such time as his natural initiative could find legitimate scope. Between 1885 and 1901 he [Fowler] published little in his own name—even his eclipse observations in 1893, when Lockyer was unable to accompany the expedition, were published in Lockyer's name.[15]

Lockyer's first experience of working with research assistants came via his cooperation with Frankland, for the latter's assistants were brought into the investigations. This was a fortunate beginning for Lockyer, for Frankland's assistants in the late sixties, Herbert McLeod and Alexander

Pedler, were very capable scientists. Both subsequently became Fellows of the Royal Society, and attained important positions in the scientific world, from which they were able, on occasion, to assist Lockyer in his plans. McLeod was appointed professor at the Royal Indian Engineering College at Cooper's Hill, where he became acquainted with Lord Salisbury. Salisbury was sufficiently interested in science to have a laboratory of his own fitted out at Hatfield, and there McLeod went at weekends to help with the experiments. These arrangements lapsed when Salisbury's appointment as Prime Minister brought some of his leisure activities to a halt, but the appointment made McLeod's friendship with Salisbury even more useful from the point of view of scientific politics. Pedler became Professor of Chemistry at the Presidency College in Calcutta, and there he helped Lockyer's eclipse expedition in 1875. More importantly, he was also appointed Meteorological Reporter to the Bengal Government where he was able to prime Lockyer with climatic data for the latter's investigations of solar–terrestrial relationships.

The first personal assistant that Lockyer appointed for himself was Richard Friswell, who, like Pedler, was born in 1849, and had also been a pupil of Frankland's at the Royal College of Chemistry. It is evident from Lockyer's laboratory notebook that Friswell was mainly responsible for the continuity in the research programme that Lockyer carried out during the early seventies. Lockyer, himself, was mainly involved in his secretarial duties to the Devonshire Commission. Raphael Meldola, who was another of Frankland's assistants, also became assistant to Lockyer in 1875. When Lockyer was unable to take charge of the 1875 eclipse expedition (because of pressure of work) Meldola took over the organisation. The relative infrequency of Lockyer's appearances in the laboratory at this time is reflected in Meldola's notebook. On 5th July, for example, Meldola noted, 'As Mr. L[ockyer] had not been down since June 22nd, [I] arranged to go to St. John's Wood on the following day.'[16] Meldola left in 1876 to take up a position in the dye industry, where his work led him to fame as an organic chemist.

Lockyer's emphasis on spectroscopic researches induced him initially to look for chemically trained assistants. He was also probably influenced by his friendship with Frankland, and by the fact that the best scientific training available at the time was in chemistry. However, Lockyer's most important assistant, Alfred Fowler, specialized in mechanics as an undergraduate at South Kensington. Fowler was, by any standard, an outstanding scientist, and his merits were obvious from early on. He was admitted into the Normal School at the age of fourteen, probably the youngest student ever to enter there. Lockyer engaged Fowler originally as a computer, in the eighties, but when the post of Demonstrator in

Astronomical Physics was created in 1888, he was the obvious candidate. From that time on until Lockyer's retirement in 1901, Fowler remained his right-hand man in both teaching and research. In the latter activity, it is a little difficult to decide who eventually was the mentor.

Lockyer's remarkable output of research papers whilst he was constantly engaged in administrative and editorial duties must be understood partly in terms of the great amount of work contributed by his assistants. This does not mean that Lockyer took no part in the research effort, for sometimes he was very closely involved, rather that he used his assistants not only for research, but also in his other duties. For example, he customarily farmed out some of the *Nature* work to his assistants. Fowler often wrote the astronomical column for him. Indeed, one of his assistants at South Kensington, Richard Gregory, became so involved in the workings of *Nature* that he was ultimately appointed assistant editor, and took over the editorship when Lockyer finally relinquished it. Similarly, Lockyer offloaded some of his administrative chores. In the latter part of the century he trained his son, Jim, in the necessary routine duties at the Solar Physics Observatory, and left the day-to-day running to him.

Lockyer was good at spotting talent when appointing assistants, and it may be that the extent of his outside interests, which left the assistants free to pursue research in their own way, actually helped to keep them reasonably content. For Lockyer's characteristic impetuosity made research when he was present a hectic procedure.

> Lockyer was always keen on the work in hand, but his impatience to see the result of an experiment when a photograph had been taken was sometimes rather trying. As soon as not, he would stand by the door of the dark room, and begin his inquiries almost as soon as the developer had been poured over the plate. When it had reached the fixing bath he would scrutinise it with a magnifying glass and when in his own laboratory would soon give an order for another experiment. His bursts of enthusiasm sometimes proved too much for some of his assistants, who especially objected to work on Saturday afternoon. I suspected that some of them had developed secret methods of bringing an experiment to an end early in the afternoon either through the failure of a battery or the cracking of a vacuum tube through overheating.[17]

If Lockyer hardly sounds an ideal scientific superior by modern standards, it must be remembered that the Victorian scientific ethos was a good deal more autocratic than ours. During the latter part of Lockyer's life, a transition towards a less distant relationship between a scientist and his assistants did begin, and the social gulf between the two diminished. During his debate with Lockyer on the endowment of research, Proctor had observed, 'A point to be most carefully insisted upon is that the head of any national observatory such as has been proposed should be one who would behave in a generous, considerate and courteous way to all who worked under him.'[18] The clear implication is that Lockyer failed to live up to this standard. Nevertheless, there are indications that Lockyer was genuinely respected by his assistants, as witness the number of them who were prepared to help him whenever possible, after they had taken up posts elsewhere.

Lockyer had undoubted powers of leadership. He never asked an assistant to tackle a job he would not try himself. In the early days of work with electric arcs, the harmful effect on human beings of the ultra-violet light emitted was not fully recognised (just as, later, the harm that could be caused by indiscriminate exposure to X-rays and radioactive materials was initially overlooked). In suffering from such effects, Lockyer took at least as much punishment as any of his assistants. After one of his visits to South Kensington, Stokes wrote to his brother-in-law,

> On Wednesday I went to the South Kensington Museum and saw Lockyer. I found him in his Laboratory with one ear, and the part of the neck near it, as red as a turkey cock. He told me that on Monday he had been working with the electric arc given by a powerful Siemen's machine. Knowing the injurious effect of the light upon the eyes, he took care not to look at it, so he stood with his back to the light about 2 feet off. Yet it affected the skin of the back of his neck which was exposed to it, producing something like erysipelas. He said that the day before the skin had been black, and that wanting to write something he was obliged to dictate it, as the inflammation had extended to his eyes. His assistants were also affected, but not to such an extent.[19]

The final word should be left with the long-suffering Fowler,

> Lockyer was not always easy to please, but my own relations with him were seldom otherwise than agreeable, and I have always felt grateful that I came under his influence.[20]

VI

WHAT IS AN ATOM?

Lockyer's early research was controversial only in the sense that it was pioneering work in a new area of science. His ideas might be queried, but his investigations were based on a set of fundamental concepts shared by his fellow scientists. During the seventies, however, he began to evolve a system of ideas about both the nature of matter and the nature of the universe which diverged increasingly from the general consensus. In his later years, therefore, he was to find himself increasingly isolated, although, being the man he was, this merely made him fight harder to establish the validity of his concepts, and so at least force his peers to try and demonstrate clearly where he was wrong. This isolation, in itself a personal tragedy, appears even more so today. For Lockyer's ideas were only partly wrong, in other ways he was dimly seeing ahead to some of the most important scientific developments of the twentieth century. But the mixture of insight and error he brought to these problems was too involved for most of his contemporaries to disentangle, and they eventually approached the solutions by other routes. In this chapter we shall consider the development of Lockyer's ideas on matter—the more important of his two major concepts—as expressed in his dissociation hypothesis. His views on the Universe—as contained in the meteoritic hypothesis—will be discussed in the next chapter.

The atomic doctrine of matter, which had been speculatively suggested in ancient times, was revived in the seventeenth century, but the billiard-ball type of atom visualised then, although acceptable in terms of physics, was not of much use chemically. It provided no scope for an explanation of how atoms could combine together to form the various chemical substances. This was not of great importance initially, but became so during the late eighteenth and early nineteenth century, when chemists began to consider the atom from a different viewpoint. They then related it to the concept of the chemical element, which was also being evolved at this time. An element they defined as any substance that could not be broken down into something simpler, and, as a rapidly increasing number of such distinct elements was discovered during the nineteenth century, each was supposed by the chemists to be reducible to a correspondingly characteristic atom. Hence, two different ideas of atoms were in use during the century: the chemical atom, of which there were as many varieties as there were elements, and the physical atom which was of one type only (all the chemically distinct elements being built up from it in various combinations).

The development of the concept of the chemical atom is usually attributed to the Mancunian scientist, Dalton, at the beginning of the nineteenth century, but it was actually some time before much value was attached to it by chemists. There were various theoretical objections, for example, that there was no proof that matter was atomic in structure—it might just as well be a continuum—or again, that a theory of matter should be expressible in a mathematical form, and the Daltonian atom was not. But the major objection throughout the first half of the nineteenth century was simply that the Daltonian atomic theory led to no new scientific results.

The basic problem during this period was that no distinction was drawn between the atoms of an element, and the fact that the element might commonly exist in nature in a molecular form, that is the atoms of a single element might join together to form a 'compound atom'.* For example, the French chemist, Gay-Lussac, reported in 1809 that when gases reacted with each other, they combined together in simple ratios by volume. At first this appeared to be excellent evidence for the atomic theory. Each atom in one volume of gas reacted with an atom of the other volume of gas to produce a simply related volume of the final compound. Unfortunately, this straightforward picture did not come out so well from a more detailed scrutiny. Thus the Daltonian theory predicted that if one volume of nitrogen reacted with one volume of oxygen to produce nitric oxide, the latter gas should also occupy one volume. In fact, it occupied two. This, and other related difficulties, were resolved quite soon by the Italian, Avogadro, who suggested that equal volume of gases under the same conditions contained an equal number of molecules, rather than of atoms, as Dalton had supposed. Unfortunately, his proposal was neglected for some time.

One result of this confusion between atoms and molecules was that the whole concept of valency—the combining power of atoms—could not be properly formulated; for that fundamental quantity, the atomic weight of the different elements, could not be determined with certainty. This was a matter of great concern in the nineteenth century. As one example (in which Lockyer was interested) we can point to its relevance for Prout's hypothesis. William Prout, a physician, had suggested in 1815 that the atomic weights of all elements were simple multiples of that of hydrogen.

* The present-day terminology of 'atom' and 'molecule' only became common after the 1830s, but we shall use it here throughout.

Good atomic weight determinations were necessary if this hypothesis was to be properly tested. Yet the matter was finally put on a sound footing at only about the time Lockyer himself was entering scientific research. In 1860, Cannizzaro managed to convince his fellow-chemists that Avogadro's old hypothesis provided the only reasonable way forward.

During the first half of the nineteenth century the chemist's concept of the nature of matter was considerably affected by the growth of a new branch of chemistry, the study of organic compounds. During the 1830s, it became apparent that certain groups of atoms could play much the same role in chemical reactions that single elements did. For example, the atomic grouping CH_3 (in modern notation) appeared to behave rather like atoms of sodium or potassium in reactions. Combinations of atoms like CH_3 were called radicals by the French chemist, Dumas, and the German, Liebig. It was an obvious extension to speculate that the elements themselves might simply be a series of 'radicals' built up in this case from some set of sub-atomic entities (indeed, Dumas proposed such a concept himself).

The analogy received further support from within organic chemistry during the same period with the discovery of isomerism—the existence of certain compounds with the same composition, but different properties. One obvious explanation of this was that it was caused by differing arrangements of the same set of atoms. Again, it could be argued analogously that elements were just different arrangements of some more fundamental matter. Dimorphism—the property that some substances had of crystallising in more than one form—which was also observed during the same period, led to the same deduction.

The chemical atom, however envisaged, could not be taken seriously until atomic weight and valency problems had been solved. The fundamental solution of these problems constituted the great advance of the sixties, and it is not surprising therefore that this was a period of most vigorous debate amongst chemists concerning atomic theory. But during the same decade, the physicists were also busy modifying their picture of the atom. The kinetic theory of gases, although suggested earlier, was firmly established at this time, especially by Clerk Maxwell. He pointed out that acceptance of the theory necessarily implied rejection of hard, impenetrable Newtonian atoms, since the new approach required elastic atoms. More important still in causing the physicist to reject the old picture of the atom was the growing interest in spectroscopy during the sixties. It was hard to see how hard, homogeneous particles could produce the extremely complex spectra that were observed. If each spectral line corresponded to a different mode of vibration, the atom must necessarily be complex in structure to oscillate in such a multitude of ways.

Thus, by the 1860s and 1870s, the chemists and the physicists were both suggesting that atoms might be more complex phenomena than had been supposed in earlier discussions. It is hardly surprising that during this exciting period Lockyer began to interest himself in spectroscopy and what it revealed of the nature of atoms. Not only was the topic in the news, but also several of his friends were deeply involved in its various aspects. At the end of the sixties, the debate between chemists over the existence or non-existence of atoms came to a head. Although, like most debates, this was inconclusive, chemists subsequently increasingly accepted an atomic theory of matter. Indeed, Lockyer noted in *Nature* that even at the time of the main debate (which occurred just as *Nature* first appeared) the majority of chemists were in favour of an atomic approach. In view of this, and of the fact that Lockyer himself always seems to have accepted the physical reality of atoms, it is interesting to note that his closest chemical associates during the period were sceptical of atomism. His collaborator, Frankland, maintained that atoms were simply useful fictions. Meldola, Lockyer's assistant when the latter first began serious research on the nature of matter, held much the same view, and continued —unusually, although not uniquely—to do so until his death in the twentieth century. Brodie at Oxford, a close friend of Lockyer in the seventies and, like him, involved in pressing for better science teaching, developed a mathematical theory of chemical reactions that did not involve assuming atomism at all.

Lockyer's adherence to an atomic theory came about partly because he approached the question via spectroscopy. Whereas it was easy to imagine that the various observed spectra were produced by the different modes of oscillation of atoms, it was difficult to envisage a method for producing spectral lines from a continuum. For this reason spectroscopists, like Lockyer or, amongst chemists, his friend Roscoe, tended necessarily to believe in the physical reality of atoms. But their atomic picture of how spectra were produced led them on to further deductions. In particular, they deduced that since each chemical atom might be expected to have a characteristic set of vibrations, the spectrum of each element should always look the same. This was the fundamental assumption on which Lockyer based his initial researches. The development of spectroscopy during the sixties and seventies, and the evolution of Lockyer's own ideas on the subject, are to a considerable extent the story of how this original viewpoint was modified.

As soon as accurate spectroscopic work got under way in the sixties, it became evident that the spectral lines from a given element were not always the same. This was noted, for example, by Kirchhoff and Bunsen in 1860. The question was whether the observed differences were genuinely due to variations in the spectral lines produced by the individual atoms, or whether they were due to differences in the total bulk of the material. Kirchhoff believed that the changes in the relative intensities of various lines which he observed from time to time were simply a result of the different amounts of emitting vapour present—the intrinsic spectrum had not changed. But it was soon pointed out (by Roscoe amongst others) that many substances were capable of giving two quite obviously different types of spectrum, one being the ordinary line spectrum and the other a 'fluted' spectrum consisting of a series of bands. Which of these two spectra appeared seemed to depend on the temperature of the emitting gas. This led to the suggestion that at lower temperatures the material was in the molecular state, giving the fluted spectrum, whilst at higher temperatures it split into atoms, giving the line spectrum. Counter-arguments were soon produced. For example, Ångström, whilst agreeing that temperature-dependent variations of spectral lines occurred, did not accept that a given element could produce two distinct spectra. He believed rather that the existence of a fluted spectrum indicated the presence of impurities, and Lockyer accepted this view initially. In any case, even if an element could have two spectra, this was not necessarily relevant to the discussion of changes in spectral lines, for it could always be argued that whilst the atom might produce one type of spectrum, and the molecule another, both were invariant when they appeared.

When Lockyer began his spectroscopic observations, therefore, it was in this same belief that atomic spectra were invariant. We must now follow the devious path of his researches by which this assumption came to be shaken. The first thing Lockyer noticed when he began looking at spectra was that different lines had their own characteristics. Some were sharply defined, some more diffuse, some were single, some proved to be two lines side by side. This led him to compare the appearance of specific lines of a given element produced by different light sources, especially the electric arc and the Sun. As his acquaintance with spectroscopic work grew, it became clear to him that spectral lines could change their appearance according to the source producing them, and in particular, they could become more or less intense. He therefore began to examine what factors might be involved in altering the intensity of spectral lines.

Lockyer always integrated his approach to observational and laboratory work. A problem or development in one would quickly affect his attitude to the other. In 1869, he extended the idea of examining localised parts of the solar surface, which had proved so useful when applied to sunspots, to sources of light in the laboratory. The two main such sources at that time were the spark and the arc, the latter being more generally useful because it gave a continuous light whereas the spark was intermittent.

The standard procedure was to pass the light from one or other of these sources directly into the spectroscope for analysis, so that the spectrum obtained actually represented an average over the total emitting volume. Lockyer simply inserted a lens between the source and the spectroscope, thus throwing an image of the source, which was usually an arc in his early experiments, upon the slit of the spectroscope. This enabled him to examine the spectra from different parts of the arc instead of looking only at the average spectrum, and led to one of his most important laboratory results. He discovered that lines from an arc could be roughly divided into two groups, long and short. The long lines were those visible over the whole breadth of the arc; the short lines were produced only in the small core region of the arc. Although something like this distinction had been noted before, for example, by Stokes in 1862, Lockyer must be given the credit for emphasizing its importance.

Lockyer and Frankland now saw that this difference detected in the laboratory could actually be reapplied to their solar observations. Lockyer in his investigations of the chromospheric emission lines sticking up all round the limb of the Sun, had soon noted that not all the lines of a given element protruded above the limb to the same extent. For example, the spectrum of magnesium contained a small, but conspicuous group of lines (which Fraunhofer had called the 'b line'). Since these lines were close together it was easy to compare their relative heights above the limb, and Lockyer, observing them in emission during a solar storm, found that they were clearly not all of the same length. Similarly, although many iron lines had been detected in the solar photosphere, only a few were found in the chromosphere. These observations suggested to Lockyer and Frankland the hypothesis that the spectra of the elements became simpler with height in the solar atmosphere, as the temperature and the density decreased.

The desire to test this started Lockyer on a prolonged investigation of long and short lines in the laboratory. The core of the arc was evidently at a higher temperature and density than the outer regions. If, therefore, the hypothesis was correct the core spectrum—that is the short lines—should be more complex than the spectrum of the outer parts—the long lines. This was precisely what Lockyer and Frankland found. In fact, they were able to go further. They showed that their laboratory division of long and short lines corresponded to a separation of solar emission lines according to their extension above the solar limb.

Thus encouraged, Lockyer began to seek a more fundamental understanding in the laboratory of how long and short lines developed. He soon realized that one key to such an understanding was the amount of the element present in the arc. As this quantity decreased the lines disappeared in order of their length: the short lines vanished first, and the longest line observed was the last to go. The long lines were thus the really significant markers of the presence of a given element in the arc. Since spectra usually appeared very readily, often only trace amounts of the element were required for spectroscopic identification of the longest lines to be possible.

This advance had a consequence of immediate importance for laboratory work in spectroscopy. One of the great problems of nineteenth-century chemistry was the difficulty of obtaining, even of identifying, chemically pure substances. Frankland, as a leading chemist, was well aware of the danger that impurities presented to spectroscopic research in particular, and conveyed the importance of this question to Lockyer (though, as we shall see, after Lockyer ceased working with Frankland, he seems to have become considerably less cautious). Now this new knowledge of how small amounts of material could affect spectra made it much easier to detect impurities. Previously, the identification of an impurity element in a spectrum had required the laborious compilation of a spectroscopic map of the suspected impurity, and then a comparison of this map with the spectrum of the substance under examination to determine whether any of the lines were coincident. Even if coincident lines were found, it had not hitherto always been clear whether the impurity was present. Sometimes certain lines from the impurity could be identified, but not others that should have been equally strong. In these circumstances, was the impurity present, or not? Lockyer's results now implied that the detection of an impurity simply required the identification of its longest lines in the spectrum. These were not necessarily the strongest lines of the element, but so long as they were observable, the impurity could nevertheless be positively identified. (In practice, the situation was by no means always as clear-cut as this description suggests. For example, Lockyer's calcium samples were contaminated with strontium, but, equally, his strontium samples were contaminated with calcium, and it required a long and detailed mapping to separate out the two sets of lines.)

Lockyer found several uses for his newly acquired understanding of impurity spectra; in particular, he realised that it might be employed to practical advantage for the assaying of metals. In order to examine in greater detail how the spectrum of an element changed as the amount present was reduced, Lockyer had begun a systematic investigation of the spectra of metallic alloys. They had the advantages that spectra could be easily obtained (they were simply formed into one of the points of a spark gap) and the relative amounts of the elements present could be determined accurately. However, Lockyer did not have the resources to prepare a series of alloy samples with small differences in the amount of one of the elements involved. He turned for assistance to the Royal Mint.

The Mint had a long reputation as a haven for scientists. Ever since the supreme master of British science, Isaac Newton, had become Master of the Mint at the end of the seventeenth century, his disciples had looked on it as a desirable appointment. At the time Lockyer began his spectroscopic researches, one of the most eminent scientists of the period, Thomas Graham, was in charge of the Mint. By the time Lockyer decided to approach the Mint, Graham had died under something of a cloud. It was alleged that his administration was far too lax, and he had been replaced by a non-scientist. The scientific part of the Mint's work (that connected with the properties of alloys, detection of impurities, etc.) had been handed over to W. C. Roberts (afterwards Sir William Roberts-Austen), who was to become a great friend of Lockyer and, later, a colleague at South Kensington. (He became Professor of Metallurgy in 1882, when Percy resigned in opposition to the transfer from Jermyn Street to South Kensington.)

Lockyer outlined to the officials at the Mint how his spectroscopic method of detecting impurities might be used for quantitative assaying of the minor components of alloys, something that the Mint found quite difficult with the standard chemical methods then employed. Roberts was deputed to cooperate with Lockyer in trying out the new method using both the facilities at the Mint and those at South Kensington. The initial results were distinctly promising. It was possible in some circumstances to detect differences as small as 0.01 per cent, and to do this using only very small samples, whereas normal assay methods required a lump weighing a gram. Lockyer, highly pleased, began to think of patenting the method he had developed for the spectroscopic comparison of samples. But further investigations showed that although the method was certainly very sensitive, it was difficult to obtain reproducible results, and the work ultimately lapsed.

The most important result of the new distinction between long and short lines from Lockyer's viewpoint was its application to the Fraunhofer spectrum. In his original attempts to deduce what elements were present in the Sun's atmosphere, Kirchhoff had asummed that the same lines would be present whether the element was studied in the Sun or in the laboratory. Identification for him meant a one-to-one correspondence of the lines. But it was soon discovered that certain lines of a given element might be seen in the laboratory but not in the Sun; whilst other lines, apparently no different, might be seen in both. It was not clear, when this happened, whether the element should be considered to be present on the Sun, or not. Lockyer now noted that the lines missing from the solar spectrum were generally his short lines. The elements in the solar atmosphere were acting like the trace impurities he injected into his electric arc in the laboratory. He therefore felt able to say with certainty that the elements were present on the Sun, even though some of the expected spectral lines were missing. As a result, Lockyer was able to add six new metals to the list of elements positively identified on the Sun—an increase of 50 per cent in the total known at that time.

These various developments were more than sufficient to convince Lockyer that the previous consensus of opinion had been quite wrong in supposing that the spectrum of any given element was invariable. He now began to examine suspiciously another accepted opinion, that the spectrum of each element was entirely unrelated to the spectrum of any other element. In particular, he began to study the possibility that some spectral lines might be common to two, or more, different elements. If this was to be tested, he would need to know the accurate positions of a large number of lines in the spectra of several different elements; so, during the early seventies, his laboratory assistants at South Kensington were kept hard at work mapping spectra in detail.

Lockyer was not the first to consider this possibility. Several of the leading spectroscopists of the time, Kirchhoff, Ångström, and Thalén, had suggested that different spectra might show coincident lines. But the significance of such coincidences was uncertain, since it might simply be due to the presence of one element as an impurity in the others. Lockyer could now eliminate this uncertainty, he believed, by considering only the short lines, for an element present as an impurity would give only long lines. Hence, he argued, if the short lines of two elements coincided, this represented a genuine coincidence of spectral lines unaffected by impurities. The only remaining problem, though this was a large one, was the possibility of chance coincidence. Since the spectrum of the average element contained a very large number of lines, there was a finite probability that two lines would coincide in position purely by accident, and not because of some physical correlation. The higher the dispersion of the spectra (i.e. the larger the scale employed) the less likely such coincidences would be. For this reason Lockyer continually tried to obtain instruments giving a great dispersion. Nevertheless, although this reduced the likelihood of chance coincidence, it could never eliminate it completely. He therefore also examined the coincident lines he found for some distinguishing feature which would definitely show that the coincidences were physically significant. He finally found what he was looking for, as he so often did, not in further laboratory work, but in his solar research.

Since he had first thought to examine localised regions of the Sun, Lockyer had spent much time comparing the spectra of the photosphere, sunspots, and prominences. He had noted that, although many of the lines visible both on the general surface of the Sun and in sunspots were widened in the latter, the degree of widening varied from line to line. Similarly, although the lines visible in prominences were bright, the extent to which they were brightened varied with the individual line. Now he made a new leap forward. A consideration of his past results suggested to him that the lines most widened in sunspots and brightest in prominences were also those that were coincident in different spectra. Here, he thought, was the clinching proof he had been looking for that coincident lines had some physical meaning. Since these particular lines seemed to occur under such a wide variety of conditions on the Sun he decided that they should be called 'basic' lines.

It should be remarked that this idea of Lockyer had to some extent been anticipated by his friend, Young, in the United States. In 1872, Young announced in *Nature* that the lines most commonly observed in prominences were also those that appeared to be common to two or more elements. Unlike Lockyer, Young felt it most probable that the coincidences could be explained either in terms of impurities in the laboratory spectra, or as chance superposition due to lack of resolution. But he pointed to the third, less likely, possibility, that there was some '. . . similarity between the molecules of different metals as renders them susceptible of certain synchronous periods of vibrations.'[1]

Lockyer was now becoming increasingly unhappy with the accepted picture of spectroscopy. The crux of the matter seemed to him that spectral lines were not behaving in a predictable, uniform manner. We have noted his observations of the way in which the relative intensities of lines varied according to the light source used. But he had found equally striking differences of another sort in the Sun, namely that the lines of an element might not all show the same Doppler effect. For example, his observations of iron lines showed that they could sometimes be divided into two groups each indicating an entirely different rate of motion. Lockyer, like virtually all his contemporaries, envisaged a solar atmosphere whose temperature diminished from the surface outwards. In such an atmosphere, the height to which a given gas extended above the surface would be expected to depend on its molecular weight. At first glance, this appeared to be true, for hydrogen extended far out into space, whilst the heavy vapours of iron floated just above the photosphere. There were, however, certain discrepancies. The 1474 line was observed at eclipses to stretch out much further from the Sun than any hydrogen line. This could be explained away by saying that the unknown element producing this line was actually lighter than any element yet detected on Earth. But some of the other observations were not so easily dismissed. Thus the lines of magnesium vapour were found to extend further from the Sun than those of sodium; yet sodium had a lower atomic weight than magnesium.

Lockyer brooded for a long time over the various difficulties which faced contemporary ideas on spectra. Whichever way he looked at it, it seemed that these ideas were vitiated by the little attention they paid to the effects of temperature. During this period, Huggins had been examining the spectra of some of the brighter stars in as much detail as possible. He had shown that the white stars, which were accepted as hot, gave very simple spectra consisting predominantly of hydrogen, whilst the red stars, agreed to be the coolest, showed the fluted bands of molecules. In between these two groups, the yellow stars like the Sun showed atomic lines due to iron and other metals. Lockyer added to this the laboratory observation that, as the temperature was raised, a molecular banded spectrum could be observed to change into an atomic line spectrum; a result which, it was generally agreed, was caused by the dissociation of molecules into atoms at elevated temperatures. To Lockyer, the conclusion to be drawn from these two pieces of information seemed inevitable. If a rise in temperature dissociated a molecule into an atom, then a further rise in temperature would dissociate an atom into something simpler still. The beginnings of this further change could be seen in the laboratory, where it would explain the odd results he had been obtaining, but it was taken much further in the Sun and in the hotter stars. From this viewpoint, indeed, the Sun and the stars simply provided a series of furnaces which extended to higher temperatures than were realisable in the laboratory. Lockyer seems to have thought of atomic dissociation first in connection with the Sun and stars, and then to have extended it to laboratory spectra.

In his physical picture of what atomic dissociation meant Lockyer was obviously influenced by his period of active collaboration in the laboratory with Frankland. The analogy that came naturally to his mind related to the branch of organic chemistry in which Frankland was an acknowledged authority. The various hydrocarbon series—such as are found, for example, in petroleum deposits—were known to be built up in a regular manner by successive additions of carbon–hydrogen radicals. Lockyer suggested that atoms might similarly be built up from smaller sub-atomic units, analogous to the organic radicals. (This suggestion had already been strongly urged in the latter part of the fifties by Dumas, the French chemist, with whom Lockyer was acquainted.) The atoms found on Earth would correspond to the higher members of the hydrocarbon series. If they were exposed to sufficiently elevated temperature, however, the atoms would break down to their simpler groupings, which would, he supposed, have simpler spectra. This analogy immediately explained the existence of basic lines. Two different atoms on Earth could, at high enough temperatures, break down to similar, simpler groupings, and so have the same spectrum. At the temperatures available in the laboratory, some of the atoms would dissociate, but most would be in their normal state. Hence the spectra observed would be mainly unique, but would include some basic lines common to two, or more, elements.

This new insight into spectroscopy was announced by Lockyer in his Bakerian lecture to the Royal Society in 1873. It is worth remarking that the research described there was in a very real sense also the work of his assistants. An examination of Lockyer's laboratory notebooks for the early seventies[2] shows that, owing to his preoccupation with the investigations of the Devonshire Commission, he was only occasionally in the laboratory. As a result, his assistants contributed not only the actual experimental and observational work, but also an appreciable amount of theoretical speculation. Thus one or two of the suggestions in the Bakerian lecture actually stemmed originally from his assistant, Friswell. In the later seventies and the eighties, when Lockyer had more time to spare for research, work on extending and perfecting the dissociation hypothesis was more evidently under his immediate direction.

In introducing the dissociation hypothesis for atoms, Lockyer was not, in fact, far in advance of contemporary thinking. The term 'dissociation' had been introduced by the chemists in the late fifties to describe the break up of molecules into atoms on heating. It had first been used in this sense by the Frenchman, H. Sainte-Claire Deville, who subsequently also pointed out the possible further extension to atoms. In 1873, the year of Lockyer's Bakerian lecture, Deville posed the query, 'What guarantee is there that hydrogen is a simple substance, and that at temperatures of millions of degrees it will not split into two elements which can combine to form hydrogen.'[3] (It should be remarked that some contemporaneous estimates actually put the surface temperature of the Sun at about a million degrees.) Public attention had already been drawn to this question. The American mineralogist, T. S. Hunt, returning not long before from a visit to Deville in France, had lectured at the Royal Institution on the chemistry of the primaeval Earth. During the course of his talk he had suggested that the intense heat of stars might be capable of decomposing the elements. Hunt had stayed with Lockyer's friend Brodie during this visit to England. Not long afterwards Brodie had lectured on a similar theme to Hunt's, citing this time Huggins' observations of stellar spectra as proof for the dissociation of elements. Despite these indications that the concept of atomic dissociation was in the air during the early seventies, Lockyer subsequently implied that he was not acquainted with either Hunt's or Brodie's speculations, though by 1874 Brodie and he were cooperating in research on the problem.

We have noted the spectroscopic problems facing Lockyer both in the laboratory and at the telescope, and his consequent belief that atomic dissociation was a necessary hypothesis to explain them. It is characteristic that he laid the main stress of his argument for the hypothesis not on these detailed problems, but on a universal principle, that of the uniformity of nature. 'The question', he argued, 'is an appeal to the law of continuity, nothing more and nothing less. Is a temperature higher than any yet applied to act in the same way as each higher temperature which has hitherto been applied has done? Or is there to be some unexpected break in the uniformity of Nature's processes?'[4]

The appeal to continuity in nature was a commonplace of the nineteenth century. It became rapidly apparent, however, that many scientists interpreted this principle differently from Lockyer. This was especially true of chemists, and, because spectroscopy and atomic theory were still predominantly seen as parts of chemistry, it was the chemists who played the major role in debating Lockyer's dissociation hypothesis. Dumas, by this time Permanent Secretary of the Académie des Sciences, wrote to Lockyer from Paris, describing the discussion that had been evoked there by a preliminary note from Lockyer on his ideas.[5] According to Dumas, it was one of the best-attended and liveliest sessions the Académie had had for some time. Besides Dumas himself, the chemists Wurtz and Berthelot and the astronomers Leverrier and Janssen had all joined in the debate. It was evident from the comments that the chemists as a body felt less need than the physicists for a unified theory of matter. Berthelot argued that the concept of atoms that could be split was itself a violation of continuity in nature. But this was an extreme position; other disputants at the Académie were more favourably inclined towards Lockyer's hypothesis.

Lockyer was greeted with a similar mixed reaction when he spoke on dissociation to the Chemical Section of the British Association in 1873. Some were impressed. Gladstone, one of the senior chemists present, was later reported as saying that,

> . . . there was no impossibility whatever in what Mr Lockyer had advanced. They [the chemists] knew that what they considered to be elements now were only elements to them. They did not know how to decompose them; but he imagined that there was no chemist whatever who imagined that they would never be able to decompose them.[6]

Others were highly sceptical. The average opinion was probably that of Roscoe, who was not prepared to reject the possibility of dissociation out of hand, but was unhappy about the number of assumptions involved. He pointed out, for example, that Lockyer's hypothesis supposed that the temperature increased in going from the arc to the spark, and further increased from the spark to the Sun and the stars. Yet, in fact, neither the laboratory nor the astronomical sources could at that time be assigned any reasonably certain temperature at all.

Lockyer, himself, was well aware that his suggestion was still highly tentative, and that he would need to present much more detailed evidence before he could carry the scientific community with him. He therefore concentrated particularly on improving his spectroscopic data. For the next few years his assistants were back at work as hard as ever, mapping the spectra of the elements at the best possible dispersion and with the greatest possible attention to eliminating impurities. In pursuit of the latter objective, Lockyer turned to those of his chemist-friends who had for various reasons been involved in the preparation of pure samples, and pursuaded them to loan or give him some of the material they had prepared. Crookes gave him some thallium, Roscoe sent some vanadium and caesium, and Roberts provided refined silver, gold and platinum.

As a result of the investigations he took up during these years, Lockyer was led to modify his concept of dissociation in certain directions. For example, one suggestion in his Bakerian lecture had been that non-metals might be more complex in their nature than metals (a speculation which probably came from Friswell originally). This seemed to fiit in with some of the current theorising of his friend Brodie. Brodie had found that in his mathematical symbolism the element chlorine had the same form as the compound hydrogen peroxide. This suggested that chlorine might not be an element in the normal chemical sense, but some new type of compound. Chlorine being a typical non-metal, Brodie's speculation had much in common with Lockyer's, and they carried out various joint experiments together in 1874 to examine chlorine in more detail.

Lockyer had speculated that what was usually referred to as chlorine was really a compound of a new element with ozone. He therefore also wrote to Thomas Andrews in Ireland, who many years before had been a student under Dumas, and who, with Brodie, was recognised as the leading authority on ozone. Some years previously, Andrews had studied the effects of passing an electrical discharge through moist chlorine to see whether any changes were induced in the chlorine. Now, at Lockyer's request, he began a new series of experiments on the question. He was forced to conclude that there was no unrecognised relationship between chlorine and ozone,[7] and Brodie and Lockyer's experiments confirmed this. Moreover, the general idea of the compound nature of non-metals came in for strong attack from Roscoe, Balfour Stewart and Crum Brown, so Lockyer allowed this particular part of the original hypothesis to lapse. There was a brief revival at the end of the seventies when some experiments by Victor Meyer in Germany seemed to suggest that chlorine might, indeed, be a compound containing oxygen. It appears that Lockyer also toyed with the alternative hypothesis that chlorine might undergo some form of allotropic modification : just as oxygen could exist as either two or three atoms linked together, so could chlorine. But an assumption of this sort brought in its train a series of others, which were unacceptable, as chemists were quick to point out.[8]

The chemists were also suspicious when Lockyer's ideas on dissociation of elements led him to question current ideas on the measurement of atomic weights, for this was one area of chemistry that was felt at long last to have been put on a firm footing. Roscoe wrote to Lockyer towards the end of 1873 suggesting that it was time Lockyer put his ideas on paper, so that his friends could consider them quietly. In the following May he wrote firmly,

> I enclose Stewart's opinion on the v.d. [vapour density] and spectroscopic question, which appears to me just and sound—and I agree with him that we cannot draw any conclusions about atomic weights from spectroscopic considerations. No doubt there is a general proposition which may be stated that the more simple spectra as a rule correspond to the simplest atomic arrangement but when you come to apply this to particular cases I cannot see that you have evidence to carry out your view.[9]

And he advised Lockyer not to publish this aspect of his work until further evidence had accumulated.

Despite these setbacks, the over-riding result of the intensive researches at South Kensington during the seventies was to convince Lockyer that his original supposition was right. Atoms did break down at higher temperatures, and the effects of this could be observed spectroscopically. By the end of 1878, Lockyer was ready to unveil the detailed version of his dissociation hypothesis before the Royal Society. Word spread of what was to come, and the ensuing meeting was said to be one of the most crowded there had ever been at the Royal Society; the chemists, in particular, turned up in full force. Since Lockyer's preliminary announcement of his idea, there had been time for the chemists to consider at least the general likelihood of dissociation. If anything their mood seems to have hardened against it. Certainly Lockyer received rather rough handling from them at the meeting. Crookes, who was sympathetically inclined towards Lockyer's speculations, told him, 'The general impression among men I have spoken to is that you were badly treated the other night by the chemists.'[10] A major problem was that Lockyer, not being a chemist, thought about the ultimate nature of matter in a manner unfamiliar to chemists, yet supported his views with evidence which appeared to be chemically based. As one of his correspondents in the United States told him,

> It is evident from your paper that the strength of your convictions is based on circumstances, of which no description can give to the reader the same

impression, which they have produced on the mind of the observer. It is therefore not surprising to me that your evidence should not convince the chemists who are pledged to a very different theory of matter, and look for a 'transmutation' of altogether another kind.[11]*

The chemists could, of course, speak with greatest confidence on the experimental laboratory evidence that Lockyer put forward in support of his hypothesis. Here, there was considerable agreement that Lockyer was not taking sufficiently seriously the possible effects of small amounts of impurities in his samples. Henry Roscoe, in particular, warned Lockyer throughout the seventies to beware of contamination. Roscoe was seriously worried that his friend's penchant for grandiose speculation might ruin his reputation. In 1874, when Lockyer was looking around for a new position as the Devonshire Commission drew to a close, Roscoe wrote to him concerning the dissociation hypothesis,

Do you really think that you may not perhaps compromise yourself by publishing views which at present are at least doubtful? If any Govt. appointment were to be in the field might not some of your *friends* (?) make capital out of your views detrimental to your prospects? I only wish to draw your careful attention to these possible chances—and both S[tewart] and I think them well worthy of your consideration.

You've done splendid experimental work, why the d——l cannot you stick to that—you will be far better repaid by so doing, i.e. in my humble opinion.[12]

(Roscoe's Manchester colleague, Balfour Stewart, added a postscript, 'Put the papers in some archives to be opened at the general resurrection of the just.') Lockyer, typically, went on his way unperturbed.

In 1878, Roscoe was again hard at work, prior to Lockyer's more detailed announcement of the dissociation hypothesis, urging him to be cautious. According to his cousin, Stanley Jevons, Roscoe was afraid that Lockyer had blundered and would have to draw in his horns.[13] During this period, Roscoe not only repeated Lockyer's experiments at Manchester, but also spent some time at South Kensington actually going over the experiments with Lockyer's assistants there. His letters to Lockyer become sharper in tone, as he accumulated evidence against Lockyer's interpretation of the laboratory results.

When you tell me that 'the *thickest lead line* is the *only* line in the spectrum of the *chemically pure zinc*', I want to know what has become of the dozens of lines mapped by Huggins, Thalén and Kirchhoff in zinc? Do you mean to say that the temperature of the arc is so much higher than that of the spark (as used by these men) that complete dissociation is set up? Most people believe that the spark is hotter than the arc. Can you say *no*? This seems to the point which you must make clear. Why do you find *all* (or at any rate very many

* The letter ends, incidentally, by pointing out to Lockyer that it is written, 'with what is called a type-writer, an ingenious Yankee machine which one plays on like a piano.'

Lockyer as a young man.
(*Photograph by courtesy of Capt. Lockyer*)

Lockyer shortly before his death.
(*Photograph by courtesy of Capt. Lockyer*)

Lockyer's first wife.
(*Photograph by courtesy of Capt. Lockyer*)

Lockyer's sons. (*From left to right*) H. C. Lockyer, A. E. Lockyer, W. J. S. Lockyer, and N. J. Lockyer. This photograph was taken c.1906 on the steps leading up to Sir Norman Lockyer's study at his house in Pen-y-wern Road. (*Photograph by courtesy of Capt. Lockyer*)

The apparatus used by Lockyer at South Kensington to compare solar and laboratory spectra.
(Reproduced from *The Chemistry of the Sun*)

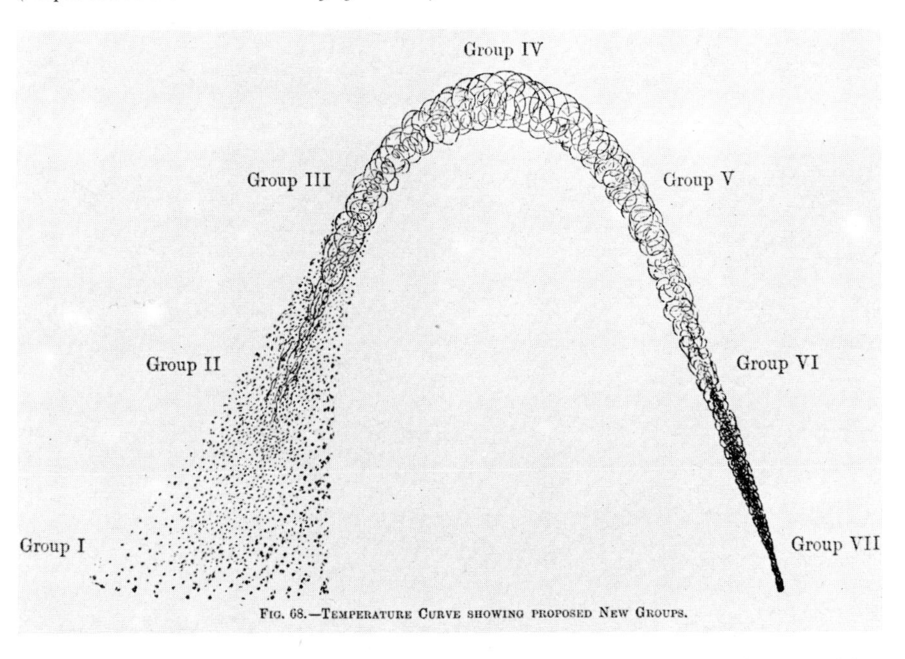

Lockyer's diagram of the curve of stellar temperatures according to his meteoritic hypothesis.
(Reproduced from *The Meteoritic Hypothesis*)

A WEEKLY ILLUSTRATED JOURNAL OF SCIENCE.

"*To the solid ground*
Of Nature trusts the mind which builds for aye."—WORDSWORTH.

No. 1.] THURSDAY, NOVEMBER 4, 1869. [PRICE FOURPENCE.

Registered for Transmission Abroad.]

PHILOSOPHICAL TRANSACTIONS.

The FELLOWS of the ROYAL SOCIETY are hereby informed that the First Part of the PHILOSOPHICAL TRANSACTIONS, Vol. CLIX. for the year 1869, is now published, and ready for delivery on application at the Office of the Society in Burlington House, daily, between the hours of 10 and 4.

WALTER WHITE,
Burlington House, Nov. 3, 1869. *Assistant Secretary R. S.*

Fourth and very much enlarged Edition, 70 Plates, 4 Coloured, 21*s*.

HOW TO WORK WITH THE MICROSCOPE. By Dr. LIONEL BEALE, F.R.S.

HARRISON, Pall Mall.

NEW WORK BY DR. BEALE, F.R.S.
Now ready, 5*s*. 6*d*.

PROTOPLASM; or, MATTER, FORCE, and LIFE.

JOHN CHURCHILL & SONS.

CROWN BUILDINGS, 188, FLEET STREET, LONDON.

SAMPSON LOW, SON, & MARSTON'S MONTHLY BULLETIN of their AMERICAN, COLONIAL, and FOREIGN PUBLICATIONS. 2*s*. 6*d*. per annum, Post free.

THE ENGLISH CATALOGUE OF BOOKS, giving the date of publication of every book published from 1835 to 1863, in addition to the title, size, price, and publisher, in one alphabet. An entirely new work, combining the Copyrights of the "London Catalogue" and the "British Catalogue." One thick volume of 900 pages, half morocco, 45*s*.

*** The Annual Catalogue of Books is with Index of Subjects. 8vo. 5*s*.

INDEX to the SUBJECTS of BOOKS PUBLISHED IN THE UNITED KINGDOM DURING THE LAST TWENTY YEARS—1837–1857. Containing as many as 74,000 references. One vol. royal 8vo. morocco, 1*l*. 6*s*.

LESSONS IN ELEMENTARY CHEMISTRY, Inorganic and Organic. By HENRY ROSCOE, F.R.S. Professor of Chemistry in Owens College, Manchester. With numerous Illustrations and Chromo-Litho. of the Solar Spectra. Fifteenth Thousand. 18mo. cloth, 4*s*. 6*d*.

It has been the endeavour of the author to arrange the most important facts and principles of Modern Chemistry in a plain but concise and scientific form, suited to the present requirements of elementary instruction. For the purpose of facilitating the attainment of exactitude in the knowledge of the subject, a series of exercises and questions upon the lessons have been added. The metric systems of weights and measures, and the Centigrade thermometric scale, are used throughout the work.

"A small, compact, carefully elaborated, and well-arranged manual."—*Spectator.*

"It has no rival in its field, and it can scarcely fail to take its place as the text-book at all schools where chemistry is now studied."—*Chemical News.*

MACMILLAN & CO. LONDON.

NEW AND EXHAUSTIVE WORK ON ORNITHOLOGY.

In Monthly Parts, **7d.** *Part* **I., November 25.**

CASSELL'S BOOK OF BIRDS,

TRANSLATED AND ADAPTED FROM THE

Text of the Eminent German Naturalist, Dr. BREHM,

BY THOMAS RYMER JONES, F.R.S.

Professor of Natural History and Comparative Anatomy;
King's College, London.

With upwards of 400 accurate Engravings on Wood, executed expressly for the Work, and a Series of exquisite Full Page Plates, printed in Colours, from Original Designs by F. W. KEYL.

*** SPECIMENS *of the* COLOURED PLATES *are on view at all Booksellers, from whom* FULL PROSPECTUSES *may be procured.*

CASSELL, PETTER, & GALPIN, Ludgate Hill, London, E.C.; and 596, Broadway, New York.

CHARLES KEMBLE'S SHAKESPEARE READINGS. Being a Selection of Shakespeare's Plays as read by him at his Public Readings. Edited by R. J. LANE. 3 vols.

This Edition is a careful reprint from the copy of Shakespeare used by Mr. Kemble at his recitals, and it is especially fitted for public and family readings by judicious omissions, and by the insertion of accents over those words which Mr. Kemble emphasised in his delivery.

THE SATIRES AND EPISTLES OF HORACE. Translated into English Verse, by J. CONINGTON, Professor of Latin in the University of Oxford.

London: BELL & DALDY.

On the 1st inst. was published, post 8vo. cloth, a Second Edition, greaty enlarged, of

CHEMISTRY: General, Medical, and Pharmaceutical, including the Chemistry of the British Pharmacopœia. By JOHN ATTFIELD, Ph.D. F.C.S., Professor of Practical Chemistry to the Pharmaceutical Society of Great Britain.

JOHN VAN VOORST, 1, Paternoster Row.

This day is published, post 8vo. cloth, 643 pp. price 7*s*. 6*d*.

THE BIRDS OF SOMERSETSHIRE

By CECIL SMITH, of Lydeard House, near Taunton.

JOHN VAN VOORST, 1, Paternoster Row.

In 5 volumes, post 8vo. cloth, 4*l*.; or with the Figures of Species coloured, 5*l*. 5*s*.

BRITISH CONCHOLOGY: or, an Account of the Mollusca which now inhabit the British Isles and the surrounding Seas. By JOHN GWYN JEFFREYS, F.R.S. F.G.S. &c. Each volume has a coloured Frontispiece and eight Plates to illustrate the Genera, and the last volume has 102 supplementary Plates with figures of all the species and principal varieties of the shells—being altogether 147 Plates.

JOHN VAN VOORST, 1, Paternoster Row.

MR. DÜRR of LEIPZIG has been appointed Agent to the Publishers of "NATURE" for GERMANY and EASTERN EUROPE. BOOKS FOR REVIEW, ORDERS, and ADVERTISEMENTS may be forwarded direct to him. Address: ALPHONS DURR, Leipzig, Germany.

The Solar Physics Observatory, South Kensington, towards the end of the nineteenth century.
(Photograph by courtesy of the Cambridge Observatories)

W. J. S. Lockyer at South Kensington preparing to observe the Leonid meteor shower (1898–99).
(Photograph by courtesy of Mrs. W. J. S. Lockyer and the Meteorological Office)

The wreck of the *Psyche* on the 1870 eclipse expedition.
(Reproduced from *The Life and Experiences of H. E. Roscoe*)

A. Fowler (on right) *en route* to the 1898 eclipse expedition in India.
(*Photograph by courtesy of Mrs. W. J. S. Lockyer and the Meteorological Office*)

The eclipse expedition in India, 1871.
(Reproduced from the *Illustrated London News*)

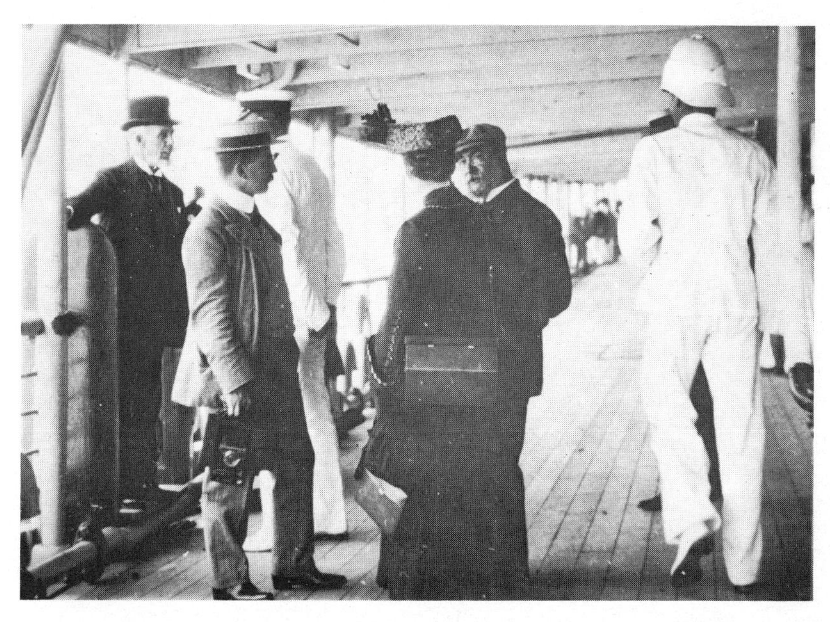

Lockyer (centre) arriving at Gibraltar during the 1905 eclipse expedition.
(*Photograph by courtesy of Mrs. W. J. S. Lockyer and the Meteorological Office*)

Inner court and sanctuary at Edfû; from a photograph taken by Lockyer during his investigations of the orientation of Egyptian temples. (Reproduced from *The Dawn of Astronomy*)

The Hill Observatory, Sidmouth, in 1915.
(*Photograph by courtesy of Mrs. W. J. S. Lockyer and the Meteorological Office*)

or the *most*) of the lines of various metals coincident or identical whilst all former observers found *none* (or next to none) coincident? They looked over the whole spectrum. You take only a part. Is your definition and your dispersion so much better than theirs? . . . These remarks are things that will strike everyone and must (as it seems to me) be met by you.[14]

Roscoe's anxiety was heightened by the fact that his researches definitely showed that some of Lockyer's supposedly pure substances were actually slightly contaminated. In particular, the zinc mentioned in the foregoing letter and the thallium that Lockyer had obtained from Crookes were found to contain small amounts of lead. This latter discovery led to some frantic correspondence between Lockyer and Crookes, for the thallium was allegedly that which Crookes had used in determining the atomic weight of that element. Roscoe's research therefore seemed to cast doubt on one of Crookes' major claims to scientific fame (a prospect which did not cause Roscoe much sorrow, for he was not very fond of Crookes). Crookes, highly agitated, re-examined the thallium samples in his laboratory with some care, and ultimately discovered, to his great relief, that it was all due to a mistake by his former assistant. It appeared that as the pure thallium Crookes had used in his own work was consumed, the assistant had added new lumps of thallium obtained commercially (and only partially purified) to the bottle. When Lockyer had asked for a sample, Crookes had sent him a piece of the commercial thallium under the impression it was purified. In his letter informing Lockyer of this, he added with an attempt at resignation,

I have written to Roscoe about the ingot of thallium. He must have put himself to great trouble to spread the knowledge he thought he had got of the great impurity of my atomic weight thallium, far and wide on Thursday, for I have heard of it in three separate quarters already! I wonder whether he would be equally keen in advertising anything greatly to my credit if he happened to find out. There's a deal of human nature in scientific men![15]

Although Crookes' reputation was thus saved, it nevertheless remained true that Lockyer had been obtaining evidence for his dissociation hypothesis from impure material. Despite Roscoe's urgent advice, however, Lockyer refused to tone down his claims, and just before the Royal Society meeting Roscoe was forced to warn him, 'I shall come up on Thursday and may feel bound to say a few words on this question of impurity, as applying to the cases which I have examined.'[16]

Lockyer does not seem to have been greatly dismayed by the criticism he came in for at the meeting, for only a month later he was off on another tack. Towards the end of January 1879, he received a letter from Stokes discussing what further experiments were necessary in order to provide clinching evidence of dissociation.

. . . the question observe is not, Are the elements compound bodies? But, has any satisfactory evidence been now obtained that they are compound bodies? You would, I imagine, find plenty of chemists from Prout downwards, who would regard it as most probable that they were compounded. I may say that, in common I suppose with multitudes of others, I have long supposed for my own part that they were. I have further speculated on the lines obtained in electric discharges as the most probable source of possible evidence of their compound nature. . . .

The fact is that spectral analysis so transcends in delicacy our chemical means of detection, that the non-detection of impurities by chemical means does not prove that the impurities are not there in quantity sufficient to show themselves by spectral analysis. The only way to answer satisfactorily as it seems to me the objection arising from the supposition of impurities, if you depend on chemical analysis, would be to purposely introduce a quantity of the suspected impurity not too small for chemical detection, but then show, supposing, which is not likely, that the fact is so, that the introduction would not account for the phenomena observed.

But if the action of the arc is continued, we might surely expect that we should be able to give some chemical evidence of a minute quantity of the products of decomposition.[17]

Henry Armstrong, too, urged Lockyer that collection of decomposed material would be the best way of swinging chemical opinion to his side. Lockyer had, indeed, already been thinking along these lines himself. Thus advised, he applied his assistants to the new task throughout the first half of 1879. By the time the British Association meeting (held in Sheffield that year) came round, Lockyer believed he had succeeded. He announced at the meeting that several elements—sodium, phosphorus, magnesium, sulphur, indium and lithium—had been partially decomposed into hydrogen. Prout's hypothesis had been vindicated with a vengeance. This result seemed to tie in very nicely with Huggins' observations of stars. The hottest stars showed hydrogen lines because the metals seen in somewhat cooler stars, like the Sun, were broken down at the higher temperatures.

But the chemists gave this new claim an even icier reception than they had accorded the dissociation hypothesis at the Royal Society meeting. Even the most sympathetic were unhappy with this development. Brodie warned of the ease with which metals occluded gases,[18] and Crookes, although he congratulated Lockyer on his wonderful new discovery, was forced to report that he could not repeat the experiments.[19] Roscoe, indefatigable as ever, devised a series of experiments similar to Lockyer's, and demonstrated quite clearly for the alkali metals that the hydrogen was either occluded in the metal, or was present as the hydroxide. In any event, it was certainly not created by decomposition of the metal on heating. As he showed, if the same lump of metal was heated strongly several times in succession, it eventually ceased to evolve hydrogen.[20] He had warned Lockyer some time before that this would be the case, and had suggested that his friend should intone the words, 'get thee behind me Sodium'.[21]

Lockyer had made a serious blunder, and he usually glossed over this aspect of his work in later years. Nevertheless, he evidently remained very attached to the idea of collecting hydrogen as a dissociation product. In 1894 he proposed that the substance giving the D lines was not sodium, but a sodium plus hydrogen combination, and hoped to relate this to dissociation. Stokes, who undertook experiments on the question, wrote to him,

> One experiment has been tried which is favourable to the supposition that what gives the D line is sodium, and not a hydride, but one or two more are projected which are likely to be decisive. I may be able perhaps to tell you on Saturday. When I say decisive, I mean as to the question, metal or hydride? Whether the so-called element sodium is or is not a compound, and if a compound whether hydrogen is one of the constituents, are speculative questions which remain as they were.[22]

The experiments did not go in favour of Lockyer's idea, but he was still toying with it in 1898, for we find Roberts-Austen then writing to him,

> Sodium is usually collected and stored in petroleum or other organic fluids. Chemists would therefore view with extreme suspicion the hydrogen extruded from sodium.[23]

The rapid deflation in 1879–80 of Lockyer's claim to have isolated decomposition products hardly enhanced the status of his dissociation hypothesis. Its acceptability was further diminished by the work of Liveing and Dewar on the coincidence of spectral lines. Some time before, after his appointment to the Chair of Experimental Philosophy at Cambridge, Dewar had taken up the study of the spectroscopic effects of molecular dissociation. He had subsequently also been appointed Professor of Chemistry at the Royal Institution, but had continued his experiments at Cambridge in conjunction with Liveing. These two first clashed with Lockyer in 1878. Lockyer then claimed he had detected the absorption lines of carbon in the solar spectrum. This, if verified, was an important observation, for it would have been one of the first identifications of a non-metal (hydrogen was considered to have the chemical properties of a metal) in the Fraunhofer spectrum. Only the year before, Henry Draper in New York had announced that he had identified bright photospheric lines of oxygen in the solar spectrum—Lockyer was, indeed, one of the first people to whom he communicated the news—but his claim was immediately disputed, though he was not conclusively shown to be wrong until some time later. Lockyer's claim, too, was not allowed to stand for long. Liveing and Dewar soon reported that his result was a misidentification. Lockyer's annoyance at this was greatly aggravated when he discovered that there was an important difference between the paper they had read at the meeting of the Royal Society, when the question of carbon on the Sun had been discussed, and the paper they had subsequently published.[24] Lockyer immediately wrote off fiercely to the Royal Society, and even more fiercely to Dewar.

The battlefront widened after Lockyer's detailed description of his dissociation hypothesis. By the middle of 1879, Lockyer was accusing Liveing and Dewar of deliberate prevarication.

> I understood you [Dewar] to say at the Royal Society that if I would point out any cases in which you had announced results which had been previously published by myself you would make matters right. I therefore prepared a list of some of the most obvious recent cases. I understand from your letter that you now decline to keep your promise. . . .
>
> I have given you an opportunity of doing justice to your fellow worker, if you are disinclined to take advantage of it it is your affair, it is certainly not mine. I have already received more credit for my work than I either desire or deserve.[25]

Fortunately, it was Lockyer's habit to show drafts of his more highly charged denunciations to his friends, who normally advised him to tone them down, advice he seems often to have accepted. In this instance, he sent a draft of the intended letter to Dewar to his friend at the Mint, W. C. Roberts. Roberts replied hurriedly, 'I think your letter to Dewar too strong, as it would be very difficult for him to make it up after having a letter like that.'[26] And Lockyer accordingly modified the wording of his letter to some extent.

The battle with Dewar nevertheless continued vigorously into the eighties. In 1881, Abney wrote from the British Association meeting in York to warn Lockyer that Dewar was going to lecture there on Lockyer's spectroscopic researches.[27] Abney urged his friend to make a special effort to attend (Lockyer was no longer a constant visitor to Association meetings), and told him that a strong army of supporters, including Armstrong and Schuster, would be there to back him. It is worth adding that Dewar's talk is one of the few items not reprinted in the British Association Reports for that year.

This controversy was a major factor in affecting the attitude of scientists towards the dissociation hypothesis during the eighties. Dewar, who was a magnificent experimenter, though by no means an easy man to get on with, became convinced that Lockyer's work on basic lines was unsound. He and Liveing therefore re-examined many of the line coincidences that Lockyer had claimed to be of fundamental importance. They pointed out, first of all, that several of these lines, when examined under a high enough dispersion, proved not to be single at all, but doublets or multiplets. In this connection, they cited work by C. A. Young in the United States. Young had, indeed, already written to Lockyer during 1879, warning him that the new, more detailed measurements of the solar spectrum being made at Princeton showed several supposedly single lines to be double.[28] It is some commentary on the pace of scientific communication at this period, that, when Lockyer asked Young whether he had read Liveing and Dewar's criticisms, his friend replied : 'I fear I shall not until we get our copies of the [Philosophical] Transactions two or three years hence.'[29]

In the next few years, this resolution of apparently single lines of the solar spectrum into multiplets was taken up by H. A. Rowland, also in the United States. Rowland, who was by training an engineer, came to be the best maker of diffraction gratings in the world, and he used them himself to map out the solar spectrum at a higher dispersion than had ever before been attempted. Later in the eighties, Lockyer acquired a Rowland grating, and it may be that the measurements he made with it helped force him to modify some of his earlier ideas on dissociation. Rowland's measurements showed that several of those basic lines not previously resolved by Young or Liveing and Dewar could be resolved by his more powerful apparatus.

There can be no doubt that this increasing ability to split lines, which he had supposed to be single, upset Lockyer. In 1882, Rowland visited England. A colleague who was with him subsequently reported back to Johns Hopkins University,

> Lockyer before he saw Rowland's results reminded me of this [drawing of crowing cock] and afterwards of this [drawing of drooping cock]. Lockyer evidently is much disturbed and was not at the meeting of the Physical Society. He does not impress me as a gentleman. Did not call on us in London.[30]

In 1887, Lockyer published *The Chemistry of the Sun*, partly to summarize his ideas on dissociation, and partly to answer the objections which had been raised in the intervening years against the dissociation hypothesis. The particular objection that basic lines could be split into more than one component he tried to counter by invoking the concept of 'line shifts' (a term he seems to have been the first to apply in spectroscopy).* This explanation depended on observations from the late seventies onwards which showed that the conditions prevailing in a light source could not only affect the appearance of a spectral line, but could also slightly alter its wavelength. Lockyer therefore suggested that since different atoms simplified by dissociation under different conditions, the dissociation products would produce the same basic line at wavelengths

* He later changed the terminology to 'line shifting', on the grounds that a shift was an article of clothing.

that were not exactly coincident, hence giving the phenomenon of multiple lines.

Lockyer's contemporaries were not particularly impressed by his new suggestion. Jewell in the United States used one of Rowland's gratings for a detailed investigation of line shifts in the early 1890s, and remarked that the result, 'effectually disposes of the necessity of any dissociation hypothesis to account for most solar phenomena.'[31] Lockyer seems to have expected a sceptical response to this line of argument. When he introduced the concept of line shifts, he added, 'I am most anxious that this should not be regarded as special pleading or as an ingenious evasion of a difficulty.'[23]

To return to Liveing and Dewar's indictment of Lockyer's dissociation hypothesis, the second major argument they advanced against it, besides multiplet blending, was the argument that Roscoe had been pressing previously, namely that several of the alleged line coincidences were due to the presence of impurities in the laboratory samples. It became clear during the eighties that Lockyer's method of eliminating impurities by looking for their long lines in the spectrum, although correct so far as it went, did not identify all impurities. In a similar way, it was shown that Lockyer's separation of long and short lines did not always lead to a certain identification of elements in the Sun. Half-a-dozen of the elements which he had claimed as definitely present in the solar spectrum were subsequently shown to have been identified on the basis of inadequate evidence.

In view of this mass of evidence against the dissociation hypothesis, it might be thought surprising that dissociation remained a live issue into the nineties. The reason was that, if the developments in chemistry seemed to go against dissociation, there were arguments from physics and astronomy which were more in its favour. The relevant development in physics had begun in the fifties and sixties, and concerned the study of electrical discharges through rarified gases. The method by which a glass tube could be exhausted of most of the gas it contained—hence allowing an electrical discharge to pass—was perfected by Geissler in Germany during the 1850s. Just about the same time, Rühmkorff in France developed an induction coil that could be used to pass such a discharge. Lockyer, himself, used a Geissler tube, but simply as a light source. The physicists, however, were mainly interested in the actual physical events occurring in the tube.

It soon became evident that, when a discharge was passing between the anode and the cathode of a tube, the latter emitted some kind of radiation. Where opinions differed was over the nature of this radiation. The belief in Germany, where much of the early work on these 'cathode rays' was done, was that they were electromagnetic in nature—like light. In England, where studies of the phenomenon were taken up in the 1870s, most physicists favoured the idea that the rays were some kind of material substance. This Anglo-German battle continued to rage until the end of the century. In 1896, Lord Kelvin, by then in his seventies and splendidly dogmatic, wrote to Lockyer, 'I am glad to see Rücker convincingly damning the torrent of nonsense about the "Kathode (undulatory) ray" made in Germany . . .'[33]

The British viewpoint was excellently stated by Crookes in the Bakerian lecture for 1879, not many months after Lockyer had addressed the Royal Society on dissociation. The lecture was important for Lockyer since Crookes proposed that the cathode rays actually constituted a new (fourth) state of matter. Ordinary matter was in some way broken down in the discharge tube. In fact, Crookes came to favour, almost as strongly as Lockyer, not only the possibility that elements were built up from something simpler, but that, given the correct conditions, they could be broken down again in the laboratory. In 1883, Crookes was again invited to give the Bakerian lecture, and this time he chose to describe his work on the substance yttria. 'No longer than twelve months ago,' he remarked, 'the name yttria conveyed a perfectly definite meaning to all chemists. It meant the oxide of the elementary body, yttrium.'[34] By extensive fractionation, however, he had—so he claimed—split yttria into at least five fractions; and, he said, '. . . these constituents of old yttrium are not *impurities* . . . They constitute a veritable splitting up of the yttrium molecule into its constituents.'[35]

Lockyer was, of course, delighted with this, though he was subsequently warned by his friend Armstrong that the rare earths formed a complex group, and could be easily misinterpreted.[36] The most important aspect from his viewpoint was Crookes' claim to have verified the existence of the yttria fractions spectroscopically. According to Crookes, the ordinary spectrum of yttria simply represented the superposition of the different spectra from the various simpler bodies into which it could be split; a view he developed in his address as President of the Chemistry Section of the British Association in 1886. This agreed very well with Lockyer's belief that the simpler the substance, the simpler its spectrum.

The lightest substance known, hydrogen, also had one of the simplest spectra. It is not surprising, therefore, that the hydrogen spectrum was much examined in the eighties and nineties in the hope that it might hold the key to the theoretical interpretation of spectra. For once the hope materialised. Huggins first pointed out, from his photography of stellar spectra, that the hydrogen lines in the visible and ultraviolet formed a series, with the lines getting progressively closer together from the red end of the spectrum to the blue. Then, in 1885, a Swiss schoolteacher, J. J. Balmer, showed that the series could be represented by a very simple empirical formula. This discovery led at once to a widespread investigation of possible numerical relationships between the wavelengths of spectral lines. As a result, a number of such relationships were brought to light, and it was found, to most people's surprise, that they existed not only between the different spectral lines of a single element, but between the spectral lines of different elements as well. This latter discovery tended to make some laboratory spectroscopists (H. Kayser in Germany, for example) more receptive to Lockyer's ideas of dissociation.

Crookes, in his speculations on matter, actually suggested that hydrogen might not be the simplest substance. He proposed instead that there might be a more primitive material—he called it 'protyle'—which possibly had half the atomic weight of hydrogen. He further speculated that this protyle might be related to Lockyer's proposed helium. In the 1890s, E. C. Pickering at Harvard pointed out that the spectra of some hot stars contained a series of lines that looked very much like the Balmer series of hydrogen, but with additional lines inserted. Lockyer took this as evidence that hydrogen could indeed break down under the extreme conditions found in hot stars, and distinguished the new spectrum as being due to proto-hydrogen. The subsequent confusions over this 'hydrogen' spectrum were not finally cleared up until the introduction of the quantum theory just before the First World War, when it became apparent that the substance involved was actually ionised helium.

Although Crookes' speculations certainly weighed with Lockyer, they were not of very great influence in the scientific world at large. The trouble was that Crookes' approach was basically that of the chemist. So while his fellow-chemists rejected his theorizing—as they rejected Lockyer's—the physicists, who might have been more sympathetically inclined, ignored it. During the eighties, Lockyer actually received the greatest sympathy for his ideas on dissociation from the astronomical community. There was very good reason why this should be so. Spectroscopic observations of the Sun and the stars were providing so many inexplicable data that a radically new approach seemed almost necessary. In commenting on the dissociation hypothesis, C. A. Young remarked in 1886,

> I suppose that at present the weight of scientific opinion is against him [Lockyer]; but for one I do not believe his battle is lost . . . our received theories so stumble, hesitate, and falter in their account of many of the simplest phenomena of the solar and stellar atmospheres, that a strong presumption still remains in favour of the new hypothesis. I am not prepared to accept it yet; but certainly not to reject it.[37]

As always, Lockyer's chief concern was to explain the spectroscopic observations of the Sun. The problems here were tied up particularly with relationships between the different layers of the Sun's atmosphere, and were therefore partly dependent on observations made during total solar eclipses. For this reason, Lockyer remained vitally interested in the organisation of eclipse expeditions throughout the whole of his life. He was already beginning to turn over the possibility of the dissociation of atoms in his mind when the eclipse of 1875 (visible from Indo-China) occurred. Although Lockyer was the obvious leader for a British expedition, he was far too much involved in the final stages of the Devonshire Commission to be able to go. But he was naturally in the thick of the preparations for the dispatch of an eclipse party. As was so often the case, these preparations were attended by various dissensions. There survives the draft of a letter from Lockyer to Hooker, then President of the Royal Society, bitterly accusing the Astronomer Royal of opposing attempts to organise an expedition (as, so Lockyer says, he had done in 1870 and 1871), and seeking the help of the Royal Society to redress the balance.[38] However, the work went ahead. Stokes and Lockyer were put in overall charge, and they selected a young Anglo-German Jew, Arthur Schuster, to command the expedition. Schuster had studied under Lockyer's old friends at Manchester, Roscoe and Balfour Stewart, as well as working with Kirchhoff and Bunsen in Heidelberg. He was thus well equipped to undertake the spectroscopic observations at the eclipse, which Lockyer wished to make a central part of the expedition's work.

Schuster, whose family had just moved to London, was living fairly close to Lockyer, and seems to have had quite close contact with him in the seventies. He offered for example during this period to translate some of Lockyer's papers into German, so that his work might be better known on the Continent,[39] and sought his advice on the possibility of writing a new text on spectroscopy.[40] Later in the century their relationship was to become more distant, and Schuster found much to criticize in the dissociation hypothesis.

The 1875 expedition was highly successful. Despite the usual delays and accidents to the transportation, and a quarrel between Schuster and Meldola over a question of authority, the party managed to carry out their entire programme without a hitch. Lockyer was mainly concerned at this eclipse with an attempt to photograph the spectrum of the chromosphere and inner corona with a prismatic camera (i.e. a camera with a prism in front of the lens). The ability to obtain a permanent record of this fleeting phenomenon would obviously be a major step forward. Proctor had violently denied the possibility of such photography, and had claimed that the whole endeavour would prove to be a waste of time. In the event some lines were recorded, with the result that, back in London, the relative heights of the emission lines concerned were measured with greater accuracy than had previously been attained. Most importantly for the dissociation hypothesis, the results confirmed that lines attributed to calcium on the basis of laboratory spectra extended to a significantly greater height above the Sun's surface than those of magnesium, although the atomic weight of calcium was considerably greater than that of magnesium. Lockyer felt that this observation could be understood only in terms of atomic dissociation. The calcium which was observed high in the solar atmosphere was not ordinary calcium at all, but a simpler dissociation product, which, weighing less, could therefore rise to a greater height.

Lockyer had hoped to extend this work at the 1878 eclipse in the United States, but the death of a son caused him to cancel his original plans, and he finally travelled out as a private individual. As it happened, he made an interesting discovery simply from naked-eye observation of the eclipse, for he noted that the shape of the corona had changed greatly since he last saw it in 1871. Since the latter eclipse had occurred near a maximum of the sunspot cycle, whereas the 1878 eclipse was near a minimum, Lockyer deduced that the appearance of the corona varied during the course of a cycle. The importance of this observation, which was initially contested, was that Lockyer now definitely began to rethink the model of the solar atmosphere he had first drawn up with Frankland in the early 1870s. His new scheme envisaged a circulation of the solar atmosphere which changed during the sunspot cycle. At maximum, streams of gas rose from the Sun's surface to form the corona, and then fell back to the surface again, their impact forming the sunspots. But at minimum, these currents were severely reduced in extent; as a result, there was little gas in the corona at this time, and so few sunspots. This hypothesis not only explained the change in shape of the corona during the solar cycle, it also explained a change in spectroscopic appearance, which other observers had detected. At the 1871 eclipse, the bright 1474 line had been easily visible, whereas, at the 1878 eclipse, only Young had been able to detect it, and he had remarked on its faintness. This could be understood, according to Lockyer, in terms of the presence and absence of gas, respectively, at these two times.

After the 1878 eclipse, Lockyer's concept of the solar atmosphere seems to have changed not only in terms of its variability during the solar cycle, but also concerning the way in which its chemical nature varied with height above the surface. He now came to view this as entirely determined by dissociation. He later contrasted his views of the solar atmosphere at the beginning and end of the seventies in the following way. His original hypothesis, he said, supposed that:

(1) We have terrestrial elements in the sun's atmosphere.

(2) They thin out in order of vapour density, all being represented in the lower strata, since the temperature of the solar atmosphere at the lower levels is incompetent to dissociate them.

(3) In the lower strata we have especially those of higher atomic weight, all together forming a so-called 'reversing layer' by which chiefly the Fraunhofer spectrum is produced.[41]

On the other hand, his new hypothesis supposed that:

(1) If the terrestrial elements exist at all in the sun's atmosphere they are in process of ultimate formation in the cooler parts of it.

(2) The sun's atmosphere is not composed of strata which thin out, all substances being represented at the bottom; but of true strata, like the skins of an onion; each different in composition from the one either above or below.

(3) In the lower strata we have not elementary substances of high atomic weight, *but those constituents of the elementary bodies which can resist the greater heat of these regions.*

Locker believed that his new model of the Sun would compel acceptance of the dissociation hypothesis, even if this hypothesis ran into difficulty in other respects. His fellow-scientists were by no means so sure. Stokes wrote to him pointing out that other explanations of the eclipse observations were equally possible.

If different elements show their bright lines extending to different heights from the sun's surface, and as a rule those of higher atomic weight extend less high, that is not, I take it, due to any effect of gravitation effecting a separation by difference of specific gravity. If you mix carbonic acid and hydrogen, and leave the mixture to itself, the carbonic acid will not settle to the bottom, and the hydrogen collect at the top. I think the most probable explanation is that the different elements require different temperatures to keep them in the gaseous form, those of higher molecular weight as a rule requiring a higher temperature; and when they are condensed into liquid or solid, forming a mist, they can no longer give out their characteristic bright lines. Then again some when they get far enough from the sun to be cool enough may enter into chemical combinations, when they would cease to give lines, or would give a different system. And even of those that remain in a gaseous state, the temperature required to make them show their characteristic lines might be different in different cases, so that on this account again there might be a difference in height at which different elements ceased to show their lines.[42]

Even those astronomers who were prepared to accept dissociation as a possible explanation of the spectroscopic observations of the solar atmosphere were taken aback by one deduction Lockyer immediately drew from his revised picture of the Sun. As we have seen, Lockyer's original theory, in common with the general belief, postulated a region of the atmosphere called the reversing layer, which lay just above the photosphere and produced the Fraunhofer absorption lines. Lockyer now believed that these lines were actually being produced throughout the whole of the solar atmosphere: those from the dissociated atoms being formed lower down than those from the ordinary atoms. He therefore had to deny that any such thing as a distinct reversing layer existed. In doing this, he seemed to run directly counter to the observational evidence, for Young at the 1870 eclipse, and other observers since, had observed the flash spectrum of the Sun (that is the sudden reversal of the Fraunhofer lines from dark to bright at the limb of the Sun near the beginning and end of a total eclipse), and this had been accepted as conclusive evidence for a reversing layer. Hence, Lockyer had to show immediately that his ideas could be reconciled with these observations. He proceeded to argue in the following way. When an observer looked at the solar limb, his line of sight intersected all the atmospheric layers one after the other, so that, even if a particular spectral line was produced only in the uppermost part of the solar atmosphere, nevertheless it would appear to project from the limb upwards. Thus, his new picture of the solar atmosphere would, in any case, produce the semblance of a flash spectrum. There would, however, be differences in detail between the appearance of the flash spectrum on Lockyer's model and on the more generally accepted model. In the latter, although the lines from the various elements would have different lengths, they would all be brightest near the photosphere, for there the density of the vapours would be greatest. On Lockyer's hypothesis, on the other hand, the short lines, since they were produced near the photosphere, would be bright; but the long lines, being produced in an outer shell of the atmosphere, would be faint over their whole length. Lockyer was thus led by his dissociation hypothesis to predict that the flash spectrum seen at eclipses would not be the exact reverse of the Fraunhofer spectrum, as astronomers generally supposed; there should instead be detectable differences of intensity between the two.

Lockyer therefore turned his attention to the detection of these predicted differences. This obviously required measurements at solar eclipses : measurements, moreover, which had to be made during a very short interval of time, for the flash spectrum was visible for only a few seconds. This, according to Lockyer, was the reason why the differences had not previously been detected. The obvious method of approach was to try and photograph the effect, but the photographic plates available were probably still too insensitive for this. Nevertheless, Lockyer decided to make the attempt at the next eclipse, in 1882, which would be visible from Egypt. This time he was able to go in person. He headed a party that included Schuster, and was to include Abney, who was put in charge of preparing the photographic plates. Abney fell ill and was unable to go, but his plates were exposed. They failed to record the flash spectrum. Lockyer was therefore forced to rely on eye estimates of the relative intensities of lines in the flash spectrum, which members of the party made. He was quite convinced that these showed significant differences, but, owing to the necessarily subjective nature of the measurements, he found it very difficult to convince his peers that he was right.

There was another eclipse in the following year, visible from the Caroline Islands in the Pacific. Lockyer was unable to go, but sent two of his assistants at South Kensington, attached to the main American expedition, with the equipment used in Egypt. Some slight advance was made, in that one or two of the main lines of the flash spectrum were recorded photographically, but there was nothing that could be used for the sort of test that Lockyer had in mind. The next eclipse expedition, in 1886, to the Caribbean produced even less, for the group of observers from South Kensington were clouded out, and failed to see the eclipse at all. Other parties were more fortunate. From Lockyer's point of view the most important result derived from the observations of H. H. Turner (who was subsequently to succeed Pritchard at Oxford). Turner examined the flash spectrum visually, and claimed that he noted differences of intensity such as Lockyer had predicted. But the scientific world remained sceptical of subjective estimates made under difficult conditions, and gave Turner's confirmation little more credence than Lockyer's original claims.

In the early 1890s, the organisation of eclipse expeditions from the British Isles, which, as we have seen, was often both *ad hoc* and controversial, was put on a more formal footing. The Royal Society and the Royal Astronomical Society agreed to establish a Joint Solar Eclipse Committee to prepare for the 1893 eclipse. This committee was subsequently continued on a permanent basis, and was responsible for organising all further eclipse expeditions during Lockyer's lifetime. Since the members of the Solar Physics Committee were also appointed to this new committee, South Kensington had a considerable say in its workings. This advantage was somewhat offset by the fact that Lockyer managed to antagonise a majority of the members within the first few years of the committee's existence. He seems to have argued that their approach to eclipse problems was too slapdash.

The 1893 eclipse, itself, still led to no further advance in the photography of the flash spectrum, though successful expeditions were sent out from South Kensington to both West Africa and Brazil. Lockyer now became determined to make an all-out attempt at the next eclipse in 1896. Photographic plates had become much more sensitive in the meantime, and the eclipse would be visible quite close to home in Northern Europe, so that Lockyer could lead the expedition himself. He selected a site on the Varanger Fiord in the far north of Norway. As a result of the new efficiency of organisation, a naval vessel was placed at his disposal. Fowler and Jim Lockyer travelled this way, but Lockyer went out on a ship of the Orient line. (Lockyer was on friendly terms with the management of this line, and they decided to put on a cruise to view the eclipse with Lockyer as their adviser.) But to everyone's disappointment, especially Lockyer's of course, this eclipse, like the last one he had attended, was obscured by clouds.

Not long before Lockyer left England for the trip to Norway, he had been approached by Sir George Baden-Powell, who offered to take, at his own expense, a small British party to observe the eclipse from Novaya Zemlya, quite close to the camp of the main Russian expedition. The Baden-Powells were a large and highly versatile family. Sir George's father had been Professor of Geometry at Oxford, and was a radical theologian who contributed to *Essays and Reviews.* One of Sir George's brothers attained fame as a British pioneer of flight, whilst another, of course, became the founder of the Boy Scouts. Sir George, himself, wrote on political and economic matters, and was at this time a Conservative M.P. He was also a keen yachtsman, and it was in this capacity that he had offered his services. Lockyer had gratefully accepted the proffered aid, and had arranged for Shackleton, one of the computers at the Solar Physics Observatory, to accompany Baden-Powell. There was no time left for elaborate preparations, so Shackleton was equipped simply with a prismatic camera that had been used in the Brazilian expedition of 1893.

Baden-Powell set off, but his yacht was involved in an accident that delayed its progress to such an extent that he could no longer hope to reach his original objective. The party therefore pitched camp at a nearer but less suitable point along the eclipse path. Fortune favoured them. The site they had hoped to reach was covered with cloud throughout the eclipse, whereas their new position stayed clear during the crucial period. Shackleton, who seems to have handled the eclipse observations with considerable competence, managed to obtain for Lockyer his first fairly detailed photograph of the flash spectrum.

The excitements of the expedition were not yet over. On the way back, Baden-Powell encountered Nansen, who was then returning from his polar journey, which had taken him, with one companion, to within 200 miles of the North Pole. Nansen had been away for three years, and there was considerable fear in Norway that he had died. When Baden-Powell returned with him, the event received great publicity. Lockyer, of course, was more concerned with the photograph of the flash spectrum. On his own return from Norway, he had felt so tired and disconsolate that he had rested instead of going to talk about the dissociation hypothesis at the British Association as he had intended. But, as soon as Shackleton returned with his material, Lockler revived, and started an immediate investigation of the photographic evidence for which he had been waiting so long. The result of the study left him in no doubt that the differences he had long expected between the flash spectrum and the Fraunhofer spectrum really existed. There was one snag; Shackleton had been equipped with one of the less powerful eclipse instruments, since his trip had been an afterthought. Hence, although Lockyer was convinced by the result, he realised that the photograph was not sufficiently well resolved to force someone less committed than himself to accept his claims without hesitation. Another eclipse was due very shortly, in January 1898, and was, moreover, going to be observable from India, which Lockyer had always regarded as a good place for eclipse observation. He decided to concentrate the resources at South Kensington on repeating Shackleton's work with more power instrumentation in India, in the hope of producing final, irrefutable evidence for his contention. He felt so certain that he was coming to the climax of this line of research that in the interim he wrote a new book, *Recent and Coming Eclipses*, to explain what information had so far been obtained with the prismatic camera, especially at the 1896 eclipse, and the major advance he hoped for at the 1898 eclipse.

Between the eclipses of 1896 and 1898, however, Lockyer's concept of dissociation underwent a radical alteration, so much so, that one can reasonably distinguish between his first dissociation hypothesis prior to this period, and his second hypothesis afterwards. But before discussing this change, we must consider how Lockyer's original hypothesis had been faring amongst astronomers during the preceding years. We have seen that astronomers were initially more favourably inclined towards his ideas than the chemists were. By the end of the nineties, this situation had changed. The accumulating evidence had convinced most astronomers that dissociation, at least as Lockyer described it, was untenable. In the first place, Lockyer's description of the circulation of the solar atmosphere became extremely suspect. No observational evidence for such a circulation could be found despite repeated attempts. The formation of sunspots by the infall of the circulating material, as Lockyer had supposed, necessarily implied that the spots were hotter than the surrounding solar surface, for they would be heated by the impact. But there was a growing consensus of opinion amongst astronomers as the century progressed that the spots were undoubtedly cooler than the surface. Moreover, Lockyer had explained the bright faculae round a spot as being the splashes thrown up by the material falling in from the corona. On this scheme, the faculae necessarily had to appear after the sunspot itself, but observations in the nineties provided good grounds for thinking that the faculae could actually be seen before the sunspot formed.

Lockyer had also argued that, if material dissociated in the hotter parts of the solar atmosphere near the photosphere, then equally it must recombine again to form ordinary atoms as it was carried by circulation currents into the cooler, upper atmosphere. He had sought for confirmation of this deduction from a detailed examination of the coronal spectrum during eclipses. His observation of the 1882 eclipse convinced him that he was right. The background spectrum of the corona, as distinct from the bright lines, did not have the continuous nature ascribed to it by most of his contemporaries, but was a highly complex mixture of lines and bands. He felt further secured in this interpretation by the results of his colleagues, Abney and Schuster, at the same eclipse, for they decided from their observations that they could distinguish the presence of hydrocarbons—low temperature compounds—in the corona. Unfortunately for Lockyer, more detailed observations at later eclipses showed beyond doubt that these preliminary results were wrong. The continuous coronal spectrum could not be attributed to the presence of molecules. Another supposed observational prop of the dissociation hypothesis disappeared.

But the worst blow to the acceptance of the dissociation hypothesis by astronomers came from one of Lockyer's favourite subjects, the study of sunspot spectra. It took some time to develop, as the particular line of research concerned was not formulated until the end of the seventies, and provided him with no results until the latter part of the eighties. What he did was to examine the width of sunspot lines over the course of a complete solar cycle to see which lines were most widened in sunspots, as compared with their width on the general surface of the Sun, and whether the degree of widening changed with time. Since there were far too many lines available for them all to be studied, Lockyer decided to concentrate on the twelve most widened lines from a limited region of the spectrum, and to examine these lines every day it was fine. Another advantage of studying only a few lines in detail was that it required only a brief spell of clear sky each day for the work to be done.

By the time that *The Chemistry of the Sun* was published in 1887, Lockyer had become convinced by the accumulated data that the lines did vary in the extent to which they were widened. In particular, the most widened lines around sunspot minimum were mainly attributable to known metals, but near maximum the most widened lines were mainly of unknown origin. This led him to conclude that the temperature of sunspots changed during the sunspot cycle. The highest temperatures occurred during the maximum of the cycle, for the unknown lines, so he thought, were produced by dissociation products, and therefore pointed to a high temperature. Similarly, the lines of the ordinary elements could be seen at sunspot minimum because the temperatures of the spots were then at their lowest. This result was highly agreeable to Lockyer, as we saw in the previous chapter, for it tied in very well with the belief he had long held that there was enhanced radiation from the Sun at maximum.

This work on the sunspot spectrum seemed to Lockyer to provide the final piece of the jigsaw. He could now explain solar activity entirely in terms of the assumed circulation of the solar atmosphere and of the dissociation hypothesis, buttressing his account at all the most important points by appeals to the observational evidence. In the final paragraph of *The Chemistry of the Sun* he asserted confidently,

> When the wonderful chemical changes which take place in the sun from cycle to cycle are more widely known, the importance of basing our views of the solar economy on both the chemical and physical facts available will be more generally recognised.[43]

Sadly for these brave words, the sunspot evidence for dissociation gradually faded away during the nineties. Lockyer had long had some contact with the observatory maintained by the Jesuits at Stonyhurst College in Lancashire, for it concentrated on studies of the Sun and of terrestrial magnetism. The first director of the observatory, Father Perry, was a member of the Solar Physics Committee for a few years prior to his death (which resulted from malaria contracted whilst observing the 1889 eclipse from South America). His successor, Father Cortie, continued the series of solar observations, and during the 1890s came to two important conclusions. First of all, he showed that the relative widening of the known and unknown sunspot lines was not specifically dependent on the solar cycle, but was determined rather by the type of sunspot observed. Lockyer's results could be understood in terms of the varying likelihood of observing a particular kind of spot during the course of a solar cycle. More importantly still, he identified the allegedly 'unknown' lines. They were actually due to vanadium and other ordinary elements, and were in no way connected with dissociation.[44]

Thus, by the end of the 1890s, Lockyer's ideas on dissociation seemed to have been rebuffed on all fronts, and very few scientists continued to regard them with favour. Not that all his efforts had been in vain. Apart from stimulating a great deal of research on the ultimate nature of the 'spectroscopic' atom, which still awaited theoretical explanation, he had pioneered the concept of studying astronomical and laboratory spectra in conjunction with each other. This mode of approach to astrophysics had caught on amongst the younger generation of astrophysicists. Indeed, it was some of their work, stimulated by Lockyer's example, which helped undermine the first dissociation hypothesis. G. E. Hale, who was to create in the United States the type of solar observatory which Lockyer had dreamed of all his life, and who was to become one of the most influential figures in twentieth-century astrophysics, was powerfully influenced by

Lockyer's approach to the subject from the beginning of his career. A long and warm correspondence developed between the two, and when Hale visited London in 1891 he noted, 'Though I took the opposite side from Lockyer in several protracted discussions with him, he certainly treated me very well, and put his whole lab. at my disposal.'[45] Lockyer's influence seems to have been particularly strong amongst the younger astrophysicists in the United States. E. B. Frost, later one of Hale's colleagues (the fact that they worked together with Gale, Professor of Physics at Chicago, on what was known—after its donor—as the Snow solar telescope led to endless jokes), has also recorded the influence of Lockyer's writings on solar physics at an early stage in his career.[46]

Despite the mounting evidence against the dissociation hypothesis, Lockyer remained wedded to it. He still felt that it provided the only way of interpreting the main spectroscopic observations, however much the details might have to be altered. Matters therefore reached something of an impasse, which was only broken, in 1897, by Lockyer's assistant, Fowler. The early work at South Kensington on comparing the arc and spark spectra of the same element in the laboratory had been predominantly visual. As sensitive photographic plates became available, this work was extended to the ultraviolet region of the spectrum, which, for many elements, was especially rich in lines. Fowler initiated a photographic comparison of the arc and spark lines of iron, for the iron spectrum contained so many lines that it contributed very largely to the spectra of the Sun and stars. His examination of the accumulated photographs led him to stress the importance of certain short lines that brightened in going from the arc to the spark. These lines had been noted before, and the name 'enhanced' lines given to them (a term that subsequently came into general use), but they had not been assigned any particularly significance. The reason why Fowler now selected them for special attention was that he realised it was precisely these enhanced lines of iron that were observed in the hotter stars. Since, according to Lockyer, the appearance of new lines in these stars was a result of dissociation at their higher surface temperatures, the natural implication was that enhanced lines were connected with dissociation. But the enhanced lines were quite distinct from the basic lines which had always been the mainstay of Lockyer's work; whereas the basic lines were, by definition, allegedly common to two or more elements, the enhanced lines were characteristic of each element individually. If, therefore, the enhanced lines were to be related to dissociation, this required a complete alteration in Lockyer's conceptual understanding of dissociation. The 'basic lines' approach had suggested that different elements broke down on heating to the same simpler substances. The 'enhanced lines' approach implied rather that each element dissociated into different, equally unique, forms of the specific element concerned.

The decision to change the whole basis of the dissociation hypothesis as Fowler proposed caused Lockyer a good deal of mental anguish. He kept Fowler's account of the new viewpoint on his desk for several months before sending it for publication. This was most unusual. As we have seen, Lockyer was more likely to send off a paper too hastily than too slowly. Fowler later declared that he could never understand why Lockyer had withheld publication for so long.[47] But it appears he privately believed that Lockyer could not make up his mind whether to publish the new idea himself, or to let Fowler publish and take the responsibility if it proved to be erroneous.[48] In the end, Lockyer decided to publish the work under his own name,* but in the form of a brief note in the *Proceedings of the Royal Society* with the distinctly unexciting title, 'On the iron lines present in the hottest stars'.

Despite this uncertain reception, Lockyer rapidly warmed to the new concept, and was soon defending the importance of enhanced lines for dissociation with much the same vigour that he had previously devoted to the defence of basic lines. Indeed, to the confusion of his contemporaries, he sometimes referred in his discussions to basic lines when he was really talking about enhanced lines. This, together with a growing disenchantment amongst his colleagues with his sweeping views on the nature of the universe, led to a cautious reception of the new dissociation hypothesis. Lockyer spelt out his revised standpoint in a book published in 1900, *Inorganic Evolution as studied by Spectrum Analysis*, which was essentially an updating of his earlier volume, *The Chemistry of the Sun.* It is completely in character that there is no clearcut mention anywhere in the book of the major change in emphasis that had taken place. Several of Lockyer's contemporaries evidently failed to note the new departure. For example, Lockyer sent the proofsheets of his book to the leading German spectroscopist, H. Kayser. The latter replied saying that the new discoveries of the physicists made dissociation seem more probable. 'The work done by Thomson and many others on the moving particles in the cathode rays, and the researches by Preston give an overwhelming weigh[t] to your idea of dissociation of different elements into identical particles.'[49]

Lockyer, in fact, remained equivocal over the nature of the dissociation products. His faith in the ultimate unity of all matter was as strong as ever, and it was certainly true that the physicists were beginning to provide experimental evidence to support the belief that the chemical

* Fowler inscribed his own copy of this crucial paper with the brief annotation, 'My work—A.F.'[48]

atom was not the final unit in nature. J. J. Thomson at the Cavendish laboratory had been pursuing the study of cathode rays for some time, and was still trying, in opposition to the Germans, to prove that the rays were particulate. In 1897, he finally managed to provide fairly good evidence that the rays were, indeed, sub-atomic particles (electrons) which were several hundred times less massive than hydrogen, the lightest known material previously. This was the first generally acceptable laboratory evidence for the division of atoms into something smaller. Moreover, the cathode rays produced seemed to have the same properties regardless of the materials forming the electrodes. These results agreed rather closely with Lockyer's original concept of different atoms forming a series of 'molecular' groupings. Thomson commented,

> The explanation which seems to me to account in the most simple and straightforward manner for the facts is founded on a view of the constitution of the chemical elements which has been favourably entertained by many chemists: this view is that the atoms of the different chemical elements are different aggregations of atoms of the same kind. In the form in which this hypothesis was enunciated by Prout, the atoms of the different elements were hydrogen atoms; in this precise form the hypothesis is not tenable, but if we substitute for hydrogen some unknown primordial substance X, there is nothing which is inconsistent with this hypothesis, which is one which has been recently supported by Sir Norman Lockyer, for reasons derived from the study of the stellar spectra.[50]

As it happened, Thomson's earliest estimates of the mass of the electron were appreciably higher than his later, more accurate values, and these incorrect estimates were quite close to values Lockyer had deduced for his dissociation products. Thomson wrote to Lockyer,

> I was much interested in the paper you sent, especially in the estimate you give of the mass of [the] smallest atom of hydrogen which is about 1/600 of that of the ordinary atom. I get for the mass of the small particles with which I have been dealing values which in different experiments have varied between 1/500 and 1/700 of that of the ordinary atom so that the two lines of enquiry lead to very concordant results.[51]

There was another important advance in physics in 1897 which seemed to fit in with the concept that similar sub-atomic particles existed in different types of atom. This was more nearly related to Lockyer's own work, for it was in the field of laboratory spectroscopy. The Dutch physicist, Pieter Zeeman, discovered that, in the presence of a magnetic field, spectral lines could be split up into two or more components. The observation was immediately explained in terms of the idea of oscillating electrons developed by his fellow-countryman, Hendrik Lorentz. Measurements of this 'Zeeman effect' indicated that the postulated electrons had properties similar to those of the cathode ray particles. Lockyer, however, was less impressed by the success of Lorentz' electron theory than by the cases where it failed. The simple theory predicted that all spectral lines should be split in the same way by a magnetic field, but the observations showed considerable variety. It was noted by Preston that, even in the spectrum of a specific element, not all the lines were affected in the same way. He remarked,

> This deviation is most interesting to those who concern themselves with the ultimate structure of matter, for it shows that the mechanism which produces the spectral lines of any given substances is not of the simplicity postulated in the elementary theory of this magnetic effect.[52]

Lockyer seized on this, and pointed out, reasonably enough, that it had been the differing behaviour of different lines of the same substance with respect to Doppler shift on the Sun which had first led him to suppose, many years before, that atoms were complex entities.

Thus, by the end of the nineteenth century, the dissociation hypothesis was in a state of transition. On the one hand, the new concept of dissociation proposed by Fowler was already clearing up some of the long-standing problems in stellar spectra. On the other, developments in physics finally seemed to be providing evidence for the similarity of the breakdown products from different elements, as had been supposed on the old dissociation hypothesis. It is hardly surprising that Lockyer was a little uncertain in his description of dissociation at this time.

The realization at South Kensington of the importance of enhanced lines occurred, as we have seen, between the total eclipse of 1896 and that of 1898. As a result, the need for definitive observations at the latter eclipse became even more crucial for the further development of the dissociation hypothesis. Prospects were good. The eclipse was due to take place during a period that generally gave fine weather, and Lockyer's party was to be assisted by his old associate, Pedler, who was still in India, and was making all the necessary arrangements on the spot. Moreover, the South Kensington group was taking out two of the most powerful prismatic cameras that had ever been used at an eclipse up to that time. It was confidently expected that these cameras would enable the positions of the lines in the flash spectrum to be determined with a much greater accuracy than had hitherto been possible, and so discover beyond any doubt whether there were any differences between the flash spectrum and the Fraunhofer spectrum. A man-of-war was being made available for Lockyer's party, as in 1896, but Lockyer, together with his son Jim, and Fowler, travelled out once more by the Orient line, and joined the naval vessel only in Ceylon. Even so, this left Lockyer with sufficient time on his hands to organise the sailors to assist at the eclipse. He was always delighted to have the chance of interesting non-scientists in science, and to show how a disciplined body of men could assist in the making of scientific observations.

The eclipse site allotted to Lockyer by the Joint Eclipse Committee was, auspiciously, in a coastal fort, similar to the site of his first eclipse observations, though this time situated to the south of Bombay. He was joined there not only by Pedler, but also by John Eliot, who had succeeded Blanford as meteorological reporter to the Indian Government. Like his predecessor, Eliot had become involved in Lockyer's continuing interest in looking for a relationship between the Indian climate and the sunspot cycle.

The tension built up as the date of the eclipse approached, but, on the day, Lockyer's optimism was justified. The weather was perfect along the whole of the eclipse track, and all the British groups obtained results. Lockyer's party experienced one or two instrumental mishaps, but nevertheless obtained some excellent photographs of the flash spectrum. When these were taken back to South Kensington and examined in detail, they confirmed beyond doubt that Lockyer had been right all along. The flash spectrum was different from the Fraunhofer spectrum. Other observers at the same eclipse (Evershed, for example) were able to confirm this finding. But the important result was not just that the two spectra were different, but that they were different in a way that Lockyer could now interpret in terms of the new dissociation hypothesis. It was found that the differences were simply due to the enhanced lines being relatively stronger in the flash spectrum than in the Fraunhofer spectrum. The implication, in terms of dissociation, was that the flash spectrum represented a higher temperature state.

This conclusion, although highly satisfactory in itself, placed Lockyer in a new quandary. Fowler's observations at the 1893 eclipse had definitely shown that the corona had a distinctive spectrum of its own. This was, indeed, confirmed by the South Kensington observations of the 1898 eclipse. Lockyer (and other observers elsewhere) obtained the first accurate measurements of the green coronal line which had long before been labelled 1474 on the Kirchhoff scale. They found that the line was actually some 14Å to the blue of its previous accepted position. It is some indication of the difficulty of spectroscopic measurements during eclipses that such a large error could have existed for so long. Whereas the 1474 line had first been identified with a photospheric iron line, and then with a chromospheric line, it was now realised to have a completely independent existence in the corona alone.

The chromospheric and flash spectra were also different from the Fraunhofer spectrum. In this case, the difference could be interpreted as indicating a higher temperature of formation for the two former. This was where Lockyer's dilemma lay. If the Fraunhofer spectrum was not a reversal of the coronal, chromospheric or flash spectra, then where in the solar atmosphere could it be produced without violating his ideas on dissociation? He argued as follows. Everyone accepted that temperature decreased with height in the solar atmosphere; therefore the Fraunhofer spectrum must be produced in a region of the atmosphere above the flash spectrum and the chromosphere. On the other hand, there seemed to be no objection to supposing that the coronal spectrum represented a lower temperature stage than the Fraunhofer spectrum. Lockyer therefore deduced that the absorption lines must be formed in a narrow region of the atmosphere between the chromosphere and the corona.

This conclusion was logically unassailable so long as the basic premises were accepted, but this Lockyer's contemporaries were unprepared to do. They argued, in the first place, that the spectroscopic temperatures he had assigned to the different solar regions were spurious, because they depended on differences observed in the arc and the spark. But these were both non-equilibrium phenomena, whereas the concept of temperature assumed that the material investigated was in equilibrium. More importantly, it became increasingly evident that strengthening of the enhanced lines could be brought about by pressure changes as well as by alterations in the ambient temperature. Hence Lockyer's observations might equally be explained in terms of the known decrease in pressure with height in the solar atmosphere, with the temperature effects having a lesser importance. The final resolution only came long after Lockyer's death, when D. H. Menzel in the United States discovered that the assumed decrease in temperature with height in the solar atmosphere did not occur. Instead, the temperature actually increased with height. Thus Lockyer was right in insisting that the chromospheric spectrum indicated a higher temperature of formation than the photosphere, but his opponents were correct when they insisted that the Fraunhofer spectrum was formed near the photosphere, not above the chromoshere. At the time, however, the general feeling was that Lockyer had again misinterpreted the observational evidence.

Finally, one general feature of Lockyer's thought that is relevant not only to the dissociation hypothesis, nor even to his astronomical researches as a whole, but to his thinking in all spheres should be emphasized. It may be mentioned at this point, since it appears most clearly in his book *Inorganic Evolution.* When Lockyer first became interested in science, there was one over-riding theme being debated—evolution. Darwin's book appeared in 1859, and its effects reverberated down the 1860s. Many of Lockyer's friends, pre-eminently Huxley, crusaded on behalf of evolutionary ideas throughout the decade. Young men, starting their careers in the sixties or seventies, in some cases had their outlook revolutionised by an encounter with the evolutionary debate in this period. For example, two of Lockyer's close friends, W. K. Clifford and G. J. Romanes, renounced their faith, the one an Anglo-Catholic, the other an evangelical, as a result of their contact with Darwinism. Lockyer was less shattered. Perhaps his Broad Church friendships helped here, though his religious enthusiasm evidently waned after his youth. But he, too, was deeply imbued with an evolutionary outlook for the rest of his life. Indeed, it became his great desire to discover the mechanism of evolution in the inorganic world as Darwin had for the organic. Both of his major scientific concepts, the dissociation hypothesis treated in this chapter, and the meteoritic hypothesis treated in the next, were expressions of this. Thus in *Inorganic Evolution* he explained to his readers,

> Just as plants and animals compose the organic or living world, so do the so-called chemical elements (either single or combined) compose the inorganic or non-living world.
>
> What we have now to consider is whether the facts . . . do or do not indicate that we must accept the chemical elements, like plants and animals, as products of evolution.[53]

By the end of the century, the younger generation of physical scientists were, perhaps, less moved by the mystique of evolution, and less likely to seek for analogies with organic evolution within their own disciplines. But to men of Lockyer's generation, the appeal was a strong one. In the mid-nineties Lockyer's old friend at the Mint, Roberts-Austen, proclaimed, 'I fully share Mr Lockyer's belief . . . and think that a future generation will speak of the evolution of metals as we now do that of animals.'[54]

According to Lockyer, only one major difference existed between organic and inorganic evolution. In the former, evolutionary changes depended on the gradual passage of time, in the latter, they depended on a gradual alteration of temperature (such as he postulated in his meteoritic hypothesis). But the essential point, he stressed, was that they were both simply aspects of a uniformly acting nature, '. . . the more different branches of science are studied and allowed to react on each other, the more the oneness of Nature impresses itself upon the mind.'[55]

VII

THE PHILOSOPHER'S STONE

The objects that interest astronomers are usually there to be studied as, and when, desired (presuming, of course, that the weather does not interfere, and that the time and place are right). Occasionally, however, there are temporary changes in the heavens which can be observed over only a restricted period of time. The year 1866 was distinguished by two events of this latter kind. In May of that year, a bright new star, or nova, flared up in the constellation of Corona Borealis, the Northern Crown. Novae formed a recognised, if infrequent, astronomical phenomenon. It was known that they brightened for a period of some months and then faded again. The importance of this new one—which came to be called in the astronomical jargon T Cr B—was that it was the first to appear since the spectroscope had been introduced into astronomy. Within a few days of the nova being reported, Huggins had examined its spectrum. He found it partly resembled that of an ordinary star, with dark absorption lines, but that superimposed on this there were a few bright lines, some of which were certainly due to hydrogen.

In November, another spectacular astronomical event occurred. There was an incredibly prolific shower of meteors, which came to be labelled the Leonids, since they seemed to radiate outwards from the constellation of Leo. This shower was not altogether unexpected, for, 33 years before, an extensive display had been seen, and later investigations had suggested that the meteors responsible were moving in a periodic orbit round the Sun, and might therefore return again. The general supposition at the time was that meteors were not really members of the solar system at all, but were merely passing through. These periodic meteors, however, obviously belonged; and were therefore assumed to have been captured. Within a few months of the 1866 Leonids appearing, an even more important link had been forged. Schiaparelli in Italy pointed out that another meteor shower, the Perseids, visible each August, was orbiting round the same track as a bright comet discovered in 1862. Almost immediately it was found that the Leonids were likewise following the same path as a comet, in this case one that had been observed earlier in 1866. Evidently comets and meteors were related; meteors must be the fragments comets left behind in their orbits.

Lockyer was at the beginning of his astronomical career when these two events occurred, and they impressed him deeply. In view of their coincident appearance, Lockyer immediately wondered whether meteors and novae might not be related in some way. This speculation, which contained the germ of his subsequently meteoritic hypothesis, was not taken further at the time.

'Meteorite' and 'meteor' were used as virtually interchangeable terms in the latter part of the nineteenth century. The only difference between them was that the former were larger blocks of material, which therefore reached the Earth's surface; whereas the latter were presumably smaller, and burnt up in the Earth's atmosphere. This position had only been reached during the first half of the nineteenth century. Previously there had been considerable dispute whether meteorites were extra-terrestrial, or not. General agreement that they were led to an increased interest in their study. Detailed scientific work on meteorites actually got under way not long before Lockyer himself first became interested in them. The importance of such investigations was obvious. Meteorites formed the only sample of material from outside the Earth that could be studied directly in the laboratory. Lockyer's belief in the unity of nature naturally led him to the view that the chemistry of bodies anywhere in the universe should be much the same. This view might be tested from a study of meteorites.

Lockyer's first work on meteorites was not, however, undertaken for this reason. We have seen that his main efforts from the late sixties onwards were concentrated on the comparison of laboratory and solar spectra. One of his problems was that individual elements seemed to give a different spectrum when other elements were present, as compared with the spectrum they gave when studied alone. If, therefore, he was going to make a proper laboratory comparison with the solar spectrum, this suggested that he must use a mixture of elements in the laboratory similar to that found on the Sun, in order that the spectral effect of the elements on each other should be identical. It seemed to Lockyer that meteorites provided precisely such a mixture. As a result, he made a certain number of laboratory observations of meteorite spectra in the early 1870s, to supplement his solar observations. In Lockyer's view, of course, it was not accidental that the Sun and meteorites could be taken as having an identical chemistry. He believed the two to be genetically related. On his picture of the Sun, the outer corona was a region of meteoric matter, which fell downwards to form the sunspots.

The real beginnings of the meteoritic hypothesis did not stem from this laboratory work. In the summer of 1874, Lockyer journeyed round Scotland and the north of England, visiting various of his friends en route. One of the friends he stayed with for a few days was Tait. Their talk naturally turned to comets, for in the April of that year Coggia in France had discovered a new comet, which had rapidly brightened. A few years before, Tait had suggested that all the extended sources of light in the sky, both the heads of comets and the nebulae, resulted from the collisions of meteoric stones, which produced highly heated clouds of gas. Lockyer had readily accepted the suggestion, though he had not tried to press it any further himself. Subsequently one of his friends in the British Museum, from which he obtained most of his meteorite specimens, drew his attention to a memoir published by the German, Baron Reichenbach, in 1858. This proposed that meteorites were the condensation products of smaller particles—the chondrules (or chondroi, as they were then called)—and that the process of condensation had taken place in the heads of comets.

The discussions with Tait brought these ideas once again to the forefront of Lockyer's mind, and he was evidently still thinking of them as he travelled southwards to stop with another friend, R. S. Newall, in Gateshead. Newall, a wealthy amateur astronomer, had ordered a 25-inch refracting telescope some years before from Lockyer's old mentor, Thomas Cooke. The objective glass, at that time the largest in the world, had been finished in 1868, though the effort of making it was said to have proved too much for Cooke, who died in the same year. The telescope was mounted by Newall at Gateshead, although the climate there hardly matched its capabilities. In fact, a major reason for Lockyer's visit at this time was to observe Coggia's comet through the telescope. As a result of his observations—it was the first comet he had ever examined in detail—Lockyer became convinced that the only possible explanation for the changes in the head of the comet was that it consisted of a large swarm of meteors in motion.

On his return to London, Lackyer began once more to study meteorite spectra, but now in the hope of learning something about comets rather than about the Sun. Although chemical and mineralogical methods of studying meteorites were naturally still being evolved in the 1870s, nevertheless some surprisingly high calibre work was already appearing. Quite good data on the chemical constituents of some meteorites were already available, and their distinctive mineralogy had been recognised. H. C. Sorby, a wealthy amateur scientist, with whom Lockyer was acquainted, had worked out the method of making thin sections of rocks, and had applied it to meteorites. But Lockyer's work formed the first extensive series of spectroscopic experiments on meteorites. Lockyer's earlier spectroscopic researches, being mainly concerned with comparison with the Sun, had been carried out at higher temperatures in the arc and the spark.

It seemed unlikely that comets were at such a high temperature, so Lockyer now extended this work to the study of meteorite spectra emitted at lower temperatures, corresponding to the Bunsen burner flame and the oxy-coal-gas flame. At the lower temperatures, molecular bands appeared —they were then called flutings—and Lockyer was able to observe how the spectrum changed from a mixture of bands and lines to lines only as the temperature increased. He could also determine how the spectral lines of different elements first became visible at different temperatures.

Lockyer's main interest was in applying these results to the interpretation of cometary spectra, but he also argued that they were applicable closer to home. The 1872 expedition of H.M.S. *Challenger*, under the direction of Wyville Thomson, was generally recognised as a major step forward in the embryo science of oceanography. Lockyer, like many other British scientists, followed the results of the expedition with interest, noting the main developments as they arose in *Nature*. One result he reported, which interested him personally, was the discovery by Murray and Renard of alleged cosmic dust in deep-sea deposits. The dust was in the form of small spherules which, as Lockyer remarked, were distantly reminiscent of meteoritic chondrules. Besides these, many much larger nodules containing a high percentage of manganese were recovered by the ship's dredge. Lockyer obtained some of these nodules from Murray, and compared their spectra with the spectra of meteorites. He claimed that the results were so similar that the nodules should be counted as extra-terrestrial material. (Murray had attributed them to volcanic activity.) The overall result was to confirm Lockyer in his opinion that material of all sizes from dust upwards was falling continually on the Earth's surface, and that, despite morphological differences, this material possessed a basic, underlying chemical unity.

A. S. Herschel, a son of Sir John Herschel and a great admirer of Lockyer's work, was also one of the leading English experts at the time on meteor observations. He was able to supply Lockyer with instances where the spectra of meteors had been observed as they burnt up in the Earth's atmosphere. These were naturally hard to obtain, since the meteor flash only occurred briefly, and its position could not be predicted beforehand. Nevertheless, Herschel informed Lockyer that the green line of magnesium had definitely been seen, as well as the yellow lines of sodium. The chemical composition this implied for meteors tied in very well with Lockyer's developing line of thought. The meteors burning up in the Earth's atmosphere threw off small incandescent particles of the same material, which formed into small round spherules. These cooled down, solidified, and fell to the Earth's surface where they were found in the deep-sea deposits.

Lockyer, having thus formulated one unifying hypothesis for the Earth, characteristically sought to extend it further. One of the mysteries of the time was the cause of the aurora. It was commonly believed to be some kind of electrical phenomenon, but attempts to simulate the observed auroral spectrum by electrical discharge through a mixture of atmospheric gases in the laboratory consistently failed to produce the desired result. But it had also been suggested by some that there might be a relationship between the occurrence of aurorae and the influx of meteoric dust (this appears, for example, in Humboldt's *Kosmos*, one of the most widely read scientific works of the nineteenth century). Lockyer went a step further still and postulated that the aurorae were, in fact, caused by heated meteoric dust. A comparison of his laboratory spectra, obtained at relatively low temperatures, with reports of auroral spectra convinced him that the latter could be explained as a superposition of lines and bands from manganese, magnesium, thallium, lead and carbon—that is much the same grouping of elements as he had already found in meteoritic bodies.

His final extension of the hypothesis was to claim that the spectrum of lightning, which was normally attributed to the excitation of the atmospheric spectrum by an electrical discharge, should really be attributed to the cosmic dust spherules. They were, he said, heated by the electrical discharge as they filtered down through the atmosphere. He proposed specifically that the spectroscopic observations of lightning made by Schuster, whilst visiting the U.S.A. for the solar eclipse of 1878, could be interpreted in terms of spectra of the same elements as aurorae.

The essential points to notice about the spectra that Lockyer used to support his hypothesis is that many of them were visual observations of rapidly varying phenomena. As a result, they could not readily be repeated or checked. Moreover, it was impossible to determine the wavelengths of the observed lines very accurately. Lockyer had only rough values available to aid him in determining which elements were present. It should be added that the consistent trend in astronomy up to the latter part of the nineteenth century was to distinguish betwen the various types of phenomena which had previously all been lumped together as 'meteors'. Lockyer's contemporaries inevitably saw his efforts as an attempt to put the clock back.

We must return now to Lockyer's prime reason for examining meteorite spectra in the laboratory, the wish to compare them with similar observations of comets. An examination of such comet spectra as had been described, mainly by Huggins, convinced Lockyer that they could be explained in a fairly direct manner as originating in the vapours given off by heated meteorites. He could quite easily identify carbon bands he had previously seen in the laboratory ('carbon' was then a fairly wide spectroscopic designation which could include hydrocarbon bands as well), together with one or two other obvious features. But Lockyer was immediately prepared to go much further. He believed that all the spectroscopic features found in the laboratory examination of meteorites could be distinguished in comets. The difficulty, he thought, was that the flutings from the different compounds overlapped, and so made interpretation of the final comet spectrum very complicated. Lockyer decided that his knowledge of spectra was now sufficiently extensive for him to undertake this task. His approach was very simple. He took the elements he had found to be ubiquitous in meteoritic material, and combined their spectra together until he obtained a satisfactory fit with the described comet spectrum.

Lockyer soon realised that the spectrum of a comet might change considerably as the comet's distance from the Sun varied, for the vapours produced would evidently be different at different points round the comet's orbit. His work on meteorite spectra convinced him, indeed, that as a comet approached the Sun, its spectrum changed to a higher temperature form. In recognising this spectral change, Lockyer was ahead of most of his contemporaries. In 1879, Young compared his own observations of the spectrum of Brorsen's comet, visible in that year, with Huggins' observations of the same comet in 1868, and was completely at a loss to explain the resulting differences. Lockyer was able to point out that it was due to the different distances of the comet from the Sun at the times of the two observations.

Once this change in spectrum was generally accepted as real, it was natural to see it as being caused by the increased solar heating when the comet approached the Sun. The higher temperature spectrum appeared as the comet neared the Sun because the comet material then became hotter. But Lockyer's interpretation was quite different. As we have seen, he pictured the head of a comet as consisting of a swarm of meteorites, and he felt sure from his laboratory experiments that solar heating would be completely inadequate to vaporize such refractory material. The necessary temperature could be generated only by collisions between the different members of the meteor swarm. Hence, the higher temperature as the comet approached the Sun had nothing to do with increased solar heating, but was due to an increase in the number of meteor collisions. Lockyer explained this increased collision rate in terms of the perturbation of the meteor swarm by the tidal action of the Sun's gravitation, which necessarily increased as the swarm approached the Sun. In a similar way, whereas most of his contemporaries believed that comets brightened as they approached the Sun because of the increased solar radiation, Lockyer ascribed this enhanced brightness, as well as the higher temperature, to the increased collision rate.

In doing so, he could claim to explain some observations that the accepted opinion could not. Thus some comets occasionally flared up for short periods, and then returned to their normal brightness. According to Lockyer, this occurred when one meteor swarm passed through another, so that there was a temporary increase in the collision rate. Similarly, it was observed that comets which passed very close to the Sun became very bright at perihelion. Lockyer explained this in terms of his theory of the solar constitution. The outer reaches of the solar corona were made up of condensed meteor-like particles; as the comet passed through this region, the collision rate therefore naturally increased.

Until 1876, Lockyer's speculations on the role that meteorites might play in various astronomical phenomena were restricted to the solar system. In 1876, however, his attention was once again turned to the possibility of changes in the stars by the appearance of another bright nova, in the constellation Cygnus. Lockyer was fascinated by the appearance of the new star, and immediately began to wonder whether it, too, might be explicable in terms of a clash of meteorite swarms. Indeed, he evidently thought, to begin with, that a nova might be something occurring quite close to the Earth, perhaps even within the confines of the solar system. He therefore asked a friend, Robert Ball, who specialised in measurements of parallax, to try and determine the distance of the nova. Ball was unable to obtain a definite result, but conclusively demonstrated that it was a distant object, nowhere near the solar system.

While Ball was coming to this conclusion, spectroscopic observations of the nova were already leading Lockyer on to a more grandiose speculation. The spectrum initially had contained several bright lines, but as the brightness of the nova declined during the following months, it was observed, by Lord Lindsay, that all these lines had faded except one. What remained was the line in the green part of the spectrum, recognised since Huggins' observations of the sixties as characteristic of nebulae. Lindsay announced that the nova had transformed into a planetary nebula.

To follow the ensuing arguments, it is necessary to understand what nineteenth-century astronomers meant by the word 'nebula'. Essentially, the word was applied to any bright extended object outside the solar system. Because such objects were faint and, at least till the advent of good photographic plates, very difficult to examine in detail, there was great uncertainty over the physical nature of nebulae throughout the nineteenth century and into the twentieth. One of the first British astronomers to mention nebulae was Newton's friend, Edmond Halley, who proposed that they consisted of some form of undefined shining fluid. In his time only some half-a-dozen nebulae were known. The real study of nebulae started with William Herschel towards the end of the eighteenth century. As a part of his campaign to map the contents of the heavens, Herschel vastly increased the number of known nebulae, so that, from being minor oddities of the universe, nebulae became an important constituent. Herschel, himself, believed initially that nebulae were simply clusters of stars, so distant that the individual stars could not be resolved by the telescopes he possessed. (In the same way, the Milky Way appears as a continuous blur of light to the naked eye, although we know it to consist, in fact, of innumerable faint stars.) From this viewpoint, Herschel necessarily came to believe in a vastly extended universe. The nebulae were scattered throughout the whole expanse of space. Subsequently, however, he saw reason to suppose that this could not be the entire explanation. Something like a shining fluid had to be invoked in some cases. He was led to this conclusion particularly from his discovery of planetary nebulae, single stars surrounded by shells of bright nebulous gas.

During the nineteenth century, opinions concerning the nature of the nebulae fluctuated considerably. About the middle of the century, Lord Rosse erected in Ireland a telescope considerably larger than anything that had gone before. It had a main mirror six feet in diameter, as compared with Herchel's maximum of four feet. Rosse found that some of the nebulae which had appeared continuous to Herschel could be resolved into stars in his larger instrument. This suggested that Herschel's original idea of nebulae as distant clusters of stars might well be substantially true. Then, in 1864, Huggins examined nebulae spectroscopically for the first time, and discovered that several of them gave a bright-line spectrum. This was clinching evidence that these nebulae could not consist of faint stars, but must be gaseous—Halley's shining fluid. The remaining nebulae, on the other hand, had spectra reminiscent of stars, consisting of a continuous spectrum which might be crossed by dark absorption lines or, perhaps, bright emission lines—the objects were too faint for certainty.

Lord Rosse had noted that one or two of the nebulae he observed through his large telescope seemed to have a spiral shape. This was difficult to discern visually, but, later in the century, the application of photography showed that a great number of such spiral nebulae existed. It was found that these spiral nebulae always possessed a continuous spectrum. Towards the end of the eighteenth century, the great French theoretician, Laplace, had proposed that the solar system originally formed from a rapidly rotating condensation of fluid. Such a condensation might well assume a spiral configuration as it spun round. It was natural for astronomers in the latter part of the nineteenth century to identify spiral nebula as being other stars in an advanced state of formation. This implied, however, that they were quite small—nothing like the vast star systems that Herschel had suggested—and therefore that they were fairly close to us in the sky. As a result, the universe accepted by astronomers in the late nineteenth century was much smaller than that envisaged by Herschel. By present-day standards, all the objects visible in the sky were our near neighbours.

This contracted view of the universe became generally established during Lockyer's lifetime, and he entirely accepted it. Where he came to differ from his contemporaries was in the emphasis he gave to the role of meteor swarms in nebulae. He set himself the task of explaining the spectra of all types of nebulae in terms of the meteoritic hypothesis he had originally put forward for comets.

Since Huggins' investigations had shown the importance of one bright green line (the so-called chief nebular line) in the spectrum of many nebulae, Lockyer concentrated first on this. Huggins had put forward various suggestions for the origin of this line. He had proposed originally that, since the line seemed to agree in wavelength with a line from nitrogen, the nebular line might represent, 'a form of matter more elementary than nitrogen, and which our analysis has not yet enabled us to detect.'[1] Then he pointed out subsequently that there was a line due to lead at the same position in the spectrum. Finally, he concluded that no element known on Earth could produce the line. In analogy with Lockyer's discovery of helium, the existence of a new element, named nebulium, had to be assumed.

Lockyer originally believed that the nebular line emanated from some hydrogen compound. But, when he began his intensive work on meteorite spectra, one of the first things he noted was the frequent appearance of a line apparently coincident in wavelength with the chief nebular line. He had been half expecting it, for Huggins had found a similar line in comet spectra. Lockyer was now thinking of nebulae as places where meteorites were colliding (Tait's original suggestion), he therefore sought the source of the line in one of the common elements found in meteorites. A comparison with known laboratory spectra indicated that there was a bright fluting caused by magnesium at about the same wavelength as the nebular line. Since magnesium was a major constituent of olivine—a common mineral in meteorites—this seemed a reasonable candidate. The problem was that in meteorite spectra it appeared as a band, whereas in nebulae the observations definitely indicated a line. Lockyer therefore carried out a series of laboratory measurements on magnesium, and found to his delight that, as he increased the temperature, the bright fluting gradually reduced in extent till it became more or less a single line. He confidently proceeded to identify the nebular line as a product of magnesium at moderate temperatures. Moreover, a closer study of the nebular line, itself, convinced him that it showed signs of this origin as part of a band. His assistants and he noticed that in the spectrum of the Orion nebula (the brightest nebula, and so the one most observed) this line had a faint fringe to the blue. This he claimed as evidence that the line was, indeed, the remnant of a fluting.

Lockyer's work on the chief nebular line brought him once again, yet more violently, into collision with Huggins. As we have seen, the two men came to be on somewhat better terms during the seventies, and this reasonable harmony continued into the eighties. Thus during the latter decade Huggins invited Lockyer to his house for a private view of photographs of the corona, which he claimed to have taken out of eclipse, before he officially released them to the public. In the event, he was shown to have been deceived by the effects of irradiation. This was probably just as well, since Lockyer had made experiments along the same lines in the mid-seventies, and there might well have been another priority dispute if Huggins' results had been confirmed. However, towards the end of the eighties the paths of the two began to diverge again, mainly over the interpretation of stellar and nebular spectra. In May 1889 there was a rapid exchange of letters over Lockyer's references to one of Huggins' papers. Huggins wrote angrily to Lockyer,

> . . . I thought the meaning [of a statement in his paper] was too clear for anyone to be misled.
>
> Nothing is so painful to me as controversy, but if the paragraph [in Lockyer's paper] is allowed to stand, there is nothing for me but to point out that the statements in it should certainly not be made and are inexcusable.[2]

Lockyer responded immediately,

> I have consulted one of my colleagues here [at South Kensington] and he agrees with me that after your *request* of withdrawal at the meeting, and after your use of the word inexcusable in your letter (which it is possible you may some time regret) I cannot withdraw the paragraph. Of course it is open to you to make any remarks you wish afterwards.
>
> I too hate controversy; it wastes too much time, but I do not shirk it with honorable antagonists.[3]

At the same time Lockyer offered to add a note to the offending paragraph which went some way towards meeting Huggins' complaints.

Huggins, mollified, immediately backed down,

> I am very much obliged for your letter. I wrote hastily to catch the post and in using the word 'inexcusable' I did not do so in a moral sense, but in a literary or intellectual one. . . .

I am much gratified that you have spared me from any controversy, but it was not from any personal feeling, but from my strong wish always to do and say what I *believe* to be the truth that I felt I could not allow the paragraph to pass without correction.

I am sorry that a difference of opinion on several points exists between us, but I am anxious that we should remain friends however much we may differ in opinion.[4]

It was a vain hope. In the next two years they were arguing more bitterly than before about the nature of nebulae and the identification of the chief nebular line. Huggins' views on nebulae had formerly been quite distinct from Lockyer's, but in the early nineties he began to veer round to a less distinguishable position, so much so that Lockyer decided Huggins had been filching his own ideas without acknowledgement.

I am now compelled to add, but I wish to make any statement with the utmost courtesy, that a complete study of the literature shows that he [Huggins] was quite familiar with my work all the time, and that while he thought fit to republish my main contention as his own on the one hand; on the other he was engaged in attempting to throw discredit on my work, and to conceal his retreat after the manner of the sepia by a great cloud of ink.[5]

But their main argument concerned the origin of the chief nebular line, which was related to the determination of its exact wavelength. The faintness of nebulae made it very difficult to observe their spectra at high dispersion, and so carry out accurate wavelength measurements. In order to establish their respective points, both Lockyer and Huggins needed better instrumentation. Lockyer therefore erected a horizontal telescope at South Kensington specially designed for the purpose. Huggins went one better, and appealed to observers in the U.S.A. to take up the question. This was an interesting move, for it clearly reveals the leading position in astrophysical research the United States had taken by the 1890s. Huggins' request was taken up by James Keeler, an expert in instrumental design, who was then at the Lick Observatory. Keeler's results indicated a small, but distinct, difference between the wavelength of the magnesium fluting and that of the nebular line. Moreover, he found no trace of the alleged hazy border towards the blue side.

Lockyer was by no means disposed to give up his identification. He argued, perfectly correctly, that since Huggins' original discovery of the chief nebular line, the measurements of its position had consistently tended to take it away from the nitrogen line Huggins had first advocated and towards the magnesium edge. Lockyer had been highly annoyed by a remark of Huggins to the British Association; 'The progress of science has been greatly retarded by resting important conclusions upon the apparent coincidence of single lines in spectroscopes of very small resolving power.'[6] Since this was an accusation frequently flung at him, and not without reason, Lockyer was delighted on this occasion to turn the tables on his accuser. He commented scathingly that Huggins' remark was 'an *apologia* of which everyone will see the propriety'.[7]

The squabble between Lockyer and Huggins over the nebular line reached such proportions that it seriously embarrassed some of their colleagues. In the spring of 1890, Liveing and Rücker were called on by the Royal Society to adjudicate between the two, as they were inundating the Society with their counter-claims. In concluding their report, the two Royal Society assessors commented,

> . . . we think it right to draw the attention of the Council to the fact that we feel that the circumstances under which we have been called upon to decide the questions at issue have not been altogether satisfactory.
>
> The notes sent in by Dr Huggins were of such a character that it would have been quite impossible to have shown them to Mr Lockyer. We have therefore been compelled not only to assume the responsibility of deciding which of Dr Huggins' criticisms were pertinent, but of ourselves stating them in our terms to Mr Lockyer.[8]

Later in the same year, Huggins complained bitterly to Christie that one of Lockyer's controversial papers on the nebular line was being distributed as if it was an official publication of the Solar Physics Committee.

> My foreign scientific friends advise me simply to ignore Lockyer's papers, but as they are published by the R.S. with the confirmation of a committee containing the official head of British Astronomy [i.e. Christie, himself, who had followed Airy as Astronomer Royal] I do not see that I can ignore them.[9]

Christie apologised to Huggins, and took the matter up with Lockyer. He admitted responsibility, but refused to apologise.

The importance of this dispute, and the reason why Lockyer defended his interpretation of the chief nebular line so strongly, was that on it hung his whole idea of the nature of nebulae, and consequently his theory of how stars developed. We must now see how these things were related. Lockyer's extension of the meteoritic hypothesis from the solar system to more distant objects was induced by his interest in novae. As we have seen, he was soon forced to discard the idea that novae were nearby, and therefore genetically related to comets. Instead, he came to believe that the relationship was one of physical similarity. Theoretical physicists, such as Sir William Thomson, had shown that there was a limit to how fast a heated body could cool : the larger it was, the longer it took. Lockyer argued that if a body like a star brightened as much as novae were observed to, it would certainly not fade in a few months as novae did. Hence, a nova must be understood not as a single, massive body, but as a multitude of small bodies heated individually, which could then cool down quite rapidly. Novae, in other words, were swarms of heated meteorites. These swarms must have been in existence before the novae occurred. Lockyer identified them with the nebulae. A comet brightened, when it neared the Sun, because of the collision of its constituent meteors. A nova brightened because separate meteorite swarms collided.

Lockyer was delighted with his extended hypothesis. He had united two hitherto separate areas of astronomy: one within the solar system and one without. He wrote post-haste to Stokes urging rapid publication of his ideas by the Royal Society. Stokes was not greatly impressed by the need for speed.

> As regards its possible utility in the mean time, in the event of a new star or comet turning up, the chances are so much against the occurrence in the six weeks that must elapse before the meeting of the Royal Society that I do not think that that need be taken into much consideration.[10]

Moreover, he had his reservations about the hypothesis itself.

> One is rather startled by having to look on meteorites as a sort of primitive condition of matter: for they contain regularly crystallised minerals: and what we know of the formation of crystals leads us I think into a difficulty as to how such crystals could be produced in such a quick process as the cooling of such small masses, after a chance collision which might have heated them.

Lockyer was in no way discouraged. He now had before him a picture that he believed could explain the whole course of stellar evolution. In 1890, he presented his ideas to the general reading public in *The Meteoritic Hypothesis* (published as usual by Macmillan).

We may start our examination of these ideas where Lockyer himself started, with the nebulae. Essentially he saw all bright nebulae as being caused by collisions of meteorite swarms. The different types of nebulae were the result of differences in the nature of the collisions. He could also introduce comets into this sequence, since they were formed on a similar, though smaller, scale. The differences in cometary and nebular spectra he explained partly by the higher temperature generated in comets, and partly by the fact that the finer dust in cometary collisions was driven away by solar radiation, whereas in the nebulae it remained mingled with the gas.

Lockyer distinguished between two extreme types of nebula. The first type was formed by more or less random collisions between meteorites, and had a fairly amorphous appearance. An obvious example was the Orion nebula. In this case,

> . . . streams or sheets of nebulous material, invisible if undisturbed, may encounter others, and in this way luminous patches of undefined shape may be produced by motions and crossings and interpenetrations, the brighter portions being due to a greater number of collisions per unit volume.[11]

At the other extreme, he placed the organised nebulae, consisting basically of a spheroidal swarm of meteorites rotating regularly round some central axis. Such a system was frequently,

> . . . invaded from various parts of space by other swarms which feel its attraction and have their paths deflected: we shall have on the general background of the symmetrically rotating nebula, which may almost be invisible in consequence of its constituent meteorites all travelling the same way and with nearly equal velocities, curves indicating the regions along which the entrance of the new swarm is interfering with the movements of the old one; if they enter in excess from any direction, we shall have the appearance of broken rings or spirals.[12]

The physical feasibility of some of the arrangements of intersecting orbits that Lockyer proposed to explain nebular forms was by no means evident. Thus the scheme he suggested to explain planetary nebulae was sent to G. H. Darwin, one of the great experts on gravitational problems in the latter part of the nineteenth century, for his approval. Darwin replied, 'Your conception of a region of greatest collisions does not commend itself to me at first sight, as I do not see how it can come about that they are more frequent at some distance from the centre than nearer in.'[13]

It should be remarked that Lockyer's attempt to classify and relate different types of nebulae was a pioneering effort, for it was only with the development of the modern type of dry photographic plate that systematic data on nebulae became available. (Faint, extended objects are very difficult to examine visually.) Although the basic principle of the dry plate was developed in the early seventies, their production on a regular basis for astronomical work only came about at the end of the decade. Hence, the detailed investigation of nebulae really began in the eighties, at just the time that Lockyer was evolving his meteoritic hypothesis. Indeed, from about 1885 the leading photographer of nebulae in Britain was one of Lockyer's close friends, Isaac Roberts. By the end of the decade, Roberts' photographs had clearly demonstrated the irregular nature of the Orion nebula, and also, for the first time, that the Andromeda nebula was a spiral.

This latter discovery, made in 1888, was especially gratifying to Lockyer. The form that Roberts discerned on his photograph seemed to correspond very closely with the conditions Lockyer had laid down a year before for the appearance of a spiral nebula. Spiral arms, along which the meteoritic material was supposed to be flowing, should lead directly into a central spheroidal nucleus. The importance of Roberts' photograph was that the Andromeda nebula was by far the largest spiral known, and a detailed examination seemed to show distinctly the lines of flow that

Lockyer had postulated. Roberts was himself much struck with Lockyer's comments.

> In presenting the photograph of the nebula in Andromeda to the R.A.S. I did not offer an hypothesis of my own in explanation of it, and only hinted that some astronomers would be tempted to appeal to it as confirmatory of the nebula hypothesis [of Laplace].
>
> Since the meeting I have had the pleasure and advantage of reading in print your remarks upon the nebula and they forcibly strike me in being the true interpretation of what we see.[14]

These ideas received further support in 1889, when two of Lockyer's research assistants, Fowler and Taylor, claimed to have discovered carbon bands in the spectrum of the Andromeda nebula. This was a visual identification; the first photograph of the spectrum was not obtained until 1899. The presence of carbon was seen by Lockyer as evidence for a low-temperature, meteoritic constitution of the nebula.

Lockyer thus agreed with his contemporaries that the nebulae were all associated with our own Milky Way. He differed from them in his belief that nebulae were cool objects, for the accepted view was that they were hot. The original reason for this supposition was that, since the stars were hot, so must be the fluid from which they formed. This requirement was gradually realised to be unnecessary in the latter part of the nineteenth century, as it was stressed, first by Helmholtz in Germany during the 1850s, and then by Sir William Thomson in Britain, that heat could be generated by the contraction process itself. A hot star could therefore be formed from a cold fluid. Nevertheless, despite Thomson's advocacy, the idea of nebulae as hot objects lingered on, perhaps because some astronomers believed that nebulae were not the beginnings of stars, but rather material left over when stars were formed, which differed from them in its nature, and chemical composition. Thus Huggins originally claimed that,

> The nebulae which give a gaseous spectrum are systems possessing a structure, and a purpose in relation to the universe, altogether distinct and of another order from the group of cosmical bodies to which our sun and the fixed stars belong.[15]

In 1891, when Huggins announced that he had changed his views on nebulae, he also admitted that his former views had been partly influenced by the religious ideas current in the sixties. This statement, which would have evoked no comment a century before, moved Lockyer to denunciation.

> It is fortunate for that large public which takes an interest in scientific matters, that the statements made by men of science—on which statements it is compelled to rely, and on which it builds its views of the universe and all it contains—are not, as a rule, thus unduly influenced by prevalent theological opinions.[16]

Lockyer's strong views on the unity of nature made it impossible for him to accept that there could be two different and separate types of material in the universe. Indeed, he believed that all nebulae, and therefore all the stars, which formed from them, contained the same elements in identical proportions. The variety of nebular and stellar spectra observed was due to the differing physical conditions. One final theoretical difficulty faced Lockyer. Laplace's nebular hypothesis had depended on the assumption that the condensing material was a cloud of gas. Would a swarm of rocky meteorites act gravitationally in the same way as a continuous fluid? For an answer, Lockyer turned again to his friend, G. H. Darwin. Much to Lockyer's relief, Darwin came up this time with a favourable reply. A swarm of meteorites would actually respond in much the same way as the molecules in a cloud of gas. Nebulae consisting of meteorites could therefore condense into stellar bodies without difficulty.

From the viewpoint of Lockyer's contemporaries, there was a much more serious problem posed by the meteoritic hypothesis. As we have noted, the nebulae were generally supposed to be hot. Stars, therefore, were hot when they first formed, and then gradually cooled down. This picture of stellar evolution could be linked to the known colours of stars, a subject that elicited a good deal of attention during the nineteenth century, for it was evident that the sequence of observed star colours—white, blue, yellow, and red—reflected differences in the surface temperatures of stars.

In the 1860s, astronomers became increasingly interested in the evolution of stars. For many of them, including Lockyer, this was a consequence of the concurrent debate over evolution in the organic world. It was natural that the colours of stars and nebulae should seem to support a theory of evolution from high to low temperature. The white stars, which were generally accepted as the hottest, were born from white nebulae, which should therefore also be hot. As these stars cooled down they became successively blue, yellow and, finally, red in colour. The important point to notice about this evolutionary hypothesis is that it postulated a linear sequence of events. Thus a white star was necessarily a young star; a red star was necessarily old. Moreover, since the cooling process was generally supposed to be accompanied by the contraction of the star, a white star would always be larger than a yellow star, which would, in turn, always be larger than a red star.

This was not the end of the matter. The question of how stars evolved came under review at just the time, in the sixties and seventies, when stellar spectra were first being surveyed and classified. It was realised very quickly that some kind of correlation existed between the colour of a star and the appearance of its spectrum, and that both were related in some way to the surface temperature. As early as 1863, the American astronomer, Rutherfurd, noted that the spectra of white stars differed from those of yellow and red stars. But the matter was really clarified by Secchi in the mid-sixties. He showed that each of the three main colour groups—blue-white, yellow, and red—had a corresponding characteristic spectrum. He subsequently pointed out that, on the basis of spectra, red stars should be divided into two groups, both of which showed fluted spectra, but in one case the flutings were sharply defined on their redward sides, and in the other, they were sharply defined towards the blue.

Secchi's work was concerned with the possibility of classifying stars into groups dependent on spectra, rather than with the creation of a spectral temperature sequence. In 1873, however, Lockyer pointed out that the spectral groups defined by Secchi confirmed that red stars were cooler than white or yellow stars. The spectra of the former contained flutings whereas the two latter types gave lines, and laboratory work clearly indicated that the transition from a fluted to a line spectrum corresponded to an increase in temperature.

In the following year, Vogel in Germany took over Secchi's classification, and turned it into a general scheme for correlating colour, spectrum and surface temperature. He redefined the three groups as follows:

Class I. White stars with spectra in which the metallic lines are faint or invisible.

Class II. Stars from bluish-white to reddish-yellow in which metallic lines are numerous and easily visible.

Class III. Orange and red stars with spectra which contain numerous dark bands (or flutings) as well as metallic lines.[17]

The surface temperature decreased regularly from Class I to Class III, and the sequence was assigned the specific evolutionary significance described above. Secchi's division of the red stars into two separate groups was modified. They were regarded rather as different sub-divisions of Class III, and were therefore labelled IIIa and IIIb. Depending on conditions, a Class II star might develop into either. Although nebulae were not specifically mentioned, it was an easy further step to insert them before Class I stars.

This evolutionary picture did not fit in with Lockyer's meteoritic hypothesis, which necessarily supposed that the nebulae were initially cool and became hotter as they condensed. Lockyer's concept of stellar evolution required a much more complicated sequence of events. He supposed that, as the meteorite swarm condensed, the rate at which the meteorites collided with each other increased. The maximum temperature was reached when all the meteorites were vaporized by the collisions. After this maximum, the mass of gas produced would gradually cool down again, as supposed by his contemporaries.

Thus Lockyer was forced to think in terms of a double temperature curve for stellar evolution, rather than a linear sequence, such as Vogel had proposed. There was an ascending temperature branch during the initial condensation, when the colour of the star changed progressively from red to white. Then, after reaching this maximum temperature, the star, still condensing, would retrace its path from white to red again. One significant point of this scheme was that the colour of a star no longer uniquely fixed its size. A red star could be either large and nebulous on the ascending temperature branch, or small and condensed on the descending branch.

In defining the different colour groups Lockyer followed Vogel's model, but since his evolutionary sequence had two branches, he naturally identified twice as many groups. First of all, there were the nebulae—which he labelled group I—at the beginning of the ascending branch. These were balanced by the ultimate, dark, and therefore invisible, condensed bodies—group VII—at the base of the descending branch. Groups II and VI were the red stars on the ascending and descending branches, respectively. Similarly, groups III and V were the intermediate stars, from reddish-yellow to whitish-blue, on the two branches, and group IV were the white stars at the peak of the curve.

Lockyer was actually not the first person to propose a double-branched temperature sequence. In 1883, a few years before Lockyer published his detailed scheme, Ritter in Germany had suggested that red stars could be divided into two groups, one with increasing temperature and the other decreasing. He also showed theoretically that a condensing sphere of gas might well first increase in temperature and then decrease. Despite this, Lockyer must be regarded as the pioneer of the idea. His scheme was far more detailed than Ritter's, and he expended a great deal of effort on searching for observational verification. His main problem in this latter respect was, of course, to find criteria for distinguishing between stars of the same colour on the ascending and descending branches. Obviously, the average density of the two types of stars would be different. But Lockyer expected this to make relatively little difference to their spectra, since he believed that the latter depended mainly on temperature. It was, indeed, the difficulty of detecting the required spectral differences which prevented Lockyer for some years from working out a complete observationally-based scheme of stellar evolution.

The obvious way of attacking the problem was to consider first the two groups that Vogel had called IIIa and IIIb. Here were stars of the same colour, but with clearly different spectra. Could one group be assigned to the ascending branch and the other to the descending branch?

Lockyer found it moderately easy to interpret type IIIb spectra, but for a long time he found it impossible to understand type IIIa spectra. None of the absorption flutings he was acquainted with seemed to fit. At last he had what seemed to be at the time a major inspiration. Like other astronomers, he had been looking at type IIIa spectra as consisting of a bright continuous spectrum crossed by dark absorption bands. In 1885, he suddenly realised that it was equally possible to interpret the observations in a directly contrary way, as being bright emission bands on a dark background. Moreover, with this interpretation, the proposed bright bands seemed to lie at much the same wavelengths as the carbon flutings already observed in comets. This identification led Lockyer to postulate that type IIIa stars were simply overgrown comets. They were diffuse clouds of incandescent meteorites. Thus they became the first stars on the ascending branch of his classification scheme, the ones he had provisionally labelled group II. Lockyer had already interpreted the spectra of type IIIb stars, since he had found them to contain normal dark flutings of carbon. (It must be remembered that any radical containing carbon was labelled simply 'carbon' by Lockyer.) The presence of these bands indicated, he argued, not only that the stars producing them were cool, but also that they were condensed bodies. Hence, they must obviously be placed at the bottom of the descending branch, in his group VI.

Besides establishing the two low-temperature points in his evolutionary curve, Lockyer had no difficulty in determining the point of maximum temperature corresponding to his group IV. He agreed with his contemporaries that there was a single group of hot stars, white in colour, with simple spectra.

As Lockyer became more deeply involved in problems of classifying stars, he built up an extensive collection of photographs of stellar spectra. Photographic plates were more sensitive in the ultraviolet than the human eye. Since the hotter stars gave out more ultraviolet light, they could be identified by the greater ultraviolet extension of their photographic spectrum. This was an accepted, if rather ill-defined, method for separating the hotter stars from those less hot. Lockyer extended the principle beyond its customary usage. He argued that the extension of the continuous spectrum into the ultraviolet varied with the amount of cool vapour present in the star. As the amount of vapour increased, so the ultraviolet extension decreased. For a given surface temperature, a star on the ascending branch, formed of vaporizing meteorites, should contain a larger amount of cool vapour than a star on the descending branch, which had already vaporized all solid particles. Hence, Lockyer began to examine the ultraviolet extension as a means of distinguishing between similarly coloured stars on the two branches of the curve.

Lockyer badly needed new criteria, for he was finding great difficulty in distinguishing between the stars of intermediate surface temperature (his hypothetical groups III and V). In the hope of finding further distinguishing features, he turned to laboratory experiments again, and examined the spectra obtained from various mixtures of meteorites at different temperatures. But the clue finally came from his favourite source—the Sun. This was not altogether surprising, for the Sun was itself a star of intermediate temperature. Moreover, it could be placed unequivocally in Lockyer's group V on the descending branch, for it was evidently a condensed star with no trace of the supposed original swarm of meteorites. Now one of the most characteristic features of the solar spectrum was a line, labelled 'K' by Fraunhofer, which was known to be associated with calcium. This line was noticeable because of its great width. When Lockyer came to examine the same line in other stars of intermediate temperature he found that, whereas in some it was as wide as in the Sun, in others it was much narrower. He decided that this difference could only depend on the degree of condensation of the stars. Group V stars were like the Sun in having a broad K line, whilst the group III stars had narrow K lines.

One of the main reasons why Lockyer had such difficulty in distinguishing between his postulated stellar groups was that, among the hotter stars, a good many of the most important lines remained unidentified. In the 1890s this situation changed radically. In the first place, helium came back into the news. We have seen that Lockyer asserted the existence of this element from the presence of a strong solar emission line, labelled D_3. Subsequently, Young in the United States found other solar lines of unknown origin, which might be related to the D_3 line, and hence be attributable to helium. But Lockyer ceased to make much reference to helium in the eighties. He carefully left the question to one side in his book *The Chemistry of the Sun*, which appeared in 1887. He was evidently influenced in this partly by the continuing failure to identify the element on Earth. More importantly, he began to suspect that helium might not be a simple element, instead, it might be—in terms of his dissociation hypothesis—a breakdown product of another element. In 1887, he wrote with reference to the D_3 line that it probably derived from a 'fine form' of hydrogen. Thus Lockyer had returned to something closer to Frankland's original position, albeit modified by the dissociation hypothesis.

The situation began to change again in 1886, when Copeland (who, in 1889, succeeded Piazzi Smyth as Astronomer Royal for Scotland) discovered the D_3 line in the spectrum of the Orion nebula. Subsequently, he detected in the same nebula some of the lines that were associated with

D_3 on the Sun. In 1890, Lockyer, himself, managed to obtain a photograph of some of these lines. Shortly afterwards, it also became apparent that several of the hitherto unidentified lines in the hotter stars coincided in position with this same set of lines (including D_3). As a result of these developments, Lockyer's attention was redirected to the question of helium in the mid-nineties.

Meanwhile, in 1888, Hillebrand, a geologist working in Washington, had begun research on an apparently quite unrelated theme, the investigation of the mineral uraninite. He found that specimens treated with sulphuric acid gave off a gas which he decided was nitrogen. In view of the sequel, it is interesting to note the evidence on which he based this view. He examined the spectrum of the gas evolved by subjecting it to an electrical discharge in a Geissler tube. He found a fluted spectrum identifiable with that of nitrogen. Now nitrogen gave two types of spectrum, this fluted one, and another consisting of bright lines only, depending on the pressure. Hillebrand's experiments were nearly always at such pressures that the fluted spectrum appeared, and this naturally obscured any bright lines from other elements present. Occasionally, however, Hillebrand's experiments were performed at such a pressure that he observed the bright-line spectrum, and then he saw several lines that he could not identify. He put this down to differences in the conditions of the electrical discharge and in the gas pressure, and rejected the possibility that a new element was present.[18] (This is worth remarking, since it has sometimes been suggested that Hillebrand failed to detect the presence of a new element solely because the needs of government service prevented him from pursuing the investigation.)

Hillebrand published his results in the *Bulletin of the United States Geological Survey*, which was well known to geologists, but reached few chemists or physicists. So, when shortly afterwards Lord Rayleigh started work on a new determination of the atomic weight of nitrogen, he was unaware of Hillebrand's discovery. Rayleigh soon found that the value he obtained for the atomic weight varied according to his source of nitrogen. He obtained one value for nitrogen from the atmosphere, and another for nitrogen from chemical reactions in the laboratory. Evidently, the nitrogen from one, or other, of the sources must be contaminated. Following up this clue, in 1895 Rayleigh announced the isolation of a new element as a highly unreactive gas. By this time the preliminary announcement of his results had attracted William Ramsay into the field, and the discovery of the element, which they called argon, was published by the two men jointly.

During the same year, Ramsay had his attention drawn to Hillebrand's work by a friend, who suggested that it might provide a new source of nitrogen. Ramsay therefore obtained some specimens of uraninite, treated them as Hillebrand had done, and examined the resulting gas in a discharge tube. The conditions he employed, however, were such as to give consistently the line spectrum, not the band spectrum. The most obvious feature in the spectrum of the new gas did not appear in the nitrogen spectrum at all, a single bright yellow line at the same wavelength as the solar D_3 line. Ramsay was well aware of Lockyer's original identification of helium in the Sun, he vividly remembered having heard Lockyer describe it in a lecture many years before, so he quickly came to the conclusion that he had found terrestrial helium. Nitrogen was, indeed, also present, but only as an impurity.

The discovery of the family of noble gases (the remaining members were tracked down without too much difficulty) provides an excellent incidental illustration of the Victorian view of scientific ideas as being in some way the property of their owner. This feeling was even more pronounced in Germany than in England, and it may be that it was reinforced in this country by the common experience that many British scientists had of training in Germany. The concept of intellectual ownership began to fall into disrepute, at least among the younger generation of scientists, by the end of the nineteenth century. It had, in any case, always come under great strain when any major discovery, such as that of the noble gases, was made. Thus Rayleigh was taken aback by Ramsay's incursion into the investigation of the peculiar differences in nitrogen samples from different sources. Having made a preliminary announcement, he expected to be left alone to complete the work. Although rumours of differences circulated, Rayleigh never complained in public. Ramsay, in turn, became exceedingly annoyed when other scientists took advantage of his own preliminary statements concerning his investigations into argon to try and anticipate his later results. He complained bitterly, and a letter he wrote to *Nature* on the subject clearly reveals the outraged Victorian sense of property.

> A habit has grown of recent years among some scientific men, which many of those with whom I have discussed the subject join with me in regretting. It is this:— After the announcement of an interesting discovery, a number of persons at once proceed to make further experiments, and *to publish their results.* To me it appears fair and courteous, before publication, to request the permission of the original discoverer. . . .
>
> An analogy will perhaps help. If a patent has been secured for some invention capable of yielding profit, and some person repeats the process, making profit by his action, an injunction is applied for and is often granted. Here the profit of the business may be taken as the equivalent of the credit for the scientific work completed; no original idea, undeveloped, is of much value; before it produces fruit, much work must be expended; and it is precisely after the publication of the original ideas, that sufficient time should be allowed to elapse,

so as to give the author time to develop his idea, and present it in a logical and convincing form.[19]

It is clear from Ramsay's covering note to Lockyer that Lockyer was in general agreement with this attitude. Ramsay remarked, '. . . lately, you will remember, you asked me about making experiments on helium before you began. Needless to say, I was only too glad to grant it.'[20]

Ramsay had sent some of his friends sealed tubes containing samples of the newly identified helium. Lockyer received one towards the end of March 1895, and immediately threw his whole laboratory resources into a study of the element. Ramsay's sample proved not to be suitable for the type of spectroscopic observations that Lockyer wished to make. He therefore obtained some pieces of uraninite—and later of clevite, another uranium mineral—and extracted his own helium. Whereas Ramsay, as a chemist, had naturally used chemical methods of extraction, Lockyer realised from his experience with meteorites that the gas could be collected by heating the mineral *in vacuo*—a much simpler method. His assistants were soon hard at work examining the spectrum of the gas, and comparing it with the unidentified lines in the solar chromosphere. Besides visual measurements, they obtained photographs in the blue, and compared the results with the photographs of stellar spectra accumulated at South Kensington. Lockyer found that several, though by no means all, of the unknown lines in the hotter stars could also be attributed to helium. The difference between the Sun and the stars was that the helium lines were reversed in the stars as compared with the Sun, that is, the lines were dark in the stars, and bright in the Sun.

This straightforward identification of solar and stellar lines with the laboratory spectrum was not accepted without query. In the first place, Runge and Paschen in Germany observed the yellow laboratory line in greater detail, and showed that it was double. Neither component had the exact wavelength allotted to the solar D_3 line. This seemed to cast doubt on the validity of identifying the solar line with the terrestrial gas. Lockyer immediately started high-dispersion observations of the solar line, and in June, he and, independently, Hale in the United States, observed that the chromospheric D_3 line was also double. Hence, what had initially seemed an obstacle actually proved to be an additional confirmation of the identification.

The next problem proved much more difficult to overcome. Ramsay had speculated initially when he studied the spectrum of argon that the gas might prove to have more than one component. This suggestion was subsequently withdrawn, but evidence meanwhile began to accumulate that, instead, helium was a mixture of gases. As early as May 1895, Lockyer was supporting this conclusion. His study of the gas obtained from clevite seemed to indicate that its spectrum changed under different conditions of electrical discharge, as he would have expected for a gas mixture. Runge and Paschen came to the same conclusion, but went further. Their detailed study of the spectrum revealed that the lines could be divided into two fairly simple groups. In one group the lines were all single, and in the other they were all double. They concluded therefore that the mixture consisted of two gases only. The name helium was retained for the gas which produced the D_3 line, whilst the other was called 'gas X'.

Lockyer pondered whether to call the new element 'X', orionium or asterium, the former because some of the spectral lines concerned had been identified in the Orion nebula, and finally came down in favour of the latter. Runge and Paschen were a good deal more cautious. Runge wrote to Lockyer,

> As to the name of gas X our opinion is this. You see the inference that there are two gases is a spectroscopical one being based on the investigation of the 'series'. Now though we think this basis quite sound, we must own that the conclusion rests on induction only and this induction is not a complete one as 'series' have not been observed in the spectra of all elements. Therefore we do not think that our claim of having shown the existence of the two gases is perfect. The final proof is given by him who first separates the gases materially. For this reason we do not want to give a name to gas X nor do we think to have any right to allow any one or to prohibit any one naming it.
>
> Professor Stoney proposed 'parhelium' which I would prefer to Orionium, the formation of the latter word being I think rather monstrous philologically.[21]

Runge's comment that the final proof of the existence of asterium would depend on its physical separation was, of course, obvious. Both he, working with Paschen, and Lockyer tried to effect such a separation by differential diffusion experiments, that is, by taking advantage of the fact that the rate of diffusion of a gas depends on its molecular weight. Lockyer, indeed, claimed that he had partially succeeded, and that his results showed the helium component, producing the D_3 line, was lighter than the asterium. Unfortunately, Runge and Paschen's experiments led them to the exactly opposite conclusion, and further attempts to separate the supposed two gases also failed to give a verifiable result. Asterium lingered on rather tentatively until the development of the quantum theory, when it was finally conclusively demonstrated that both sets of spectral lines were produced by the single gas, helium.

Although the discovery of the new element, or elements, explained the origin of several of the hitherto unidentified lines in the hotter stars, many other lines remained unidentified, especially those in Lockyer's intermediate stellar groups III and V. When, in 1897, the importance of enhanced lines was realised at South Kensington, it became possible to track down the source of many of these lines too. It was found that they consisted of those metallic lines, especially of iron, which were enhanced in going from arc to spark spectra. Thus, by the end of the nineteenth century, Lockyer had a whole range of known lines in each of the spectral groups he had previously defined. This enabled him, in turn, to rank stars in order of the surface temperature far more certainly than he could do before, and even to make quite good guesses at the actual values of the temperatures. In order to distinguish the atoms producing the enhanced lines from those producing the ordinary spectrum, Lockyer introduced the prefix 'proto' (thus he spoke of the iron spectrum and the proto-iron spectrum). The proto-element was essentially a breakdown product of the ordinary element, the distinction between what we would now call the neutral atom and the ionised atom.

These various advances in spectroscopic interpretation convinced Lockyer that the ascending and descending branches of his evolutionary curve could be distinguished by means of differences in the appearance of individual lines, as he had formerly asserted. Now, however, the differences covered hydrogen, the metals and the proto-metals, as well as calcium. At the end of the century, Lockyer drew up a final scheme of evolutionary classification to take account of the new data that had been acquired since the meteoritic hypothesis had first been formulated. From quite early on, he had realised that there were recognisable sub-groups within his major groupings, and had distinguished these by adding a Greek letter to the Roman numeral denoting the main groups. Thus group III was divided into three sub-groups, IIIα, IIIβ, and IIIγ. But he preferred ultimately to call each group by a characteristic name, basing his usage on geological nomenclature. In this, he was following a suggestion made to him early in 1899 by T. G. Bonney, the Professor of Geology at University College, London.

> You want, I imagine, to get a name to express the 'spectral' presence of a certain group of lines in a star so as to be a symbol of an epoch in the history of condensation or of cooling—Thus, as 'Cambrian' expresses a very early period in geological history, so would X-ian express the same in that of a star.
>
> Now the best principle of nomenclature in regard to geological formations (there are some bad ones) is to name them from places where the formation is largely and typically exposed. Thus Cambrian, Silurian, Neocomian [i.e. lower Cretaceous], are good.
>
> Suppose then you have a star which is a good example of a particular stage in history why not form the name from it, whenever possible, and I should think it generally would be, e.g. Sirian, Cygnian, Vegian?[22]

In any scheme of classification there are always a few objects which do not seem to fit. The success of a theory of evolution can often be judged by its ability to account for these misfits. We must therefore consider how successfully Lockyer treated these in his final spectral classification. The first type he had to deal with consisted of stars with bright lines, instead of the more usual absorption lines. Lockyer was convinced from the beginning that these stars must be related to the nebulae. His earliest measurements on rather poor material did not provide him with much evidence that the same bright lines appeared in both, but subsequent work at South Kensington and elsewhere did indicate the presence of the same lines in some cases. More than this, he became convinced that bright-line stars showed bands of carbon similar to those found in comets. He therefore asserted that bright-line stars were simply a more highly condensed form of a meteorite swarm. They formed a grouping between the nebulae and the red stars on the ascending branch of his evolutionary curve. If he was right, it followed that the bright-line stars should appear as diffuse patches of light on photographs—something between the extended images of bright nebulae and the point images of stars. He could, in fact, point to at least one case where this seemed true. The brightest stars in the group known as the Pleiades all had bright lines. In 1886, Roberts took a long exposure photograph of these stars and showed conclusively that each was surrounded by a bright nebula. Lockyer noted,

> The principal stars are not really stars at all; they are simply *loci* of intercrossings of meteoric streams, the velocities of which have been sufficiently great to give us, as the result of collisions, a temperature approaching that of α-Lyrae [one of the hotter stars].[23]

Lockyer, looking round for another way of testing his explanation of bright-line stars, lighted on a type of star which had been discovered by two Continental astronomers, Wolf and Rayet, in the late sixties. These Wolf-Rayet stars were easily distinguished by their peculiar emission-band spectra, and seemed excellent candidates to examine for a semi-nebulous appearance. Lockyer therefore asked Roberts to take a long-exposure photograph of a region of the sky where some of these stars occurred. Robert reported discouragingly,

> There is no appearance of a nebula or even a trace of nebulosity, but the plate is almost covered with stars. . . . Will this fact not account for the remarkable spectra you refer to? You of course get the *combined* spectra of a multitude of stars which if examined separately would give a different result.[24]

Huggins, who was completely sceptical of Lockyer's explanation of bright-line stars, seized on this absence of nebulosity as clinching evidence of its failure. But Max Wolf, on the Continent, subsequently photographed the same region under more suitable conditions and with better equipment, and found that the stars Lockyer had specified were, indeed, surrounded by faint nebulae. Lockyer's claim thus seemed to have been fully justified.

Huggins also clashed with Lockyer over the nature of the spectra of bright-line stars. Huggins (working, as usual, in conjunction with his wife) stated that he could find no evidence for the bright carbon bands Lockyer claimed to be present. This particular argument proved to be inconclusive at the time, since some evidence seemed to support Lockyer's contention, and some Huggins. In one important respect, however, Lockyer was forced to change his views of bright-line stars. Further study of their spectra during the nineties by means of photography revealed that in many cases the bright lines were superimposed on underlying dark absorption lines. This observation necessarily indicated the presence of something like a normal star in addition to the nebulosity. Lockyer therefore finally pictured a bright-line star as being what he called a 'gaseous star', consisting of a central condensation raised to a high temperature by the infall of a surrounding nebula.

Another important class of peculiar star that Lockyer had to consider consisted of those stars whose brightness varied with time. The existence of such variable stars had been recognised since at least the beginning of the seventeenth century, but it was only during the nineteenth century that such variations were realised to be of moderately common occurrence. Some 400 variable stars were known by the 1890s. It was evident from early on that not all light variations could be explained in precisely the same way. Some stars varied regularly, and some irregularly, with time. During the nineteenth century, different theories were accepted to explain the different kind of observed variation in light. Thus, some of the regular variations could evidently be attributed to the components of a double star eclipsing each other in turn, as they moved round their orbit. Not all regular variations could be explained in this way, however, and a popular alternative explanation was based instead on what was known of the Sun. The idea was that some stars might accumulate more spots on their surfaces than the Sun did, and that these spots might accumulate preferentially in one part of the star. Then, as the star rotated on its axis, its dark and light faces would be alternately exposed to an observer on Earth, thus leading to a periodic change in brightness.

Lockyer was unhappy with the general nineteenth-century tendency to accept *ad hoc* theories of stellar variability. He felt that any consistent theory of stellar evolution should at the same time explain the existence of light variations. He was therefore concerned that the meteoritic hypothesis should be capable of measuring up to this standard.

As we have seen, Lockyer believed from the start that novae resulted from a clash of meteorite swarms. It was therefore natural that he should try to explain all other light variations in the same way. Clearly since the types of light variation differed considerably, so must the nature of the collisions. He divided regular light variations into three types, and assigned a slightly different cause to each. In the first type, it was supposed that one swarm of meteorites moved in an elliptical orbit round another swarm. At the point of closest approach the two meteorite swarms collided, and there was therefore a periodic increase in light each time this occurred. Next, a meteorite swarm might be moving round a body that was already fully condensed. The gravitational tidal action of the condensed body could then induce individual meteorites in the swarm to collide with each other in a particular part of the orbit only, so producing regular light variations. Finally, if two condensed bodies moved round each other, they could, given a favourable inclination of the orbit, appear as an ordinary eclipsing binary. Lockyer explained irregular variability, on the other hand, as dependent on one of two causes. It resulted either from the presence of more than one swarm of meteorites orbiting the central body, or from the collision of independent meteorite streams moving through interstellar space.

Since the only thing that distinguished nebulae from stars on Lockyer's theory was their lesser degree of condensation, it seemed to him reasonable that nebulae whose light varied should exist equally with variable stars. He was able to point out that variable nebulae had, indeed, already been identified (for example, one was discovered by Hind in 1852), and that his was the first theory to relate such variations to those of stars. He also drew attention to the fact that, if light variations were caused by meteorite collisions, then it would be expected that variable stars would have bright emission lines in their spectra, at least for a part of the time. Again this was a conclusion borne out by most observations. Lockyer could quite reasonably claim therefore that his theory of variable stars unified the observations in a way never hitherto achieved. Despite this, even Lockyer's friends were a little unhappy with some of his ideas on light variation. They objected, for example, to the requirement that all variability must result from the interactions of two, or more, distinct bodies.

Novae not only focused Lockyer's attention on the meteoritic hypothesis initially, they also seemed to provide the best evidence for his concept of stellar variability. When, in 1892, a nova flared up in the constellation of Auriga, astronomical photography had just reached the stage where it was possible to photograph the spectrum (previous novae spectra had been observed visually). Lockyer and his assistants used the instrumentation at South Kensington for this purpose. A detailed examination of their photographs immediately showed that both bright and dark lines were present. The dark lines were adjacent to the bright lines, and always occurred towards the blue side. The observation was confirmed by other observers, including Huggins, and there was general agreement that this peculiarity of the spectrum must be interpreted as being produced by two objects, one corresponding to the dark lines and the other to the bright. From the displacements of the two sets of lines it was possible to estimate, *via* the Doppler effect, the relative speeds of the two postulated bodies. It came out to 600 miles per second—high, but not excessively high by astronomical standards. Of course, Lockyer saw this result as confirming his picture of novae.

> . . . the spectrum of Nova Aurigae would suggest that a dense swarm was moving towards the earth with a great velocity, and passing through a sparser swarm, which was receding. The great agitation set up in the dense swarm would produce the dark line spectrum, while the sparser swarm would give the bright lines.[25]

Lockyer's contemporaries, however, did not feel compelled to accept that the two colliding bodies were necessarily meteorite swarms. Another suggestion, originally proposed in the 1860s, that received support, was that two stars had made a close approach to each other, and had been violently disturbed by their mutual gravitational interaction, so producing an outburst of light. This possibility seems to have been favoured by Huggins, though his views on the matter were by no means settled. Lockyer observed sourly, '. . . all suggested explanations of the phenomena of new stars which have been put forward, except my own, have appeared to find favour with him in turn.'[26] Another explanation, suggested by Seeliger in Germany, was that interstellar space contained clouds of cosmic material. A nova occurred whenever a star encountered one of these clouds, and was heated by the consequent friction.

It looked at first as though there was no obvious method of choosing between these different theories. But the nova itself provided a new pointer. After its initial burst of light, it faded quite rapidly, and within a month of its first appearance it had become too faint for the spectrum to be observed at South Kensington. But six months later it suddenly flared up again, and when the spectrum was now examined it was found that it indicated the presence of nebulosity. The new flare-up seemed definitely to dispose of the idea that a nova was simply due to the close passage of two stars, for it was most unlikely that such an event would occur twice in quick succession. On the other hand, it appeared to vindicate triumphantly Lockyer's opinion, since it was an integral part of his hypothesis that the collisional interaction of two meteorite swarms should produce a nebula. Lockyer might reasonably have hoped that this new development would induce his peers to look more favourably at the meteoritic hypothesis. This did not occur, and we must now examine why.

During the nineties, several novae were discovered. This jump in the discovery rate was almost entirely due to the efforts of the Harvard

College Observatory. As we have seen, Lockyer's friend, Henry Draper, was a pioneer of the photography of stellar spectra in the United States. After his early death in 1882, his widow presented his telescope and a sum of money to Harvard with the request that they should be used for a survey of stellar spectra. The project was to be called the Henry Draper Memorial. E. C. Pickering, the director of the Harvard College Observatory, was delighted with the gift, as he and his brother, W. H. Pickering, were already experimenting in this field. Mrs Draper's gift was made in the mid-1880s, and by the end of the decade a photographic survey of northern hemisphere stars had been completed. In the 1890s this was complemented by a similar survey of the spectra of southern stars from an out-station in Peru. The spectra obtained were examined and classified at Harvard, and ultimately published in catalogue form. (The first version of the Henry Draper Catalogue, in 1890, contained somewhat over 10,000 stars, but this number was subsequently greatly increased.) The novae were picked up as a by-product of the survey, being identified on the plates by their spectral peculiarities.

It was found that whenever the spectrum of a nova could be examined in detail, the dark lines were always towards the blue side of the bright lines. It therefore became fairly certain that the displacement of the two lines could not be explained in terms of random collisions, for, in that case, the bright lines should have been sometimes to the red and sometimes to the blue of the dark lines. So, at the end of the century, explanations of novae in terms of two bodies, such as was a vital part of Lockyer's thesis, began to fall into disrepute, and theories involving a single body became more popular.

This work on novae at Harvard told against Lockyer's ideas. More importantly, the spectral classification developed at Harvard was so detailed and extensive that it eventually swept all other classifications into oblivion. In particular, Lockyer's classification was submerged by the Harvard system. The women astronomers employed at Harvard for spectral classification, Mrs Fleming, Miss Maury and, especially, Miss Cannon, divided up the stars into various groups according to the similarities of their spectra. Lockyer and his assistants at South Kensington were doing much the same thing; the major difference was the vastly greater number of spectra available for classification at Harvard. The instrumentation at Harvard was better for this task than Lockyer's, and, even more importantly, the seeing conditions at South Kensington gradually deteriorated during the whole of Lockyer's stay there. As might be expected, their mutual interest in spectral classification led Lockyer and Pickering to a long exchange of correspondence. The sad recurring theme of Lockyer's letters concerns the way his work on stellar spectra was held up by bad weather. He naturally took a warm, if slightly envious, interest in the Draper Memorial project from its inception. In an early letter to Pickering, he remarked, 'I saw a great deal of Mr and Mrs Draper when I was in America and I looked at the time upon Draper's death as a personal loss.'[27] He then went on to offer some of his own photographs for the Draper Memorial work, if Pickering so wished.

The Harvard spectral classification distinguished each group by a letter of the alphabet, and was to some extent a temperature sequence, for the hotter stars tended to be denoted by letters earlier in the alphabet, the cooler stars by later letters. There were exceptions to this rule; for example, stars with emission lines were mainly grouped together, and, in any case as the system evolved, letters were added or dropped. Although on the surface the Harvard system contained many more groups than the South Kensington classification, as a result of the ultimate sub-division of the groups in the latter there was in several cases a virtual identity of grouping. For example, the Harvard criteria for their type B stars were almost the same as the South Kensington criteria for their sub-group IIIγ. The interchange of information between Harvard and South Kensington almost certainly helped to some extent in producing this similarity; nevertheless, some differences remained. The most important of these concerned Lockyer's group II (red stars with some bright lines and often variable in light). The South Kensington sequence placed these just after the nebulae, whereas the equivalent Harvard class (group M) was placed next to stars like the Sun.

It will be noted that this difference concerns the relative positioning of the classes. This, indeed, reveals the most important distinction between the two schemes of classification. The South Kensington classification, as we have seen, was intimately involved with Lockyer's meteoritic hypothesis, and therefore with a double-branched temperature curve. The Harvard classification, on the other hand, was not specifically linked to any theory of stellar evolution, though it seemed to be best fitted to a linear sequence. Some astronomers regarded the Harvard approach as preferable because it was empirical. It was still uncertain at the end of the nineteenth century whether temperature could be uniquely correlated with spectrum. Whereas some agreed with Lockyer that temperature was the main factor determining the type of spectrum produced, others, including Huggins, believed that the density of the stellar atmosphere played a significant part. This idea was linked closely with the concept of a linear evolutionary sequence, and of a star as a contracting sphere of gas. As the sphere contracted, the surface temperature would certainly change, but so would the density at the surface, for the star was becoming more highly compressed. Hence, in a linear evolutionary sequence, there was as close a link between density and colour, as there was between temperature and colour, and the spectrum might therefore depend on either. Because of this uncertainty, there was felt to be some advantage in using a classification which was as non-committal as possible.

The main reason, however, why the South Kensington scheme was rejected was not because it committed itself to a double-branched evolutionary curve, with the concommitant assumption that temperature was the main determinant of spectra, but more specifically because the scheme was dependent on the meteoritic hypothesis. By the end of the century, so many objections had accumulated against this hypothesis that it retained few supporters. We have seen the difficulties arising from the study of novae spectra in the 1890s; but the objection that probably ranked the highest in people's minds, perhaps because of the long public battle between Huggins and Lockyer, was the identification of the chief nebular line. By the end of the nineties, almost everybody except Lockyer regarded this dispute as resolved. The differences in wavelength between stellar and laboratory measurements no longer made it possible to regard the line as a remnant of the magnesium fluting. This, in itself, was a body-blow to Lockyer's hypothesis, but there were other objections almost equally insistent. For example, Lockyer's habit of finding carbon flutings in stars when other people could not. Or the feeling that, since so many meteorites were made of iron, a star in its initial stage should show predominantly iron lines in its spectrum, which seemed not to be the case. Or, again, the increasing doubts of theoretical astronomers whether two or more interpenetrating streams of meteorites, such as Lockyer postulated for variable stars, could be gravitationally stable.

One final reason for opposition to the meteoritic hypothesis is worth mentioning separately, for it reflects yet again one of Lockyer's main virtues and vices as a scientist, his desire to unite all aspects of nature into one picture. In pursuit of this ideal, as well as for perfectly practical reasons, Lockyer consistently tried to inter-relate his meteoritic hypothesis with his dissociation hypothesis. In this he was highly successful, but the result was that evidence against one hypothesis therefore also affected the status of the other hypothesis. We have seen that the dissociation hypothesis was under fire towards the end of the century, this, therefore, reflected adversely on the meteoritic hypothesis. For example, Pickering accepted Lockyer's identification of the enhanced lines, and used them in the Harvard classification, but he was not prepared to commit himself to the dissociation hypothesis. Lockyer wrote reproving Pickering for always referring to the enhanced lines of metals, whereas on Lockyer's scheme they were due to proto-metals.[28] But Pickering's usage merely reflected the opinion of most astronomers at the time. When Lockyer and his assistants produced new observational results, his contemporaries used them, but they were not prepared to swallow the accompanying theoretical propositions.

The initial enthusiasm for the meteoritic hypothesis, which certainly existed in some quarters, died away in a little more than a decade. After Lockyer had revealed the details of his hypothesis in 1887, Airy wrote to him,

> I think (but I am only an outsider) that the connexion established between the various astrometrical and spectral phenomena is the greatest step that has been made—lately or ever (so far as I know)—in celestial physics.[29]

Another outsider, Tennyson, was impressed in another way.

> Must my day be dark by reason, O ye Heavens, of your boundless nights,
> Rush of Suns, and roll of systems, and your fiery clash of meteorites?[30]

But the insiders, the astrophysicists, were less impressed, and by the early twentieth century, even a sympathetic writer on Lockyer was forced to observe, '[One astronomer has concluded] that "the weight of evidence is against the truth of the Meteoritic Hypothesis"; and this is the opinion of the majority of modern astronomers.'[31]

Perhaps we can end with part of *Punch*'s commentary on the meteoritic hypothesis. One would suspect that several of Lockyer's scientific peers read it with relish.

CHAPTER I

MR. NORMAN LUCKIER, the eminent astronomer, was walking in his garden. Suddenly he was staggered by a sharp blow on the head. Something fell at his feet. It was not his head. He picked it up. It was a meteoric stone. This set him thinking.

'Here,' said he, as he rubbed his newly-acquired phrenological development with one hand and held the meteoric stone in the other, 'is a solid, ponderable body, which I can handle, examine, and analyse, and it comes to me,' continued the eminent scientist, extending his arms and looking round him, then directing his gaze upwards, his eye dilating with the grandeur of the discovery,—'it comes to me direct from the Cosmos!'

CHAPTER II

THERE was a chuckle from behind the neighbouring hedge, and, as the Philosopher returned to his sanctum to write a paper on the 'Spectra of Meteorites,' a small boy stepped cautiously out into the road, and hurried down the lane.

'Ooray!' muttered the small boy to himself; 'the old gent don't know my name. What did he say about "Crismas"?' And he vanished into space.

CHAPTER III

THE Philosopher, with aching head, sat down to write, and penned these words,—

'Cosmical space is filled with meteorites of all sizes, flying about with immense velocities in all directions.'

'Good Heavens! or, rather, Bad Heavens!' exclaimed a simple-minded visitor, to whom he read this statement, 'why, "Cosmical space" must be uncommonly like a proclaimed district in Ireland, or Trafalgar Square during a Socialist riot.'

The Philosopher perceived that he was not in the presence of a sympathetic mind, and regretted having invited the visitor to lunch.[32]

VIII

FAMILY AND FRIENDS

Lockyer's first wife was a fecund bearer of children; during the first fifteen years of their marriage, nine children were born. The first child, called Joseph Norman like his father, died young; so the second child, another boy, called Norman Joseph, who was born in 1860, became the eldest. After him there was a regular succession of children every two years until 1873. Only two of these were girls; the elder, Rosaline Annie, born in 1864, seems to have inherited something of her father's stubbornness. Lockyer was moulded somewhat in the popular image of a Victorian father, and although intent on helping his children, was noticeably authoritarian. It appears that his children, even when grown up, found him a little overawing. Rosie, however, when she was 23 insisted on marrying a man of whom her father strongly disapproved. Lockyer's prognostication that the man was a waster may not have been entirely incorrect, for, in the nineties, Rosie had to go on the stage to make ends meet. The second daughter, Winifred Lucas, on the other hand, never married. She was the last member of the family to arrive, and, after the death of her mother, acted as companion first to her father and then to his second wife. She also seems on occasion to have acted as a moderating agent between her father and her brothers.

Of Lockyer's other sons, Alexander Edmund and Ormonde Hooley Spottiswoode, the two youngest, both later emigrated, the former to the United States, and the latter to Australia. Hughes Campbell entered the Royal Navy, and was therefore away from home for long periods of time. The only member of the family who shared his father's interests was William James Stewart Lockyer, nearly always referred to as Jim, who was born at the beginning of 1868. It should be remarked that Lockyer's children were generally named after one of their godparents. Thus Alexander Edmund was called after Alexander Macmillan, and Hughes Campbell after Tom Hughes. The 'William James' part of Jim's name came, of course, from his mother's father, but the 'Stewart' was in honour of Balfour Stewart, who was a godparent.

Despite the death of his first child, Lockyer and his wife were not greatly afflicted by serious illness or death amongst their offspring by Victorian standards. It is true that Jim had bouts of illness, which continued throughout his life, but they seem to have been annoying, rather than seriously worrying. One child, who was a constant concern to the Lockyers in this respect, was Frank Ernest, the third son. It has been mentioned in an earlier chapter that Lockyer suffered from a nervous breakdown in 1877, and went to recuperate in France. He was forced to cut short his stay abroad when Frank fell ill again. To Lockyer's great distress the boy failed to rally, and finally died in the summer of 1878, when he was in his mid-teens.

This was the year when there was a total eclipse visible in the United States. Lockyer had intended to lead an official party to observe the eclipse, but his son's illness forced him to withdraw. After Frank died, and partly for relief from the strain he had been under, Lockyer decided to go out to the United States after all, and observe the eclipse as a private person. He was invited to attach himself to various of the American eclipse parties, and finally decided to observe with Henry Draper's party. At the last minute, he actually separated from the main group, and observed at a station further along the line of the Union Pacific Railway. Lockyer helped finance the trip, as was his custom, by acting as correspondent for one of the London papers, in this case the *Daily News.*

He returned to London in a somewhat happier mood, and was further cheered by his election to the Athenaeum early in 1879. He had been put up long before but had not been elected. Many of Lockyer's friends had been members for years; for example, Huxley had been elected in 1858 and Frankland in 1860, though Lockyer's sparring partner, William Huggins, was also only elected in 1879. But in September, disaster again struck his family life, even more seriously than before. We have seen that Lockyer's wife was an important, if unobtrusive, influence on his life. J. W. Draper, writing home to his son, Henry Draper, in 1870 remarked, 'He [Lockyer] is such another young fellow as you and has a wife (so Mr de la Rue told me) who takes the same interests in his pursuits that Anne does in yours.'[1] But now, after only a short illness, she died, and was buried as Frank had been in Brompton Cemetery. Lockyer was left with seven children to care for, the eldest 19, and the youngest 6 years old. It is hardly surprising that there was a noticeable lull in his scientific activities during the early eighties.

In this crisis Lockyer's friends rallied to his aid. Donnelly, for example, made arrangements to go abroad with him for a while to help distract his attention. During this period Lockyer cemented several friendships that were to last a lifetime. By no means all of these were connected with astronomy or with South Kensington. Thus one person of whom he saw a good deal at this time was George Romanes, the biologist, who, like Lockyer, was an enthusiast for the widespread application of evolutionary ideas. After Lockyer had unveiled his meteoritic hypothesis at the Royal Society in 1888, Romanes wrote to him, 'You were always a man in whom I believed (astronomically speaking) as likely to prove one of the greatest men of our century; and now I am glad to see that my faith will not be in vain.'[2] Another of the friends whom Lockyer cultivated in the early eighties was William Black, the Scottish novelist. Black had moved to London from Glasgow during the sixties, and during the next decade had established himself as the writer of travel novels, such as *The Strange Adventures of a Phaeton.* Black, incidentally, was for a while sub-editor on the *Daily News* for which Lockyer also occasionally wrote.

In 1882 there was a total eclipse visible from Egypt, and Lockyer organised an expedition to observe it. He took the opportunity to plan excursions to the major Egyptian antiquities at the same time, and therefore invited several of his non-astronomical friends to join the party. Black was one of those who accepted, for he was not only a keen traveller, but could also use the experiences in his novels. One who had to refuse was Lawrence Alma-Tadema, the painter. Alma-Tadema was extremely interested in archaeology, a fact that appears in his paintings, and would have liked to have gone, but 1882 was a busy year for him in London. He was holding his first retrospective exhibition, and was also involved in the Royal Academy exhibition.

Lockyer had many artist friends, and was a regular attender at the Royal Academy, but was by no means an uncritical admirer of the pictures displayed there. He was, in particular, frequently irritated by inaccurate representations of nature in the paintings, and for some time seriously considered writing a book on science for artists. Although this never materialised, he did lecture and write on the subject, a series of articles appearing in *Nature*, for example.

The most important Victorian artist with whom Lockyer was on close terms of friendship was the pre-Raphaelite painter, William Holman Hunt. Lockyer probably met Hunt through Tom Hughes, though he might also have met him through Rossetti (who wrote for the *Reader*) or through Woolner, both of whom were members of the pre-Raphaelite group. One of the interests that Lockyer and Hunt shared in common with Hughes was an attachment to the Volunteers. Although Lockyer was closer to Hunt than to any other artist in the later years of his life, during the eighties he associated more with the Scottish artist, William Orchardson. (It is worth noting, incidentally, how many of Lockyer's friends were Scottish. When circumstances allowed, he dearly enjoyed a trip north of the border.) The main link between the two was that Lockyer built a seaside house at Westgate-on-Sea in the 1880s, which was almost next door to Orchardson's country house. When they were down there together, Orchardson often invited Lockyer along to his tennis parties. One of Orchardson's daughters later wrote, 'I remember nothing of him [Lockyer] except that he teased little girls and wore a perfectly disgraceful straw hat; when I grew up, however, I found he was a very witty talker.'[3]

By the time his Westgate-on-Sea house was established (about 1884), Lockyer had regained his usual ebullience, and was once again scientifically active. Appropriately enough, his renewed interest was triggered off by one of the largest volcanic eruptions ever. In August 1883, Krakatoa, a volcanic island in what is now Indonesia, blew itself to pieces, and threw material high into the atmosphere. In the ensuing months, extraordinarily colourful sunsets were seen all over the world, and the Sun temporarily turned blue. There was general agreement that these effects were due to dust particles in the upper atmosphere, but there was dissension over the source of the particles. Lockyer was a strong supporter of the idea that the dust was debris from the volcanic eruption; but in Germany, as one of his friends told him, '. . . they are all agreed that we have got caught in a comet, and that the dust is all of meteoric origin.'[4] A reason for this attitude was that it seemed remarkable that the atmospheric effects should be so prolonged. Lockyer thrived on this new argument; indeed, it got him back into the habit of popular lecturing again. He had tended to let this side of his activities lapse after his wife's death. Early in 1884 the Royal Society set up a small committee to examine what had been discovered concerning the Krakatoa eruption and its aftermath. As a result of the interest Lockyer had shown, he was appointed a member of the committee. It was thought initially that the work of the committee would soon be over, but when it got down to a detailed study of the problem, a surprising amount of information was found to be available, and the committee was not able to issue its final report until 1888.

Not long after the Royal Society committee was established, Lockyer set off on another of his jaunts to the Continent, visiting France and Italy, and seeing several of his astronomer friends there. After his return, he became involved with Canon Samuel Barnett in the establishment of Toynbee Hall, with the intention of involving university graduates in the problems of the London slums. In the early days the lecturers included, besides Lockyer, several of his closest friends, such as Tom Hughes and Romanes. (Another of the lecturers was Sir William Besant who recruited Lockyer, in 1884, to be a vice-president of his Society of Authors.) Barnett had started, in 1881, an annual exhibition of art for the benefit of inhabitants of the East End, and Lockyer was involved for several years with this, persuading various of his artist friends to lend their paintings. It is clear that, although Lockyer was no longer the ardent churchman of his youth, he still retained much of the social conscience which his association with the Christian Socialists had stimulated. His name frequently appeared in connection with liberal movements, especially where they were related to education. Thus, in 1889, he was invited to be President of the Sunday Society, which had been set up to obtain the opening of museums, art galleries, libraries and public gardens on Sundays for the recreation especially of the lower classes.

Lockyer's selection of Westgate-on-Sea as the site for his country house reflected his attachment to this part of the Kent coast. His last holiday with his wife had been taken just a little further along the coast, at Herne Bay. Westgate-on-Sea was rapidly growing in popularity as a seaside resort at the time (this expansion was later halted by a diphtheria outbreak which was alleged to be due to bad drainage), but the atmospheric pollution which faced Lockyer in South Kensington was, of course, quite absent. It is not surprising, therefore, that he soon conceived the idea of using the place for astronomical observation. Early in 1888, he started work on the construction of a branch observatory there. During this period photography was becoming an essential part of astronomy, but the photographic emulsions were still fairly insensitive. Hence the photography of faint objects, as was necessary in astronomy, required long exposure times, and this, in turn, required a clear atmosphere.

Lockyer's friend, Isaac Roberts, was looking round for a new observatory site in Sussex at about this time, having finally given up hope of obtaining reasonable results in the vicinity of Liverpool. He wrote to Lockyer, '[I] have been driven almost to distraction in watching the sky closely 365 nights each year in order to secure say half a dozen intervals suitable for photographic work.'[5] His advice on the need for a dark sky for photographic work was reinforced by that of A. A. Common, who was a fairly frequent visitor at Westgate-on-Sea. As a result, Lockyer, who had already erected a refractor in the grounds of his house there, decided to install a much larger telescope for photographic work by the sea. He persuaded Common to grind him a 30-inch mirror for a new reflector, and obtained a small sum of money from the Government Grant Committee, which he used to build a canvas and wood structure to house the telescope. At the same time, he built himself a larger house, which could hold the observers as well as his own family. (Lockyer often brought down assistants from South Kensington to help with observations at weekends.) This 30-inch reflector was eventually transferred to the main site at South Kensington, but during its years at Westgate-on-Sea it provided a significant proportion of all the spectra studied there.

During the latter part of the 1880s, Lockyer's social life became a good deal fuller. In 1886, he was elected to the Royal Society Club, and, early in 1887, this was followed by election to the committee of the Athenaeum. His colleagues on the latter included Edward Cardwell, who, as Secretary for War in Gladstone's first ministry, had been partly responsible for deciding Lockyer to leave the War Office. Apart from these official activities, Lockyer also took part in organising a variety of other social events; for example, a dinner to honour Tyndall which was attended by most of the leading scientists in Britain. The *Nature* editorial on this event is interesting, as it indicates very clearly Lockyer's preoccupation with the need for scientists to have a higher professional status. On such occasions, he said, 'Men become more vividly conscious that though students of Nature are excluded from the State recognition which is extended to the Church, to medicine, and to the law, they too are members of a great profession.'[6]

Lockyer could afford to devote more time to social activity now, for his children were all grown up. He was no longer involved in their immediate care, but rather in helping them along in their various careers, in which his wide range of acquaintances proved helpful. He was especially interested in Hughes Campbell Lockyer's naval career; for, like any right-thinking Victorian, he was an ardent advocate of the importance of sea power in maintaining the British Empire, in which he strongly believed. Although not a conspicuously good sea traveller, Lockyer delighted in voyages, and indulged in many, of various lengths, besides those forced on him by his work. One of his friends, J. G. S. Anderson, was a director of the Orient line, and arranged for Lockyer to have several free trips aboard boats belonging to the line. One especially memorable excursion was a visit to the Naval Review at Portsmouth during the Queen's Jubilee. Another of Lockyer's acquaintances was in the P. and O. line, and this brought him further offers of sea excursions. Moreover, besides his eclipse expeditions, which were usually organised in collaboration with the Royal Navy, Lockyer's naval friends often provided him with other opportunities for short sea voyages. In turn, Lockyer was interested in the application of science to naval problems, and, when the opportunity arose, gave what support he could to members of the navy who were trying to make such applications.

At the beginning of the nineties, Lockyer took up the cudgels on behalf of a naval lieutenant, R. H. Bacon, who was developing a new type of glass for use in binoculars. Bacon was later responsible for the design of the dreadnoughts, and ended his career as an admiral, but at this time he seemed to be making little headway against the apathy of the Admiralty. As he sadly wrote to Lockyer,

> The Navy is a heartbreaking service, you cannot get men brought up all their lives to trusting and acting on their own opinions to carry out or even witness

trials of any sort without jumping at conclusions. Halfway through any trial, the authorities say 'I think this or that' and stop them.—As if what they 'jumped at' had anything to do with the *real truth* of the facts. I cannot help growling. I had a really rather interesting set of experiments squashed in this way just as we were really getting some results. Now we are as much in the dark as ever![7]

The son in whose career Lockyer took most interest was naturally Jim, for it was intended that he should follow his father into astronomy. Jim studied both at Cambridge and South Kensington, but his main introduction to astronomical research was in Germany, whither he was dispatched by his father in the early 1890s. Germany was the mecca of aspiring research workers in the latter part of the nineteenth century. They went there to study their trade; then, returning to their own countries, they both practised what they had learnt, and simultaneously tried to disseminate the gospel of the pre-eminence of research. This custom of going to Germany for a period had spread throughout the scientific community in Britain, though it was especially noticeable amongst the chemists, and increasingly the time was used to read for a Ph.D., a degree that had not then been established in Britain. In the nineties, a year or two in Germany was a normal part of the training of many astronomers, from countries as far afield as the United States. Indeed, Germany played the same role in the nineties that the U.S.A. has played during the last decade or so. In view of the prestige that attached to a German Ph.D., it was a very reasonable move for Lockyer to dispatch his son to Göttingen, one of the most reputable German universities. Moreover, for the benefit of Jim's future in astronomy, it was probably as well that he should have some qualification not too closely related to his father's work at South Kensington, for the general suspicion of the latter would certainly have harmed his prospects. (This does seem to have happened later, in fact.) In Germany, Jim would certainly obtain an independent qualification, for Lockyer's approach to astronomy was regarded with considerable suspicion there. As one of the visiting Americans later recalled,

. . . in my studies at Potsdam, Germany, under H. C. Vogel and J. Scheiner, we looked somewhat askance at Lockyers' tendency to build up a theory like his Meteoritic Hypothesis on inadequate observational data, and the same thing applied to his Hypothesis of Dissociation.[8]

Lockyer, of course, had to bear the expenses of Jim's sojourn in Germany, and, in typical Victorian fashion, continually suspected that his son was squandering his opportunities and spending money with insufficient justification. Jim, on his side, was continually appealing for more cash.

I have been expecting a letter nearly every day from you enclosing some money as at present I am sponging on Payn [with whom he was staying in Leipzig] here. I don't quite know what gives you the idea about my wasting my time and being in a fool's paradise: directly I wrote about the trip I made, I said to myself that you would think that I had been doing something of that kind the whole time, so you see I was not far wrong.[9]

From his letters home, Jim does seem to have had quite a pleasant time in Germany, though interrupted by one of his bouts of illness. But his father also obtained a certain amount of direct benefit from his son's time abroad, for he used him as the *Nature* representative in Germany during this period.

Both father and son had been hoping that the experience of working in Germany would enable Jim to win one of the few scholarships available for research in astronomy on his return to this country, but in November 1891, he heard that he had not been nominated. He wrote with a tinge of bitterness to his father,

I am awfully sick about it and if not for my own sake, I should liked to have got it for yours.

All along I have had an inkling that I should not get it, not from the point of view of whether I was worth it or not but simply from the fact that even if there was a benefit of the doubt to give, another fellow would be bound to have it, as I *might* seem to be more highly favoured, which of course would not do.[10]

Jim was eventually awarded his Ph.D. and returned to England, where he became his father's chief assistant at South Kensington. For some time he was on the same basis as he had been in Germany, with his father paying his salary, but the reorganisation of the observatory towards the end of the century put his appointment on a more permanent footing. His salary became a charge on the increased Treasury grant to the Solar Physics Observatory. By the beginning of the twentieth century, Jim was recognised as being in charge of the day-to-day running of the observatory, though his father still retained very tight control. For example, the following letter is typical of the type of communication that passed from Jim to Lockyer when the latter was away from the site (as he frequently was). Note that Lockyer's signature was required before even such minor items as bolts could be acquired. Jim begins,

I received your letters and wire this morning. The pouch etc. have been sent on to you, also Heis' atlas (later).

All the slides are being made as fast as they can. I have had the up and down stellar genera table made out and photographed. The classification (photographic) I will have done tomorrow.

The 6-inch is working badly again and two photographs recently taken show those step-marks in the line. There seems to be lack of rigidity between the wooden tube and the mounting of the prisms—I will have it seen to as soon as I can. Butler [one of the assistants] has been down to the coelostat most of the day rating etc. The clock goes very well and runs for 2h. 10min.

> Please sign the enclosed req[uisition] as these bolts are kept in stock and are wanted at once.
> I have written to Fowler, and will get the 'definition' table copied out and photographed as soon as he answers. I will look at [?] Gould's paper tomorrow as I could not get to it today as there were so many things to start.
> Have commenced inquiring into any relationship between Rydberg's series and Mendeleeff's periodic law, but have not as yet struck oil.[11]

Throughout the eighties and nineties, much of Lockyer's time continued to be taken up with matters related either directly or indirectly to *Nature*. By this time, it had firmly established itself as the most important journal dealing with science in general. A convincing example of its prestige appears from a letter that Crookes wrote to Lockyer in the summer of 1895.

> I have been working night and day to get in type a paper on the spectrum of helium, before my holidays. I will send you an early proof tomorrow, and I should much like to see it in 'Nature' if you can see your way to insert it. It will appear in the 'Chemical News' on Friday, but my circulation is not to the same class of researchers as that of 'Nature', and having taken a great deal of trouble about it I want the results to get to the right people.[12]

Unfortunately, high prestige did not mean high profits. *Nature* consistently failed to come out on the right side of the ledger throughout the century, the first profit margin occurring in 1899. Yet Macmillan continued to give it support. In November 1894, the firm organised a dinner at the Savoy to celebrate the twenty-fifth anniversary of *Nature*, and the publication of its fiftieth volume. Fifty people attended, including several of Lockyer's old backers. Huxley was there (he died in the following summer). Lockyer persuaded him to write a brief retrospective note for the anniversary number of *Nature*. Huxley also spoke at the dinner, and praised Lockyer for his success.

> Mr Lockyer was to his contributors not so much the Editor as the friend. This was no easy task; it was often better to be an architect building a house under the dual control of a husband and wife, or a member of the London School Board, than an editor.[13]

Another old member of Macmillan's staff who was present told the assembled diners that *Nature* was, '. . . the most influential journal of science in Europe'.[14]

The variety of men and topics that Lockyer had to deal with in *Nature* continued to be vast. Indeed, as new branches of science appeared and grew, the range of the journal became increasingly complex. As an example of such a developing science we can take one of Lockyer's early interests, anthropology. As we have seen, Lockyer was an eager listener at the debates of the 1860s, which really marked the arrival of anthropology as a separate entity.[15] Another person who attended meetings of the Anthropological Society of London during that period was Colonel Lane Fox, later to become General Pitt Rivers. Lockyer may have encountered him there, or they may have been introduced by Lubbock. Pitt Rivers had been drawn into anthropology initially by his interest in the evolution of the musket and of fortifications. When, in 1882, Lubbock (who was then M.P. for London University) succeeded in pushing through Parliament a Bill for the better protection of ancient monuments, Pitt Rivers was appointed to be the first official Inspector of Ancient Monuments. (A couple of years later Lubbock married his daughter.) Subsequent to his appointment, we find Pitt Rivers instructing Lockyer during the eighties about the current state of anthropology, for the benefit of the readers of *Nature*. (It was about this time that the anthropological section of the British Association was set up.)

> I forgot to tell you another reason when I met you yesterday why I had called my collection anthropological. One reason was as I said because evolution was not understood by the public in relation to the arts, but now the term evolution at any rate, if not the science of it, has been made more familiar to the general public than anthropology. But I also wished to promote the use of the word Anthropology in its widest sense as we understand it in this country and not in its restricted sense as Physical Anthropology only, as it is used in France. The history of the rise of anthropology in France and England is different. In France it grew out of the Faculty of Medicine and was composed chiefly of medical men. Hence physical anthropology became the leading study with French anthropologists, and anthropology came to be understood as the science relating to the constitution and races of Man, but this is really only a part of it, and the part of it which might if necessary be relegated to zoology. Anthropology taken in [its] broadest sense is the study of the 'human period of evolution' and is characterised and distinguished from all other periods by the rise of the arts. Social anthropology rather than physical anthropology has always been the leading study in England. Anthropological science with us grew out of Geography, the medical profession not standing so high with us as in France. Our extensive colonies, our commerce, the whole genius of our race has contributed to give to anthropology a wider significance with us . . .[16]

Pitt Rivers and Lockyer also shared an interest in museums. Pitt Rivers suggested to Lockyer that the science collections at South Kensington should be reorganised into a museum of technical evolution (i.e. to show how the different arts and sciences had evolved).

Pitt Rivers was not Lockyer's only adviser on anthropological matters. Anthropology in the nineteenth century, like several other areas within the humanities, relied considerably on philological studies, and for advice on this Lockyer relied on the leading expert in Britain, the German, Max Müller, who ultimately became Professor of Comparative Philology at Oxford. Not long after he received the foregoing letter from Pitt Rivers, Lockyer received a letter from Müller concerning a festival, which had just taken place, honouring the centenary of the birth of Jacob Grimm.

Müller, in explaining to Lockyer why he had not written anything for the festival, made the following claim, 'I believe I erected myself the most lasting memorial to Grimm by calling the law regulating the changes of consonants in Sanskrit, Greek, Latin and German, "Grimm's Law". I used this name first.'[17]

A completely different area of science which burgeoned during the latter part of the nineteenth century, and in which Lockyer, and therefore *Nature*, showed an especial interest, was marine biology. Of the many scoops that *Nature* made in this field, one of the most exciting was contained in a letter Lockyer received on the last day of 1895. It described

> . . . the very remarkable results obtained by the Prince of Monaco from the capture of a sperm whale under his eyes in the Azores and from the contents of [its] stomach. It appears that this animal feeds . . . on gigantic cephalopods of a size and power undreamed of even in poetry or fiction.[18]

It was an eyewitness account of what became the first full-scale investigation of the giant squids.

The biologists, and especially those concerned with the progress of evolution, continued to be the most belligerent group of writers to *Nature.* Ray Lankester, one of the more easily aroused biologists, was particularly liable to set Lockyer editorial problems. On one occasion, for example, Lankester sent Lockyer a letter for publication in *Nature* which violently attacked Lockyer's friend Romanes. Lockyer managed to gloss over this by showing Romanes the letter, but not publishing it. Lankester, who was watching suspiciously for any attempt to disguise the point at issue, finally accepted this judgement.

> You are quite right not to print my letter about Romanes, as it is not argumentative but purely denunciatory. I am glad he has seen it, as he will now know what a humbugging piece of foolery I consider his attempt to say 'Darwin-and-I' and 'the Darwin-Romanes theory', is. It is time that he knew that I consider him a wind-bag.[19]

The religious controversy that had still surrounded Darwinian evolution when *Nature* first appeared, was much less obtrusive in the late eighties and early nineties. Huxley was still battling stoutly, though now mainly in opposition to the ideas of eminent laymen on miracles. His two chief antagonists were Gladstone and the eighth Duke of Argyll. Huxley clearly relished these encounters. He wrote to Lockyer in the summer of 1887 saying that he was in bad health and spirits, and adding, '. . . if things go on in this way I shall have to write another letter to the "Times" or criticize the G.O.M.'s [Gladstone's] last literary effort.'[20] This controversy, especially the various debates with the Duke of Argyll, spilled over into the pages of *Nature*, though it was a little difficult to know how much space should be devoted to a matter which most scientists regarded as old hat. Thiselton Dyer, who was both Hooker's successor as the Director of Kew and his son-in-law, was persuaded to write a reply to some of the Duke's pronouncements for the sake of the lay readers of *Nature*, but he told Lockyer,

> I hate this sort of job and after struggling with it have produced a very indifferent result. The Duke's letter is such infernal rot that replying to it is like beating a bladder. If I were to prick him as he deserves you would say I was too rude. In the effort to be moderately civil I have only succeeded in being dull.[21]

Of course, Lockyer's first love, astronomy, also figured largely in his *Nature* correspondence. By the latter part of the century much of the more interesting material in this field came to him from the United States. One of his frequent correspondents there was his old friend Samuel Langley, from 1887 Secretary of the Smithsonian Institution in Washington. The year after his move to Washington, Langley wrote to Lockyer regarding the English edition of a book previously published in the U.S.A. This was called *The New Astronomy*, a name Langley introduced to describe the new science of astrophysics, and one which subsequently achieved some popularity. Langley explained that the English edition was in preparation.

> . . . but in the meantime, I see that a gentleman whom I may surely call our common friend—need I say that I mean Mr Proctor?—has laid hands on my title, and is publishing or advertising to publish 'The Old *and* New Astronomy'! Under these circumstances, if you care to give 'The New Astronomy' a review in 'Nature', I would rather like to see it done a little in advance of the placing on the market of a new and revised English edition.[22]

In fact, Proctor tried to effect a *rapprochement* with Lockyer at about this time, and although the bitter memories remained, the fire of their controversy had died away.

In the latter part of the century, the work Langley initiated on studying the value of the solar constant was of continuing interest to Lockyer; for, if variations of the incoming flux of radiation from the Sun could be positively identified, it might be possible to correlate them with climatic variations on Earth. However, the most exciting letter that Lockyer received from Langley was not concerned with astronomy at all. Langley had been carrying out experiments on the possibility of building heavier-than-air flying machines. In 1896, he managed to make one of his large steam-powered flying models fly for over 4000 feet, and thus encouraged he wrote to Lockyer.

> I have been for some years experimenting on mechanical flight, but have never said so publicly, and such references as have been made to my supposed work in this direction, by the press, have been unauthorised by me.

Now, I am still far from my goal, but having attained a definite, if not a complete, success I have decided to make it public.

I have accordingly sent a letter to *Nature* . . .[23]

By no means all letters coming to *Nature* were taken up with new discoveries. A much more time-consuming problem was correspondence with authors over material which for some reason or other seemed unsuitable for publication. Examples of contentious correspondence have already been given, but not all the difficult correspondence fell into this category. Thus a continuing problem of a different kind was presented by J. J. Sylvester's pressing requests to have his poetry published in *Nature.*

I am very glad you kept my sonnet back as it is vastly improved since I sent it into Nature.

But now it has received the very highest polish of which it is susceptible. It is a fraction reduced to its lowest terms—i.e. it has been reduced to the most simple form of expression capable of conveying the meaning—it is now *simple* and *direct* as if it had come from the hands of Dante who I think would not have been ashamed to claim it as his own.

Dante whom I have only quite recently begun (and only *begun*) to study in earnest is the great master of style—*simplicity, directness* and dignity are his chief characteristics. The language of poetry is one in whatever *tongue* it may be written and Dante is the great physician for all of us who aspire to write in that language, 'in the language of the heart.'

Oh! Swinburne, Oh! Arnold, Oh! Tennyson—what puling, maudlin, affected, screeching owls of night ye are, viewed in the light of example of that heaven-inspired prophet and singer whose doctrine you have desecrated or whose example you have been too puffed up in the pride of your own little minds to profit by. Adieu Nature! and believe me to be an aspirant to your communion.[24]

Editorial policy remained as live an issue as ever; for example, the question of signed reviews continued to be source of controversy. As one reviewer wrote to Lockyer,

I have initialled this but I am bound to tell you that I object altogether to signed reviews. They lose all their force. One doesn't care twopence for the individual opinion of W.H.P. or J.T. The most disparaging review of one individual would not cause me a moments discomfort but the anonymous strictures of an unknown and mysterious editor carry weight and produce effects. . . . A reviewer does not like to speak the truth when he criticizes a friend's book over his signature whereas he can speak out when he is shielded by the aegis of an irresponsible editor.

I could say much more against your practice (which I regard as a sop to fashion) had I time.[25]

It might well be that Lockyer had no desire to shield his reviewers. As it was, he had to take responsibility for all the leading articles, though, in fact, only a minority of them were his own work. During the eighties, Romanes and Geikie both wrote more leading articles than Lockyer, and he recruited his colleagues at South Kensington to write many of the remainder.[26] These leading articles by other hands more than once landed him in hot water. But, as the century drew to its close, and *Nature* became increasingly prestigious, attacks on its editor required a correspondingly greater nerve. As Huggins wrote to G. E. Hale in the nineties, 'the scientific men here, for the most part, are too much afraid of him to do anything.'[27]

Lockyer's world-wide correspondence was, of course, by no means restricted to scientific matters. On the one hand, his friends kept him up to date with many of the ordinary happenings in their own place of work, or country of residence, as, for example, in the following letter from a friend in New York early in the nineties.

> For the first time in many years I have had the opportunity to witness a political contest in my native country. It was not a pleasant sight, and has greatly strengthened my feeling of loyalty to good King Umberto, and my satisfaction with the simpler and honester methods of Italy. The animosities, the accusations of fraud made by both sides, the utter lack of anything like political principles, the want of ability in all the leaders and candidates, the basest conceivable exhibitions of partisanship, and the puerility and scurrility of the press, furnished a thousand arguments against that irretrievable blunder of the century—the extension of the suffrage to the ignorant and irresponsible classes.[28]

On the other hand, the success of *Nature* as the mouthpiece of science resulted in its editor being regarded by many non-scientists as a spokesman for his scientific peers. Partly in consequence of this development, Lockyer was frequently approached for his support in non-scientific matters. For example, at the end of the century his old Christian Socialist friend, J. M. Ludlow, wrote to him enclosing an address of congratulation to Captain Dreyfus' French supporters, asking that he should add his own signature, and obtain the signatures of some other representative British scientists.

Lockyer's writings were not, of course, limited to the pages of *Nature*. Apart from publishing several books in the latter part of the century (mainly, though not entirely, with Macmillan), he also contributed occasionally to various newspapers and periodicals. He could have written much more frequently for the latter had he wished. We find Edmund Gosse pressing him for articles in the late eighties,[29] and Frank Harris in the early nineties.[30] Some of these outside articles involved Lockyer in unexpected complications. For example, he contributed some of the astronomical articles to the famous ninth edition of the *Encylopaedia Britannica*, which led to him receiving, several years later, the following letter.

> I dare say you noticed the paragraph in today's Times about the supplement to the 9th Edition of the Encylop. Brit.? It would be *very* desirable to get Creighton's most mischievous article on vaccination corrected . . . It is difficult to exaggerate the harm Creighton's article has done and is doing.[31]

The writer was Elizabeth Garrett Anderson, one of the small number of women doctors, exerting her energy to persuade Lockyer, as an eminent scientific contributor to the *Encylopaedia*, to correct one of its more glaring errors. A reputable epidemologist, Charles Creighton had been recruited to write the article on vaccination, but had then proceeded to deny that vaccination had any value at all. Unfortunately, his article had appeared at just the wrong time, so far as supporters of vaccination were concerned. Vaccination had been made compulsory in England during the 1850s, but a strong anti-vaccination movement had appeared, and Creighton's article provided it with valuable ammunition. In fact, the agitation became so fierce that eventually, in 1898, just before the announcement of this supplement to the *Encyclopaedia*, conscientious objectors to vaccination were given legal exemption. This was why Elizabeth Garrett Anderson was so disturbed. She feared that any influential anti-vaccination writing might cause large numbers of people to take advantage of the new escape clause. Lockyer actually required little persuasion, for he was firmly convinced of the value of vaccination. He once remarked that one of the great changes he noticed, as compared with his youth, was the disappearance of pock-marked faces.

Lockyer, himself, continued to consult Lauder Brunton concerning his personal health problems. He still had his ups and downs. Thus, in 1899, he wrote to his South Kensington colleague, Rücker, saying, 'I must go away for a few days or I shall break down.'[32] Nevertheless, he seems to have been less affected by serious breakdowns than in his earlier days. On the other hand, as time passed he became increasingly stout, and began to run into problems with high blood pressure. As a result, he made a habit in the latter part of the century of visiting various spas to take the waters.

Lockyer's main form of exercise was golf, of which he was very fond. One of the joys of Westgate-on-Sea, from his point of view, was that observing sessions there could be combined with a round of golf at the St George's Golf Club in Sandwich. Golf had established itself in England during the nineteenth century as a particularly health-giving pastime, especially since, following the example of the Royal and Ancient at St Andrews, the best golf links were thought to be those constructed by the seaside. As one writer observed,

> It is a manly and eminently healthful recreation, pursued as it is mostly amid the fresh sea-breezes; while, as exercise, it has this peculiar merit, that, according to pace, it may be made easy or smart at pleasure, and thus equally adapts itself to the overflowing exuberance of youth, the matured and tempered strength of manhood, and the gentler decays of age.[33]

There was, however, a continuing confusion about the rules of the game. Although the code in use at St Andrews was generally agreed to be pre-eminent, not only were there local deviations, but even the St Andrews' code itself was open to varying interpretations. Lockyer therefore combined with the honorary secretary of St George's Golf Club (a barrister, W. Rutherford) to compile *The Rules of Golf*, which appeared early in 1896. This codified the St Andrews' rules with explanations of difficult points. It was a small book which, it was intended, should be slipped in the pocket and referred to on the course. To Lockyer's delight, and surprise, the book proved to be highly popular, and was of some importance in enforcing a uniform set of rules at all clubs throughout the country.

Many of Lockyer's scientific acquaintances were equally keen golfers. P. G. Tait in Scotland was, indeed, a far greater fanatic. During the summer months his letters to Lockyer were nearly always written from the clubhouse at St Andrews, where he often started a round at 6 a.m. During the early nineties, Tait devoted considerable thought to the theoretical aspects of golf. Tait's son, Freddie, was one of the leading amateur golfers in the latter part of the nineteenth century, and he took part in his father's investigations. One story has passed into golfing folklore. It is recounted that Tait calculated the maximum distance a golf ball could be struck; then his son went on to the thirteenth fairway at St Andrews, and hit one sixty yards further. Unfortunately, the story is apocryphal, for what Tait calculated was how far a ball would carry if no spin were imparted to it. What his son's drive showed was that a spinning ball carried much further than a ball that was not spinning. Having confirmed the effect of spin, in fact, Tait went ahead with the design of a golf club with a specially grooved head to impart maximum spin. The ultimate form of the club was quite complicated. The specially shaped head could be keyed to the shaft at varying angles. As Tait reported to Lockyer, according to the angle selected, the club could be used as a driver, a spoon or a cleek.[34] Freddie Tait refused to use his father's universal club, but a few experimental models were made. Tait told Lockyer that they cost about 10s 6d each without the shaft.[35]

Tait was rather far afield for Lockyer to play with frequently, but there were plenty of scientist-golfers closer to hand. Two astronomers with whom he played were A. A. Common and H. H. Turner. According to a letter to Lockyer from Turner, in 1890, Common was the better golfer, 'Common won the monthly medal : he is getting an awful swell.'[36] This letter from Turner was actually mostly concerned with describing the formation of a new national astronomical society, the British Astronomical Association. He assured Lockyer that the new society was not being started aggressively (presumably relative to the Royal Astronomical

Society); but he continued, 'As a professional astronomer I shall have nothing to do with the new one which I think is really meant for the education of amateurs.' It is interesting to note here the extent to which astronomy had become professionalised since Lockyer entered the field in the 1860s. In the former period, anyone with an amateur interest in astronomy would naturally have joined the Royal Astronomical Society and might, indeed, have expected to play a major part in the Society's affairs, if his inclinations so led him. Since that time, the role of the professional had, for a variety of reasons, been gradually advanced at the expense of the amateur. One of the factors was actually the success of the early pioneers of the new astronomy, such as Lockyer himself. The resultant increasing need for better instrumentation, larger telescopes, and a better knowledge of physical science, tended to separate out those who could devote most of their time and energy to astronomy from those who could only work at it in their spare time. At the end of the nineteenth century the distinction was still far from clear cut. As a result there were occasional overlappings of functions between the Royal Astronomical Society and the newly formed British Astronomical Association, as some observers had feared. For example, both were involved in the organisation of eclipse expeditions. But such major problems as the Royal Astronomical Society had, remained internal. Compared with the seventies and eighties, the nineties were a relatively peaceful time in the Society's history, although disturbances did still occur. Lockyer's old opponent, Ranyard, tried on four separate occasions either to stop the Royal Astronomical Society from awarding a Gold Medal altogether, or, at least, to make sure that only foreign astronomers were eligible to receive it. He never succeeded, though he came quite close on one occasion, and the dispute finally lapsed on his death in 1894.

Lockyer, himself, was involved in various new controversies with the Society, though most of them were minor. Thus, in 1892, there was a dispute, in which he figured, over whether the Society should become a centre for the dispatch of astronomical telegrams to alert observers to recent new discoveries, for example, of comets. It is, indeed, evident that he remained at odds with various fellows of the Society into the nineties. Pritchard wrote to him in 1893 complaining that the Royal Astronomical Society had completely ignored the measurements of stellar parallax which were being carried out at Oxford. (Whereas the Royal Society, encouraged by Lockyer, had just awarded Pritchard a Gold Medal.) Pritchard remarked, 'We who know the constitution and habits of the R.A.S. Council need feel *no surprise*.'[37]

Lockyer had always gravitated towards the Royal Astronomical Society Club whose conviviality he enjoyed, and whose members generally included those Fellows of the Society with whom he was on most cordial terms. In 1896, Lockyer was elected a Vice-President of the Club. H. H. Turner subsequently observed that the Club records for the occasion noted, 'All [the Club Officials] were duly elected and returned thanks in very nice modest speeches.'[38] He added, 'It is significant that the character of the speeches seemed worthy of record, and we are driven to the conclusion that it was a unique occasion.' During the following year, however, Lockyer finally decided to resign from the Society, and his resignation was reluctantly accepted in 1898. This resignation meant that Lockyer also left the Club, though he subsequently attended at intervals as a guest.

During the eighties and nineties, Lockyer was actually more involved with controversies within the Royal Society than in the Royal Astronomical Society. Thus a major dispute of the eighties into which he was drawn was started off when Stokes, then President of the Royal Society, decided to stand as the parliamentary candidate for Cambridge University. Huxley was immediately up in arms. He wrote to Lockyer, 'So now we are to have the Presidency of the R.S. made a party matter and all the dirt of politics imported into science.'[39] Huxley had carefully refrained from publicly supporting any political movement whilst he had been President, and he was convinced that this political aloofness should be maintained. He assured Stokes that his opposition was entirely on this matter of principle. In his letters to Lockyer, however, he revealed that his worries were not entirely unrelated to Stokes' political views. 'The prestige of the Presidency of the Royal Society will now be at the back of an M.P. who is certain to give his best assistance to everything churchy and reactionary.'[40]

Lockyer was not initially perturbed by Stokes' decision to stand. As he told Huxley, '. . . the general idea was that Stokes was entering the House not on party lines'; but he went on '. . . this idea if anybody ever had it, and I for one had, apparently on Stokes' authority, has been entirely shattered by his address.'[41] In the nineteenth century, unlike today, it was not regarded as particularly odd for a scientist, even if still active in scientific work, to enter the House of Commons* Indeed, that House contained greater scientific talent in the latter part of the century than at any time since. Thus Lubbock, Playfair, and Roscoe served for many years. All were friends of Lockyer, and all stood in the Liberal interest. It was not therefore Stokes' standing for Parliament that brought in Lockyer on Huxley's side, but a fear that the office of the President of the Royal Society, who was by this time generally supposed to be the

* The existence of University seats (now abolished) made this easier.

leading spokesman of science to the public, might be associated with an ultra-conservative political and theological viewpoint. Lockyer inserted an editorial drawn up by Huxley in *Nature* alerting the scientific community to the danger. Stokes, however, refused to withdraw his candidacy, and to Huxley's and Lockyer's annoyance was, in due course, elected. As it happened, the irritation did not last for long; Stokes ceased to be President of the Royal Society in 1890, and left Parliament in 1891.

Early in the twentieth century, Lockyer became entangled in another quarrel, this time not an internal squabble within the Royal Society, but one involving other parties. A year or two earlier, a joint committee of the Royal Society and the Royal Geographical Society had been established to organise a National Antarctic expedition. The intention of the expedition was that it should be primarily devoted to scientific investigation (this was why the Royal Society was involved), and a group of civilian scientists under J. W. Gregory, at that time Professor of Geology at Melbourne University, was organised. At the same time, the President of the Royal Geographical Society, Sir Clements Markham, successfully pressed for Captain Robert Scott, who had no previous experience of polar exploration, to be appointed naval head of the expedition. The general understanding was that Scott's job would be to ensure that the scientific objectives of the expedition were met, and therefore Scott would be subordinate to Gregory so far as general policy questions were concerned. However, when Gregory arrived in England in January 1901, he was informed by Markham and Scott that the latter was to be sole leader of the expedition, with the ultimate say in all matters of policy. Indeed, it was proposed that the civilian scientists should be required to sign ship's articles, so that Scott's authority could not even be questioned.

Up to this point, Lockyer had not been personally involved, though he had been an interested spectator, for his involvement in that peculiarly Victorian science of physiography, and his position as an expert on the organisation of expeditions, gave him a recognised position in discussions of geographical exploration. It is not surprising, therefore, that Gregory now appealed to Lockyer for his aid in re-establishing the scientific basis of the expedition. Lockyer took up the cause immediately with the Royal Society. To such effect that when Gregory returned to Australia in mid-February, he wrote to Lockyer in the following terms,

> The R.S. has taken so firm a line over the Antarctic and the Geogr. Soc. advocates appear so hopelessly divided, that Antarctic affairs have progressed favourably. . . . You have strengthened the R.S. attitude greatly. At first the R.S. representatives did not appear to realize fully what the position was. Your action was exactly at the right time and had, I believe, a great influence on the result.[42]

Unfortunately, these beneficial developments, from the point of view of the scientists, were not maintained. Gregory, far away in Melbourne, could not effectively counter Sir Clements Markham's influence in London, and Lockyer, having no official standing with the expedition committee, could not intervene directly. In May, Lockyer received another letter from Gregory.

> I sent you a cable this morning to tell you that I had resigned the Antarctic. The Joint Committee has abandoned the points in its programme which it considered in February essential to the scientific part of the programme. As altered they give no power to the head of the 'civilian scientific staff' to secure that the scientific work will have a fair chance, and accordingly I do not feel it right to accept the post. If the scientific work can be done without hampering anything else, it will no doubt be done: but if it threatens to delay the 'farthest south' part of the programme, then it will probably be sacrificed.[43]

Gregory's assessment was essentially correct. Scott's first antarctic expedition proved mainly memorable not for any scientific achievement, but for having penetrated to within 750 miles of the South Pole.

In the latter years of the nineteenth century, and the earlier years of the twentieth, the Royal Society was involved in a series of disputes mainly caused by the concern of many of its members with questions of the professionalization and specialization of scientific activity. In 1893, the feeling of several Fellows that the Society should be strictly for professional scientists came to a head. In that year Sir Henry Howorth, a well-known writer and *Times* correspondent, was proposed for election. A group of Fellows, headed by Armstrong, Lankester and Romanes, joined together to oppose his nomination on the grounds that he had not contributed to the progress of science. Lockyer trod a middle course in this controversy, possibly because he had friends on both sides. Thiselton Dyer, for example, wrote to him in horror at the proposed opposition. A few years later Lockyer did involve himself in a related type of dispute. In 1898, the Treasurer of the Royal Society resigned, and a virtually unknown Fellow, a Mr Kempe, was proposed to fill the vacancy. Several Fellows of the Society believed that the main offices in it ought always to go to eminent scientists, and a group led by Crookes, Hooker and Lockyer therefore protested against Kempe's nomination. They first approached Lubbock to stand as an opposition candidate, and then, when he declined, proposed Lockyer instead. One of the main reasons given for selecting Lockyer was that he had a larger circle of acquaintances within the Society than any other Fellow. However, in the ensuing election Lockyer was defeated for the Treasurership and resigned from the Council.

Lockyer was actually more concerned at this time with another issue involving the professionalization of the Royal Society, the question of what areas of knowledge should be represented in its ranks. The Royal Society's original charter of the seventeenth century had covered a wide swath of interests, but, as time passed, the Society became increasingly restricted in its scope. This aspect of the Society's development particularly affected Lockyer in the 1890s for he became involved in archaeology during this decade, and this led him to press for the election of archaeologists to the Royal Society. During most of the nineteenth century, the Society had been prepared to consider the election of Fellows in fringe areas, for example, Pitt Rivers who worked in anthropology and archaeology was elected, but, as the century drew to a close, the Royal Society became increasingly attached to the exact sciences only. As a result, Lockyer found it impossible to get some of the leading archaeologists in the United Kingdom elected to the Society.

The question of the Society's scope came to a head in 1901, when there was a meeting of a newly formed International Congress of Academies, which Lockyer attended as the Royal Society's representative. The continental academies had not become specialised like the Royal Society, but had retained responsibility for a much broader range of subjects (such as philosophy, history, philology, etc.). Hence, when the Congress was divided into two sections, literary and scientific, this was a straightforward step for most of the participating academies, but the Royal Society found itself in difficulties. Ironically enough, the Royal Society had been partly responsible for the setting up of the Congress. Talks were therefore started within the Society to decide what steps ought to be taken so that the United Kingdom should have an adequate international representation in non-scientific subjects.

There were two possibilities: either the Royal Society should be expanded to have the same breadth as the continental academies, or, alternatively, a separate body should be established to provide for those subjects which the Royal Society excluded. When the Society sounded out the opinion of leading academics in non-scientific disciplines, they expressed a definite preference for inclusion in an expanded Royal Society. This was the course which Lockyer favoured strongly, and he proceeded to campaign vigorously for its acceptance. Like many of his fellow scientists, he firmly believed in the necessary unity of all knowledge, and had expressed this belief for many years. For example, when the Yorkshire College of Science, which later became Leeds University, opened in 1875, Lockyer wrote in *Nature*, 'We do not want a Yorkshire College of Science, but a Yorkshire College in which science will be found in its proper place.'[44] But the Council of the Royal Society, at a special meeting in May 1901, decided that the Society should only include scientists in the restricted sense of the term which had come to be accepted in Britain. Representatives of other subject areas, thus excluded, were therefore forced to band together to form a new body, the British Academy for the Promotion of Historical, Philosophical and Philological Studies.

Lockyer was extremely annoyed at this outcome, and made his irritation quite clear in the columns of *Nature.* This brought him the following letter of agreement from Sir John Evans, who had been elected an F.R.S., as an archaeologist, in 1864.

> I have just seen this week's Nature. *If* the view taken on p.104 of the action of the Royal Society be correct* is it not time for some of us to enter an indignant protest against the action of the Council in limiting the scope of the Society far within that contemplated in the Charter? The British Academy at present has no prestige and so far as I can see has no bond of cohesion among its members. Its future is so doubtful that I for one declined to be an original member, and to say that the R.S. should refuse such men as Layard—were he alive—Budge or my son Arthur and refer them to the heterogeneous mass that calls itself the British Academy is to my mind little short of an insult. I do not believe that more than a tenth of the R.S. is in favour of restriction—Even those who (saving your presence) regard the Society as a sort of Trades Union of Professors do not object to a certain amount of leavening from an outer world whose scope is as scientific as their own and equally likely to conduce to Natural Knowledge.'[46]

But the protests were in vain. The Royal Society continued on its chosen path, and the British Academy gradually acquired its own prestige as the years passed.

Another question involving the Royal Society, where Lockyer again proved to have more liberal opinions than the general run of Fellows, concerned the admission of women. This problem agitated several of the scientific societies from the latter years of the nineteenth century onwards. The older societies had been established at a time when it was natural for many of them to have an all-male membership, though women might occasionally be elected to some form of associate membership. With the growth of agitation for female equality, it was inevitable that the election of women on an equal standing with men should become the centre of debate. Thus, in 1892, the Royal Astronomical Society, one of the all-male societies, refused after discussion to elect three women astronomers as Fellows. Women did not, in fact, become fully eligible for membership of this society until 1915.

Lockyer's view of the status of women in science was probably similar to that of his friend, Huxley. There was no reason why women should

* *Nature* had reported: 'By its action, the [Royal] Society limits its sphere of activity to that of the experimental sciences, and dissociates itself from the scientific study of archaeology, philology, philosophy, political economy and similar branches of knowledge.'[45]

not engage in science, but some separation of the sexes in teaching and research was preferable. When he visited Paris in 1872 he was amazed to find men and women medical students studying together.[47] Round the turn of the century, he became recognised as one of the leading advocates in this country of women's rights in science. One example of this was his support of Hertha Ayrton, the wife of his friend, W. E. Ayrton of the City and Guilds Institute, for admission to the Royal Society. W. E. Ayrton was one of the leading exponents of electrical engineering in Britain, and his wife worked in the same field. In 1901, she read a paper to the Royal Society summarising her work on the mechanism of the electrical spark. Afterwards a group of Fellows, including Lockyer, Abney and Meldola, got together to press for her election as a Fellow of the Society. Their attempt was unsuccessful, and the first women F.R.S.s were not elected until 1945.

Hertha Ayrton became one of the militant suffragettes in the period before the first World War, but this evidently had no effect on her friendship with Lockyer. In 1911, Lockyer threw the weight of *Nature* into advocating the election of Madame Curie to the Académie des Sciences, another all-male enclave. Hertha Ayrton wrote to thank him, saying,

> I know that you have always consistently held the views put forward there [in *Nature*], and I have not forgotten your most kind backing of my nomination for the Royal Society, therefore I was not surprised, only very very pleased to see the generous acknowledgement made in Nature of the great debt that Science owes to Mme. Curie, and the plea for equality of treatment of intellectual work without regard to the sex of the workers.[48]

However, in this, as in much advocacy of women's rights in science prior to the war, no headway was made. Madame Curie was not elected.

By no means all of Lockyer's activities within the Royal Society were controversial. As an example of a major undertaking which was not, one can instance his work in the nineties on a proposed international catalogue of scientific literature. In the 1850s, the Royal Society had started compiling a catalogue of all the papers that had appeared in the main scientific journals since 1800. This compilation had subsequently been extended up to 1883. It was felt in the nineties that the project should be completed to the end of the century, and a subject index provided, but that this would be too expensive a task for the Royal Society to tackle alone. Accordingly, an international conference was convened in London to consider the project. Lockyer and Michael Foster attended as the two Royal Society representatives. Lockyer turned up at the conference with a complete subject division for astronomy he had worked out. (Christie, the Astronomer Royal, had suggested a very similar scheme, but gave to navigation the space that Lockyer devoted to astronomical spectroscopy.)

The conference agreed that the Royal Society should advance the money for the project, and that this would subsequently be repaid by the various co-operating national bodies. The work was commenced and continued until the first World War. After that upheaval, few of the national bodies on the Continent could pay off their share of the debts, and this admirable endeavour foundered, having cost the Royal Society far more than it had bargained for.

Lockyer was in frequent demand during the last years of the nineteenth century and the early years of the twentieth century as an official, or semi-official, adviser. Thus he was one of the British scientists approached to act as a nominator for the newly instituted Nobel prize in Physics. He proposed Lord Rayleigh in 1904 and J. J. Thomson in 1905. The former received the prize, but Thomson was passed over in 1905, receiving it instead in 1906. Lockyer's advice was especially sought, however, with reference to scientific developments in India. As we have seen, he was involved earlier in the attempt to establish astronomy, and particularly solar physics, on a firm basis in India. When he visited India again, to observe the 1898 eclipse, the India Office asked him to visit all the Indian observatories by this time in existence, and suggest what changes, if any, were desirable. Lockyer took his son, Jim, on this tour of inspection, and also John Eliot, one of his main scientific correspondents in India, and director-general of Indian observatories. They travelled through the length of the country from the observatory at Dehra, in the foothills of the Himalayas, to that at Madras on the south-east coast, visiting others at Calcutta, Bombay, and Poona on the way. Lockyer concluded that the observatories were, for the most part, reasonably active, but the work they carried out was almost entirely routine, and mainly of restricted interest (e.g. the provision of local time services, tide tables, and weather forecasts). He suggested that the great need was for some form of central direction, and uncovered the odd situation that the Indian Government believed the Astronomer Royal in England was in overall control of the Indian observatories, whereas the Astronomer Royal was equally clear that he had no such responsibility. Lockyer took the opportunity of this report to emphasize that work such as his own on the relationship between solar physics and weather prediction indicated that a solar observatory at Kodaikanal, which had been mooted many years before, should be completed as a matter of urgency.

When Lockyer visited the observatory at Poona, he was delighted to find one of his former students at South Kensington, a Mr Naegamvala, in charge. After his return to England, however, he heard that Naegamvala was under heavy criticism for failing to produce worthwhile results, and that this was primarily because he was trying to hold down a second job at the University of Bombay. The problem was one that Lockyer had already noted in his report.

> . . . the system at present adopted by the Indian Government of treating its scientific servants on a different principle from that adopted in other departments is not really conducive to the best interests of the State. The chief disadvantages under which scientific men now labour in India are want of promotion and of graded increases of salary throughout their service. Men of science are after all men, and are no more likely than others to work heartily without any hope of increased pay or advancement especially when they are reminded by the promotion and increased emoluments granted to those in other branches of the same state service of their own waterlogged condition.'[49]

As Lockyer knew, the status of scientists in India had long been depressed. In the 1870s, one of his scientific correspondents in India had remarked to him, '. . . in High Indian circles men of Science are considered as loafing impostors, who trade on the general ignorance at home.'[50] But things began slowly to improve in the twentieth century, and for this Lockyer was partly responsible. For example, during the period 1908–09, his advice was sought on possible ways of improving science teaching in Bombay, and he helped draw up plans for a science institute there. Sir George Clarke, the Governor of Bombay, obviously took a quite different view of science from many of his predecessors.

> I am simply appalled at the science teaching in India. It is mostly contemptible and it produces no results. I am trying to make changes; but the difficulties will be very great. This University [Bombay] is about 40 years out of date, and education generally in India is unsound in method and shallow in character.[51]

Clarke had lectured at the Royal Indian Engineering College in the seventies, together with Lockyer's former assistant, McLeod. The College had been set up in London to provide engineers for India, since the methods of training of engineers in Britain were then regarded as inadequate. The College had a high reputation as a teaching institution, and Clarke naturally became acquainted with the people at South Kensington.

Lockyer's range of artistic and literary acquaintances continued to expand throughout the nineties and into the twentieth century. During this period he came to be on friendly terms with such writers as Rudyard Kipling, Rider Haggard, and Arthur Conan Doyle. The link was Lockyer's preoccupation (which we shall examine in a later chapter) with various aspects of the imperial future. Another, perhaps less obviously compatible, literary acquaintance was Mark Twain. The two met at intervals when Clemens was in England. For example, he stayed with Lockyer in 1907, and they subsequently went on together to Oxford, where Mark Twain received a D.Litt. and Lockyer received a D.Sc. Lockyer had been given a D.Sc. at Cambridge three years before. This was his first academic degree, and was obtained only after he had retired from university teaching.

In the last decade of the nineteenth century Lockyer lost his most treasured friend in the literary world when, in 1892, Tennyson died. Some years later, Lockyer provided the following brief description of his contacts with Tennyson.

> I soon found that he was an enthusiastic astronomer and that few points in the descriptive part of the subject had escaped him. He was therefore often in the observatory. Some of his remarks linger fresh in my memory. One night when the moon's terminator swept across the broken ground round Tycho* he said, 'What a splendid Hell that would make'. Again after showing him the clusters in Hercules and Perseus he remarked musingly, 'I cannot think much of the county families after that'. In 1866 my wife was translating Guillemin's *Le Ciel* and I was editing and considerably expanding it; he read many of the proof sheets and indeed suggested the title of the English edition, *The Heavens*.
>
> In the seventies, less so in the eighties, he rarely came to London without discussing some points with me, and in these discussions he showed himself to be full of knowledge of the discoveries then being made.
>
> Once I met him accidentally in Paris; he was most anxious to see Leverrier and the Observatory. Leverrier had the reputation of being *difficile*; I never found him so, but I certainly never saw him so happy as when we three were together, and he told me afterwards how delighted he had been that Tennyson should have wished to pay him a visit. I visited Tennyson at Aldworth [Tennyson's home] in 1890 when he was in his 82nd year. I was then writing the 'Meteoritic Hypothesis' and he had asked for proof sheets. When I arrived there I was touched to find that he had had them bound together for convenience in reading, and from the conversation we had I formed the impression that he had read every line. It was a subject after his own heart . . . One of the nights during my stay was very fine, and he said to me, 'Now, Lockyer, let us look at the double stars again', and we did. There was a 2-inch telescope at Aldworth. His interest in Astronomy was persistent until his death.
>
> The last time I met him (July 1892), he would talk of nothing but the possible ages of the sun and earth, and was eager to know to which estimates scientific opinion was then veering.[52]

As a footnote to this, we might note that Tennyson, writing to Lockyer in 1890 to thank him for the complimentary copy of *The Meteoritic Hypothesis*, told him, 'I thank you heartily for your splendid hypothesis and in my anthropological spectrum you are coloured like a first-rate star of Science.'[53]

Lockyer's admiration for Tennyson in no way decreased after the latter's death, and in later years he combined with his daughter, Winifred, to compile a book on the scientific aspects of Tennyson's poetry. It was intended particularly to show the range and accuracy of Tennyson's knowledge of science and natural history. This book appeared in the same year, 1910, that Lockyer's oldest and closest acquaintance in the artistic world, Holman Hunt, died. Lockyer was one of the pall-bearers at the

* A large lunar crater named after the Danish astronomer, Tycho Brahe.

memorial service in St Paul's. Hunt, like most of Lockyer's artist friends, turned to him on occasion for scientific information, but their friendship had essentially been on a social level, buttressed by joint interests such as their membership of the Athenaeum. Hunt had been elected in 1868, together with Clements Markham, David Masson and Herbert Spencer.

The Athenaeum then, as now, was thought to be a happy hunting ground for machinations behind the scenes. The elections to membership were correspondingly fiercely contested. Thus, in 1888, the year after Lockyer became a member of the Athenaeum, he joined with Huxley to get their joint chief, Donnelly, elected. More importantly, they persuaded R. E. Welby, the permanent secretary of the Treasury, to join them. Welby was on bad terms with Donnelly, for he regarded the expenditures of the Science and Art Department as entirely excessive, and was trying to restrict them. Lockyer and Huxley were, of course, both affected by this quarrel, and they hoped the election might act as a means of reconciliation. Immediately after the election, Lockyer wrote in triumph to Huxley.

> You will be rejoiced to hear that you and Welby between you have scored a magnificent success. There were 25 of the Committee present and on the first ballot (the only one) D[onnelly] pulled off 25!!! and so came in easily as first favourite.
>
> I am more pleased than I can say as this ought to bring D[onnelly] and W[elby] closer together—and this way science lies—and I have begged D[onnelly] to write to W[elby] to thank him because W[elby]'s proposal was I think of great importance.[54]

Huxley replied immediately.

> It is very desirable that Welby and Donnelly should be good friends and I hope this occasion may bring that result about. D[onnelly] is hot but he is very generous and can 'wipe the slate' at need.[55]

This was actually a good year for the election of Lockyer's friends and colleagues, Rücker and Lauder Brunton were elected as well as Donnelly.

Lockyer naturally used the Athenaeum for giving dinners for his friends on important occasions, for example, he gave one there in 1895 to honour the visit of a leading French scientist, Cornu. (Lockyer, as a well-known Francophile, was often involved in the entertainment of visiting Frenchmen.) At another level, Lockyer continued to provide open evenings at the Solar Physics Observatory for friends and acquaintances. They could view the stars in fine weather, or if, as frequently happened, the clouds prevented this, they could drink tea out of laboratory beakers, and feel that they were seeing behind the scenes of the scientific world.

For some fifteen years from 1890 onwards, Lockyer's position at South Kensington involved him in a persistent controversy over a question of social status. Since the professors at South Kensington came under the Science and Art Department they had applied through Donnelly for the right to wear Civil Service uniforms on official occasions. This was refused to all of them except Lockyer, who was granted the privilege as Director of the Solar Physics Observatory. Lockyer's gratification did not last long. Five classes of uniform existed, graded according to rank within the Civil Service, and he had been assigned to the fifth, and lowest, class. He was horrified. Like any good Victorian, he had a very clear feeling for status, and the Chemist at the Mint, whose post he regarded as being certainly no more important than his own, was entitled to wear a fourth-class uniform. He immediately requested that he should be upgraded to the fourth class, but to no avail. For years the correspondence went backwards and forwards. In 1905, he was summoned to a Royal Levée, and forthwith presented an ultimatum to the Lord Chamberlain.

> I have the honour to acknowledge the receipt of a card conveying the summons of their Majesties to accompany my wife and daughter to the Court on the 29th. instant. I beg you will convey to the Lord Chamberlain my desire that I may be excused from attending because, since the office I hold, that of Director of the Solar Physics Observatory, has had a civil uniform attached to it, it would not be decorous to come in ordinary levee dress, and because also the grade assigned to the office by the Lord Chamberlain's department—the fifth—is in my opinion not one of which I ought to avail myself. I yield to none in my respect and loyalty to the King, and I write this letter because I am anxious that the true reason of my recent absence from levees and of my desire to absent myself from the only Court ceremonial to which I have been summoned may be known to their Majesties. I make no appeal against the decision of the Lord Chamberlain, nor do I wish to compare the practice of other European Courts with regard to the position accorded to men of science employed in State service, but I feel that I should be doing wrong to my successors to acquiesce in a view which equates the Director of an important institution known all over the world, who must be a man of scientific eminence, with private secretaries and other similar officers of the lowest grade.[56]

But it was no use; the Lord Chamberlain remained untouched by all Lockyer's arguments.

In the matter of social status generally, however, the nineties were a period of considerable advance for Lockyer. His first civil honour had come as early as 1873, when the Emperor of Brazil, Dom Pedro II, who was greatly interested in science, had made him a Knight Officer of the Imperial Order of the Rose. It may, perhaps, have abated his pleasure a little that Huggins was similarly honoured. In the New Year's Honours List just over twenty years later, Lockyer obtained his first British honour, being made a Companion of the Bath. He received letters of congratulation from all sides, but there was also in informed quarters an undercurrent of surprise. As Donnelly wrote to Huxley,

> I dare say you saw with as much surprise as I did that our Astronomer had been made a C.B. before the Astronomer Royal! I had just written to congratulate him (Lockyer!)—I could honestly do that for it will sweeten a somewhat acid body for some time—when in came Acland, much disgruntled. He had not [?] been consulted. Lockyer had asked him before to get him an honour and he had refused. He would never have thought of recommending him without consulting me! and now he felt himself in a very awkward position. . . .
>
> Apparently Lord Kelvin had written to Gladstone and one [?] of his Secretaries had bungled. At least Lord K's name was the only one mentioned by the *junior* private Secretary! who wrote to Acland, in reply apparently to a protest.[57]

(Sir Arthur Acland was Donnelly's direct superior as vice-president of the Committee of Council on Education, with a place in the Cabinet.)

If Lockyer's C.B. caused a certain amount of comment, his further elevation to K.C.B., in 1897, was straightforward enough. Lord Salisbury wrote and told him that it was in recognition of his services as Professor of Astronomical Physics at South Kensington.[58] This was the Queen's Jubilee Honours List, and a wide variety of people were honoured in it; Crookes, Frankland and Huggins were all awarded K.C.B.'s with Lockyer. It is likely that Lockyer had some hopes of further distinction, especially of the O.M.,[59] after all, his old duelling partner Huggins was one of the first scientists appointed to the Order, but the K.C.B. remained his highest honour.

IX

A NEW ORIENTATION

Early in 1890, Lockyer went with his friend Anderson of the Orient Line on a trip to Greece and Turkey. Like any good Victorian tourist, he tramped round the monuments and buildings; but he took with him his scientific curiosity, too, and this was soon excited by what he saw.

> . . . one day when we were visiting the ruins of the Parthenon, and again when we found ourselves at the temple at Eleusis, [my friend] lent me his pocket-compass. The curious direction in which the Parthenon was built, and the many changes of direction in the foundations at Eleusis revealed by the French excavations were so very striking and suggestive that I thought it worth while to note the bearings so as to see whether there was any possible astronomical origin for the direction of the temple and the various changes in direction to which I have referred. What I had in my mind was the familiar statement that in England the eastern windows of churches face generally—if they are properly constructed—to the place of sunrising on the festival of the patron saint; this is why, for instance, the churches of St. John the Baptist face very nearly north-east. This direction towards the sunrising is the origin of the general use of the term *orientation*, which is applied just as frequently to other buildings the direction of which is towards the west or north or south. Now, if this should chance to be merely a survival from ancient times, it became of importance to find out the celestial bodies to which the ancient temples were directed.[1]

On his return to England, Lockyer enquired of his archaeological acquaintances what measurements had been made of the orientation of ancient temples, and found that the question had hitherto attracted little attention. Some general statements could be made—such as, for example, that Egyptian pyramids were usually oriented east and west—but detailed studies were lacking. Lockyer therefore began a search for survey data on ancient temples, so that he could compile his own list of orientations. He found that most of the information available referred to Egyptian temples, which had been surveyed by French scientists in 1798 during Napoleon's Egyptian campaign, and by the Russians in the 1840s. So he concentrated his attention initially on Egypt.

At this period, the art of scientific excavation was in its infancy. Indeed, the examination of monuments still standing above ground left a good deal to be desired. The Frenchman, Mariette, who was recognised as the leading investigator of Egyptian temples during the period 1850–80, had not hesitated to blast his way into them with dynamite. It is hardly surprising, therefore, that Lockyer found even the best existing surveys of temples inadequate for his purpose. Just at this time, the man who had replaced Mariette as the leading expert in Egyptian archaeology, W. M. Flinders Petrie, was back in England. One of the first acts of the Egypt Exploration Fund, on its establishment in 1883 (the year after Franco-British control over Egypt was established), had been the dispatch of Flinders Petrie to Egypt to start excavations. Now, in the early 1890s, Flinders Petrie was collecting together the knowledge he had acquired during the previous decade with the intention of writing it up in book form (the result—*Ten Years' Digging in Egypt*—appeared in 1892). Lockyer naturally turned to Flinders Petrie for advice, and the latter responded cordially, remarking on the difficulties involved.

> I have not entered on the question of orientation, apart from the exact polar orientation of the earlier structures. In the case of temples the first difficulty will be to get correct data. All that we know depends (I expect) on magnetic bearings, which are probably reduced with no correct knowledge of the variation in Egypt . . .
>
> Another serious consideration is how far local configuration influenced the positions. In some cases the hills were such that no other arrangement than the existing one could have been made. As an instance, the skew causeway from the 2nd pyramid at Gizeh down to the granite temple, which determined the skew entrance passage of that temple, is clearly fixed by a ledge of rock. If it had been made due E. it would have needed a built causeway about 50 ft high and ¼ mile long; and as Egyptians then worked, that was a far more serious matter than 10 mile of earthen railway embankment.[2]

In November 1890, the month after he received this letter, Lockyer gave a series of lectures at the Museum of Practical Geology in Jermyn Street under the general title, 'On some points in the early history of astronomy'. During the course of these he emphasized that, if the orientations of Egyptian temples did indeed have an astronomical basis, then measurements of the orientations might be used to establish the dates of building of the temples. (These lectures, which were originally given as a course for working men, were revised by Lockyer early in 1891, and published in *Nature* during the summer of that year.) A member of Lockyer's audience was able to inform him that some work on temple orientation had been published a few years before in Germany. Lockyer immediately wrote off for the relevant papers, but he had already made up his mind to go out to Egypt himself, partly to see if he could obtain some measurements of his own, and partly to see if anybody on the spot knew anything of astronomically oriented temples.

Lockyer soon found on his arrival in Egypt that not only had the significance of temple orientations not been discussed there previously, but most people were not even prepared to believe that there was anything significant in the matter to discuss. However, one of the acquaintances he made out there did become interested in the possible relevance of astronomy to Egyptian monuments, and began to search through the hieroglyphic inscriptions which had been found in ancient temples, to see if any reference to astronomical orientation could be found. To Lockyer's delight, he finally found an inscription describing the foundation of the temple at Edfu which seemed entirely to vindicate Lockyer's speculation. The relevant part of the text ran as follows, 'I cast my face towards the course of the rising constellations; I let my glance enter the constellation of the Great Bear . . . I establish the four corners of thy temple.'[3] At the same time, another friend of Lockyer, an American then in Egypt, put his river steamer at Lockyer's disposal, so making it possible for him to cruise up the Nile measuring the orientation of major temples en route. Lockyer's belief in the existence of astronomically significant orientations rapidly hardened.

The investigations in Egypt took longer than Lockyer had expected, and he found that his position with the Science and Art Department required him to return to London before he had finished. Fortunately for him the Lord President of the Council, who had the final say in such matters as leave in the department, became intrigued with his new line of work, and intimated that he could apply for a further extended period abroad. As a result, Lockyer was back again in Egypt by the end of 1891, his main purpose being the collection of as large a number of accurate orientations as possible. Evidently he needed assistance, and therefore turned for help to the Public Works Department in Cairo. Here he was in luck, for the official concerned, Sir William Garstin, proved sympathetic to his proposals, and detached a Royal Engineers officer, Captain H. G. Lyons, to survey Egyptian temples under Lockyer's general guidance. (Much later, in 1912, Lyons returned to England and took up a post in the Science Museum at South Kensington, eventually becoming its director.) Garstin was an influential man in Egypt. He was one of the few who co-operated happily with Kitchener after the latter was appointed Commander-in-Chief of the Egyptian Army in the spring of 1892. Nevertheless, he and Kitchener were soon in competition for funds; Kitchener wanting it for war against the Sudan, Garstin for the construction of the proposed Aswan Dam. The latter project gave rise to a major controversy, and Lockyer soon found himself drawn into it.

The problem was that the construction of the dam would flood a considerable stretch of the upper Nile, and, in the process, drown several ancient temples and monuments for six months each year. This possibility led to immediate protests from England, especially from the Society for the Preservation of the Monuments of Ancient Egypt, one of those finely esoteric groups the Victorians delighted in establishing. An action subcommittee of this society was set up, with Lockyer, representing the Royal

Society, as a leading member. Another member of the sub-committee was Sir George Clarke—the meetings were usually held at his house—and it seems to have been their contact during this period which first led Lockyer to develop a close friendship with him. (This, as we have seen, was put to use when Sir George Clarke went to India in the following decade.) The sub-committee decided that the first necessity was a detailed survey of the area. Lockyer, who had taken on the responsibility of communicating decisions to Egypt, wrote to Garstin. The situation was potentially delicate, since the sub-committee was essentially opposing Garstin, whilst Lockyer wished to remain on good terms with him. Lockyer successfully steered a middle course in this potentially delicate situation. Later, in 1894, after the required maps of the area had been prepared, Garstin wrote to him.

> I was exceedingly glad to read your opinion as to our proposals for Assuan. If we have you for an ally, I do not care much what . . . others may say or think. No one can be more averse than I am to the removal of Philae temple, if it can be avoided. I do not myself think that it can be avoided, if a dam is to be made at all.
>
> . . . in my heart I am convinced that Assuan is immeasurably the best site . . . I think the good of Egypt must be allowed to weigh in the scale against the sentiment connected with the Temple.[4]

Under Lockyer's guidance, the sub-committee proposed that the dam should indeed be built at Aswan, and not elsewhere as critics had suggested, but that it should be smaller than had originally been intended, so that most of the ancient buildings would remain above the water level, and the remainder could be protected by building watertight coffer dams. This proposal was accepted in Cairo, at least as an interim measure. A decade later, Sir George Clarke was in touch with Lockyer again to tell him that Garstin had decided the Aswan dam should be raised to retain a depth of some 20 feet more water.

Captain Lyons was meanwhile hard at work measuring the orientations of temples in the Aswan neighbourhood, and evidently enjoying himself. He wrote to Lockyer, 'This life on a native sailing boat with Azimuths and Magnetic Obs[ervations] to pass the time and a few inscriptions is delightful.'[5] In fact, Lockyer's encouragement of surveying in Egypt had long-term effects both for science in Egypt, and for Lyons himself. Towards the end of 1895, Lyons wrote joyfully to Lockyer,

> The 'Science' Bacillus really seems approaching cultivation in Egypt since we are to have a Geological Survey or as I would prefer to describe it Reconnaissance for Egypt. You can't *survey* deserts in any accurate sense of the word. I feel as though someone of more standing than myself should have charge of it, but naturally I am delighted to do it, and of course it would take a new man some time to know the country and especially the desert tracts. I believe you were one of those mainly responsible for the first inoculation of Egypt with the Scientific Bacillus![6]

Lockyer, apart from his visits to Egypt, had also been busy reviewing all the literature he could find relevant not only to the question of astronomical orientation of monuments, but to the much wider problem of the origin of astronomical observation in the ancient world. In 1894, he announced his main conclusions in *The Dawn of Astronomy* (published for once by Cassell, and not Macmillan), which was given the sub-title, 'A study of the temple worship and mythology of the ancient Egyptians'.

Archaeology was not a clearly defined subject in the nineteenth century, it was rather a mixture of what we would now call archaeology together with anthropology and philology. The last subject was in many ways the predominant one, for the discovery and translation of inscriptions was still regarded as the single most important objective of archaeology. In this atmosphere, it was natural that Lockyer should extend his studies to a general investigation of how astronomy first came into being in the ancient world using inscriptions as well as the buildings themselves. He decided that there had been three distinct stages in the development of astronomy.

> . . . in the first stage, wonder and worship were the prevalent features; in the second, there was the need of applying the observation of celestial phenomena in two directions, one the direction of utility—such as the formation of a calendar and the foundation of years and months; and the other the astrological direction.[7]
>
> Only more recently—not at all, apparently, in the early stage—were any observations made of any celestial object for the mere purpose of getting knowledge.[8]

Before considering orientation, therefore, Lockyer tried to work out how the first stage he had suggested came about in Egypt. He investigated, in particular, whether an astronomical basis could be assigned to the pantheon of Egyptian gods.

When Lockyer turned to the question of significant alignments of temples, he concentrated initially on possible alignments towards the Sun at solstices or equinoxes, since these were the times of most obvious significance. He noted that several temples at Karnak were oriented towards either the rising or setting solstitial Sun, whilst the pyramids and temples at Giza were oriented towards the equinoctial Sun. This, and other evidence, led him to decide that two schools of astronomical thought could be traced in ancient Egypt which were associated with different religious tendencies. There had been a southern Egyptian worship of the solstitial Sun opposed to a northern Egyptian worship of the non-solstitial Sun. He tried to trace how these two schools of astronomy had arisen and how their fortunes had fluctuated with time, basing his discussion both on the orientation of buildings and the evidence from inscriptions. It seemed to Lockyer that his investigations indicated changes in the patterns of Egyptian worship which must reflect either invasions of the country from outside, or major internal disturbances.

The basic premise in these deductions was that the temples had acted as astronomical observatories, and were used by the priests in determining the calendar. He explained at some length how this had worked for the temple of Amen-Ra at Karnak. Its layout was similar to that of a horizontal telescope pointing towards the setting solstitial Sun, for it was so built that only at this one time each year could the Sun cast its light down the entire length of the temple. When this occurred, the priests would know that a new solar year was beginning. Lockyer next pointed out that, because of a slow change in the obliquity of the ecliptic,* a temple which was correctly oriented towards the solstitial Sun some thousands of years ago would no longer be correctly oriented in modern times. He compared the current position of the solstitial Sun on the horizon at Karnak (determined at his request by the Public Works Department in Egypt) with the direction indicated by the alignment of the temple. The corresponding difference in the obliquity of the ecliptic (about one degree) could be used to determine the date at which the shrine of Amen-Ra had been built; according to Lockyer's calculations this was about 3700 B.C.

An advantage of studying Egyptian temples in the nineteenth century was that they were one of the few types of ancient monument then known for which even approximate dating was possible. The numerous deciphered inscriptions had allowed time periods to be assigned to successive dynasties, and, hence, allowed tentative dates to be allotted to any building with a dynastic inscription. Of course, the further back in time one went, the more the errors accumulated. As it happened, Mariette had investigated the inscriptions on the temples at Karnak, and it appeared from his work that the oldest dateable parts went back to 2400–3000 B.C. Lockyer was well pleased with the agreement between his astronomically determined date and that derived from the inscriptions. He believed that the extensive temple investigated by contemporary archaeologists was actually the final building. It had probably grown slowly on the site and along the orientation of an originally much humbler shrine.

Besides the question of orientation towards the Sun, Lockyer also investigated the possibility of buildings being oriented towards the positions of some of the brighter stars at their rising and setting. This had already received some attention earlier in the nineteenth century, especially from the French scientist, Biot. It seemed most likely that stellar

* The inclination of the equatorial plane of the Earth to the plane of the Earth's orbit round the Sun.

orientation would be related to the heliacal risings or settings of the stars.* But just as the rising and setting positions of the Sun on the horizon changed over a long period of time, so did the positions of the stars, due to the precession of the equinoxes†. However, as Lockyer emphasized, the change in the position of stars on the horizon caused by precession was much greater than the change in position of the Sun resulting from the variation in the obliquity of the ecliptic. Whereas the latter changed by about one degree in 6000 years, the former could change by the same amount in less than 300 years; so orientation of buildings towards stars could provide, in principle, a much better dating mechanism than solar orientation.

Before trying to apply this mechanism, Lockyer examined what was known of the constellation patterns amongst the Egyptians. His major conclusion here was that the Egyptian religious myths probably represented descriptions of astronomical events. For example, he interpreted the story of the god Horus avenging the death of his father, Osiris, on Typhon, the god of darkness, as simply expressing the observed fact that the rising Sun blotted out the circumpolar stars. Similarly, he explained changes in the form of the myth at different times as due to changes in the circumpolar stars owing to the effects of precession.

Lockyer's interpretation of the religious myths as relating to celestial events provided him with a good reason for supposing that different temples might be oriented towards either Sun or stars. He followed this up by pointing to certain Egyptian temples so aligned that the Sun could never have shone along their length, yet they seemed to be built on the same 'horizontal telescope' principle as the solar-oriented temples. The problem was to find positive evidence that such temples had a stellar significance. Lockyer suggested one possible way in which this might appear. The position of a star on the horizon would be changed appreciably by precession over the course of a few centuries. If any temple with a stellar orientation was in use for a longer period than this, then its structure would have had to have been modified, and a new orientation imposed. He suggested that temples where this had occurred could, indeed, be found, and instanced the temple at Medinet-Habu, where the central axis of the earlier part of the temple was inclined at an angle of a few degrees to the later part.

* A star rises heliacally when it appears in the sky just before the Sun at sunrise; similarly, it sets heliacally when it disappears below the horizon just after the Sun at sunset.

† The precession is a slow motion of the Earth's axis of rotation, such that the celestial poles seem to trace out circles against the background of stars approximately once every 26,000 years.

The examination of the various temple sites, mainly at Lockyer's behest, soon showed that temples at quite different sites were definitely not oriented towards the Sun, but were, nevertheless, all oriented to very much the same point on the horizon. Lockyer felt that the only reasonable interpretation of this was that they had all been aligned on the same bright star. This led him on to consider which stars might have been chosen for this purpose. He had a star globe specially made for himself which could be used as a computing device to show the positions of the stars in the sky at any date in the past. It gave alignments accurate to one degree, which was about as precisely as most temple ruins could be surveyed. Where more accurate alignments were required, Jim Lockyer was called into service to make the necessary calculations.

Lockyer's investigations showed that temples on the same site could be oriented towards different stars. At Karnak, one of his favourite sites, he determined that one temple was aligned on the northern hemisphere star, γ-Draconis, whilst another was aligned on the southern hemisphere star, Canopus. The star which he saw as the most likely target of temple orientation was neither of these, however, but the Dog Star, Sirius. Not only was Sirius the brightest star in the sky, but it was also already known to have played a major part in Egyptian astronomy, since it had long been used as a guide to the time of the annual Nile flood. Sure enough, temples oriented towards Sirius were detected, one such being the temple of Isis at Denderah.

Lockyer, as usual, sought confirmation from Egyptian mythology. In this case, he identified Isis as certainly being Sirius. This led him to interpret the inscriptions at Denderah, which had been recorded by Mariette, as indicating that the temple had been oriented towards the cosmical rising of Sirius, which confirmed his astronomically based conclusion. Finally, from a comparison of the orientation of the temple with the current position of Sirius, Lockyer deduced a date of 700 B.C. for the construction of the temple. Using the same mixture of data from orientation and from the astronomical interpretation of inscriptions Lockyer deduced dates for the alignment of other Egyptian temples for which information was available.

This work naturally forced Lockyer deep into the interpretation of Egyptian mythology, for in using the inscriptions he had to decide which star was related to which god. This led him far afield. For example, he investigated the Egyptian calendar at some length, mainly in order to study what relationship there was between feast days and individual stars. As we have seen, his investigations induced him to believe that all Egyptian mythology had an astronomical basis. Not only were gods identifiable with stars, but the way in which a god was described indicated the type of observation being made of the corresponding star. Thus, when a god was represented as mummified, this meant that the corresponding star was to be observed as it set. Lockyer's attempts at an interpretive approach came in for considerable criticism from his contemporaries, and are recognisably the most speculative part of his book.

The final chapter of *The Dawn of Astronomy* turned away from Egypt to the place where Lockyer's archaeological researches had started, Greece. But his main interest had now shifted, he wanted to show not simply that Greek temples had astronomically significant orientations, but that they had been constructed under Egyptian influence. Much of the chapter was based on work by Lockyer's old friend, F. C. Penrose, who had made a detailed study of Greek temples.

Penrose was some twenty years older than Lockyer. He had been at Cambridge in the 1840s (where, as one of the leading oarsmen, he had first drawn up the type of chart to record the results of bumping races which is still in use today), and had there become a close friend of Charles Kingsley and J. M. Ludlow. He had thus become one of the early members of the Christian Socialist network, and had, indeed, subsequently settled in Wimbledon. After leaving Cambridge, he became an architect, and in the early 1850s was appointed surveyor of St Paul's cathedral. However, he came into extended conflict with the dean and chapter, and, perhaps as a result, turned to astronomy as a peaceful hobby. (He had first become interested in astronomy at Cambridge, where another of his friends had been J. C. Adams, then involved with his calculations concerning the possible existence of a new planet in the solar system). Penrose devoted most of his energy in the fifties to examining the theory and practice of constructing sundials, work which was later of considerable value to Lockyer, since it required consideration of the way in which buildings were oriented relative to the Sun. In the 1860s, at much the same time as Lockyer, he began serious astronomical observations, and joined the Royal Astronomical Society. His subsequent work was of sufficient importance for him to be made an F.R.S. in 1894. But Penrose had also always been interested in archaeology. He designed the British School in Athens in the 1880s, and became its first director, though only for a brief period. (The British School ran into difficulties in the next decade, and Lockyer was roped in by G. A. Macmillan to help organise a petition to keep it open.[9])

Penrose thus formed an ideal aid for Lockyer in his work on orientation. Indeed, he seems to have been thinking about the possible significance of the orientation of buildings even before Lockyer approached him. Moreover, his interest in Greek archaeology had led him to make a special study of Greek temples. It is hardly surprising that Lockyer commented on his good fortune in finding someone who was, '. . . an astronomer as well as an archaeologist (for, alas! they are not, as I think they should be, convertible terms).'[10] In the early nineties, Penrose carried out several detailed surveys of the orientation of Greek temples, and his analysis of the results led him to become one of the most enthusiastic supporters of Lockyer's ideas.

Although Penrose was more immediately involved in Lockyer's work than any other archaeologist, other professionals were quite prepared to take his ideas seriously. Thus Müller at Oxford helped Lockyer with the interpretation of inscriptions. He was particularly interested in Lockyer's belief in the central importance of the Sun in Egyptian religion, for he himself was the leading exponent of the solar myth theory in England. (This was the idea that myths were ultimately based on a description of events in nature—especially those connected with the Sun.) Similarly, abroad, Gaston Maspero, the leading French expert on Egyptian archaeology, who had himself spent much of the eighties in Egypt, followed Lockyer's work with interest, and frequently provided him with advice. (Lockyer sometimes stopped with him in Paris during expeditions to and from Egypt.) But, Penrose apart, Lockyer's leading exponent amongst archaeologists was certainly E. A. T. Wallis Budge of the Department of Egyptian and Assyrian Antiquities at the British Museum. After *The Dawn of Astronomy* appeared Wallis Budge urged Lockyer to extend the work to other areas of the Near East besides Egypt.[11] Indeed, he proceeded to make some of the necessary investigations himself. In 1899, he wrote triumphantly,

> Lockyer's theory is proved up to the hilt by me as concerns pyramids and temples in Sudan as far back as XIIth. dyn[asty] B.C. 2500. It is more than probably right for temples and step pyramids up to B.C. 3500, but to make this certain *archaeologically* we must have excavations.[12]

Wallis Budge's attachment to Lockyer was not only due to the latter's archaeological ideas. In 1893, a Mr H. Rassam, who had formerly worked at the British Museum, brought an action for slander against Wallis Budge. Rassam had previously, in the 1870s, been in charge of excavating Babylonian sites on the British Museum's behalf, but, according to Wallis Budge and other British Museum employees, the material he sent back was mainly rubbish. During the same period, however, collections of Babylonian tablets of considerable value, which the British Museum felt bound to purchase, began to appear on the London market. These continued to appear throughout the eighties, though Rassam, himself, returned to England in 1882, and the British Museum, finally becoming suspicious, despatched Wallis Budge to the Near East in 1887 to try and discover where the tablets were coming from. He found that the agent, whom Rassam had appointed to protect the British Museum sites from pilfering, was actually rifling them, and sending the best material on to London to be sold. In reporting this to his superiors at the Museum on his return to England, Wallis Budge also suggested that there was a strong indication of negligence, or even collusion, on the part of Rassam, but the British Museum trustees, although they dismissed the native agent, took no action against Rassam. News of the report got about, and led to Rassam bringing a slander action. His action was successful, and Wallis Budge was ordered to pay £1000 damages.

It was at this point that Lockyer came into the affair. The British Museum trustees refused to help Wallis Budge pay anything towards the damages and the cost of the action, though many observers felt that theirs was the ultimate responsibility for the case having been brought at all. Wallis Budge immediately appealed to Lockyer, asking him to take the matter up in *Nature*, because, 'It seems to me that I have been betrayed all round, although I have only done my duty and what I was told to do.'[13] Lockyer responded with an article strongly slanted in favour of Wallis Budge, which almost certainly helped raise contributions for a fund, started by Wallis Budge's colleagues at the British Museum, to defray his costs. Wallis Budge, needless to say, was always grateful to Lockyer for this action, though Rassam complained bitterly that Lockyer was prejudiced.

If Penrose and Wallis Budge became fervent exponents of Lockyer's ideas, most archaeologists were much more cautious in their reactions. Thus Maspero, who, as we have seen, was a friendly observer, told Lockyer that he had greatly enjoyed reading *The Dawn of Astronomy*, but, 'I am still not sure what is my final opinion: the application of mathematics to archaeology gives an appearance of rigour and certainty which always make me mistrustful.'[14] Where Maspero suspended judgement, others happily attacked Lockyer's thesis with all the vigour they could muster.

> In discussing Professor Lockyer's work we propose to confine ourselves almost entirely to an examination of its leading theory, a somewhat more general form of which has been already maintained, with greater learning, though perhaps not greater success, by Herr Nissen, that the Egyptians orientated their temples to the rising or setting places of stars. We take this course for two reasons—first, because it is this part of the work which has alone attracted general interest, as containing perhaps the only one of his theories which its author has himself understood sufficiently to make it intelligible to others; and, secondly, because, if we were to attempt a refutation of all Mr. Lockyer's arguments that seem to us unsound, and all the statements which are demonstrably incorrect, it would be necessary for us to write a book nearly as long as his own.[15]

Lockyer was, of course, quite capable of caring for himself in such arguments. He extracted particular pleasure from the lack of skill evidenced by some of his archaeological critics in the type of surveying and calculation on which orientation measurements depended. 'They [the archaeologists] have all been making a great fuss about the explanation of the fine temples at Der-el-Bahari, and it now turns out that rough measures were made, and they don't know whether the given bearing is true or magnetic.'[16]

He also noted with relish that careful archaeological surveys relying on magnetic bearings had been carried out in places were the magnetic variation was quite unknown, so that the orientation relative to true north could not be determined.

Values of magnetic variation were often quite poorly determined in the nineteenth century. When Lockyer later became interested in surveying ancient monuments in the British Isles, he found that the magnetic variation was not even very completely determined for this country. Being Lockyer, he immediately persuaded the Admiralty and Ordnance Survey to prepare an up-to-date map of the magnetic variation for the whole of Britain and its coasts. He took the failings of archaeologists as surveyors sufficiently seriously to try and educate them in better methods. In 1909, he published a book on *Surveying for Archaeologists*, with the assistance of his son, Hughes Campbell Lockyer, who had carried out considerable surveying operations in the navy, basing it on a series of articles previously contributed to *Nature*.

The attacks on Lockyer's archaeological ideas were inspired by the customary mixture of motives, including the identification of genuine errors in his original discussion, but it was especially important that the climate of opinion amongst archaeologists at this time was decidedly antagonistic to his method of approach. In the relatively limited discussions of the orientation of Egyptian monuments prior to Lockyer's studies, most attention had been directed towards the pyramids at Giza. The most important early survey of these was by an army officer, Colonel Vyse, in the 1830s. His report, which was published in 1840, had a note appended by Sir John Herschel, who pointed out that the sloping passage leading down into the Great Pyramid would have pointed in previous times to the bright star α-Draconis at its inferior culmination.* Twenty years later, Vyse's results were taken up by another man, John Taylor, who became obsessed with the idea that the various measurements of the Great Pyramid indicated that it was a monument of especial religious significance. He believed that it had, indeed, been built under the direct guidance of divine inspiration, perhaps by Noah. The Astronomer Royal for Scotland, Piazzi Smyth, was fascinated by Taylor's

* Inferior culmination occurs when a star passes through the observer's meridan below the pole.

suggestion, and went out to Egypt to make his own measurements of the Great Pyramid. Henceforth, he pressed the belief not only that it had been constructed under divine guidance, but furthermore that its measurements, if properly interpreted, had a prophetic significance. He proceeded to write a book in an attempt to validate this claim. The book went through several editions, and by the eighties, the fad of pyramidology, as it came to be called, was well established.

One result of this spurious science was to cause many of those seriously interested in archaeology to look with suspicion on anyone who alleged that measurements of ancient buildings had a religious significance, and this naturally reflected on the question of temple orientations. Lockyer was well aware of this difficulty. After referring to Vyse's work in *The Dawn of Astronomy* he added, 'Much that has been written has been wild and nonsensical'.[17] Actually, at an early stage in his own investigations, he had written to Piazzi Smyth to query a point concerning Sirius which had arisen from the latter's work in Egypt. Smyth in his reply had sadly signed himself, 'Yours in retirement and penury'.[18]

Because of the pressure of his other work, Lockyer was unable to pursue his interest in astronomical archaeology much further in the latter part of the nineties, though he continued to correspond about it. There was, for example, an exchange of letters between himself and Max Müller concerning the interpretation of Near and Far Eastern mythologies during this period. As Lockyer approached retirement from his Chair at South Kensington, however, he began planning a major new effort. In *The Dawn of Astronomy* he had briefly mentioned Stonehenge as an example of a structure which was generally supposed to be oriented towards the rising Sun at the summer solstice. He now determined to apply the ideas on orientation, which he had worked out for Egyptian temples, to Stonehenge, and furthermore to see whether the variety of ancient stone monuments which were known to exist in Britain and in Brittany had any astronomical significance. This was a question already under discussion. In explaining his reasons later, Lockyer emphasized the continuity between his Egyptian and his British archaeological investigations.

> In continuation of my work on the astronomical uses of the Egyptian temples, I have from time to time, when leisure has permitted, given attention to some of the stone circles and other stone monuments erected, as I believed, for similar uses in this country. One reason for doing so was that in consequence of the supineness of successive Governments and the neglect and wanton destruction by individuals, the British monuments are rapidly disappearing.[19]

There was, however, also a very good scientific reason for trying to extend the work to Britain. Owing to its more northerly latitude, changes in the obliquity of the ecliptic would produce greater changes in solar orientation in Britain, over a given period of time, than in Egypt, and should hence make it possible to date solar-oriented structures more accurately.

In 1901, the year of his retirement, Lockyer, together with Penrose and others, such as Fowler, went down to Stonehenge, and carried out a detailed survey of its visible structure. (The most detailed survey previously had been made by Flinders Petrie in 1880.) Lockyer wrote to his old acquaintance Sir John Lubbock (now Lord Avebury*) early in 1901.

> I am getting together the orientation information regarding Stonehenge and am arranging to send down to get some fresh measures. About 2000 B.C. seems the most probable date so far as we have gone.[20]

From their new survey, Penrose and Lockyer finally managed to produce a date for the origin of Stonehenge that was much more precise than this preliminary estimate. They fixed it at 1680 B.C. ± 200 years. Lubbock subsequently wrote to Lockyer, 'Stonehenge I still consider to belong to the Bronze Age, but probably to the earlier part. This would correspond fairly well with your date.'[21] Flinders Petrie had also attempted to calculate a date for Stonehenge from his earlier survey, but, according to Lockyer, his estimate went badly astray owing to an error in the calculations.

During the early years of the present century, Stonehenge came to be at the centre of a major controversy. This was sparked off by the blowing down, during a storm on 31 December 1900, of one of the outer trilithons and its lintel, an event which focused public attention on the need to prevent further decay of Stonehenge. The structure was evidently undergoing a rapid deterioration, partly due to the effects of weather, but mainly due to the attentions of visitors.

> . . . the real destructive agent has been man himself; savages could not have played more havoc with the monument than the English who have visited it at different times for different purposes. It is said the fall of one great stone . . . in 1797 was caused by gypsies digging a hole in which to shelter, and boil their kettle; many of the stones have been used for building walls and bridges; masses weighing from 56 lb. downwards have been broken off by hammers or cracked off as a result of fires lighted by excursionists.[22]

In 1882, Lubbock had managed to push through parliament an act for the protection of ancient monuments; but, because there was a lack of adequate supervision, this act was not as effective as had been hoped. Public agitation, such as that which now arose over Stonehenge, led to further acts in 1900 and 1910, and finally, a consolidating act in 1913.

* Lubbock bought Avebury ring in 1871 to save it from destruction, when it was proposed that the ground should be used for building purposes.

Lockyer strongly supported the extension of State powers in this area, though ironically, one of the first results of the new seriousness towards ancient monuments was that Lockyer could not get permission to open some long barrows he was interested in.[23]

Stonehenge was in private hands at this time, and Lockyer, as a result of his surveying work there, came into close contact with its owner, Sir Edmund Antrobus. Lockyer found that Antrobus wished to make arrangements for the repair and future preservation of Stonehenge, and proceeded to give him as much support as possible, observing in passing that if Stonehenge had been situated on the Continent it would have been cared for by the State long before. Arrangements were made for Professor Gowland, an associate at South Kensington, to supervise the re-establishment of the so-called 'leaning stone', which was obviously about to fall. In fact, Gowland's work proved to have a much wider significance than just repair of the fabric, for his excavations brought to the fore once again the question of where the stones at Stonehenge had originated.

It was already known that two types of stone were to be found at Stonehenge, but most people, including Lockyer, seem to have accepted the contention of the latter's geologist colleague at South Kensington, Judd, that both types had been deposited on Salisbury Plain by a retreating glacier at the end of the previous Ice Age. This meant that the builders of Stonehenge had simply collected together material which was close at hand. It was now proposed, instead, that the stones might have come from great distances, with the implication that the society which had built Stonehenge was capable of a high degree of organisation. This, of course, provided some indirect support for the belief that Stonehenge, itself, could be a quite sophisticated solar temple. Moreover, Gowland's excavations revealed a fair number of implements which had presumably been left behind by the builders of Stonehenge, and could be used to help determine the archaeological dating of the structure.

Besides arranging for the repair of Stonehenge, Antrobus, acting on the advice of various interested societies, proceeded to fence off the site, and impose a charge of one shilling, for entrance, partly as a method of limiting the number of visitors, and partly to recompense himself for the cost of the repair work. It was this action which led to the dispute. Even some members of the societies advising Antrobus were unhappy at this move. Lubbock, who was on the advisory committee of the Society of Antiquaries, resigned from it because he thought that the rights of the public were being ignored. Nor was he the only one of this opinion; Antrobus wrote to Lockyer,

> My Father was constantly abused for not taking sufficient care of Stonehenge now the other side are annoyed that I am taking the only steps I could as a

private person and owner take to preserve it from maltreatment by idle and thoughtless individuals . . . I think Mr. Flinders Petrie must have been in a bad temper when he wrote his letter [to *The Times*], and he really should not talk of my private property as 'our great monument', and what does he mean by 'other societies with as good a right to be heard were curtly refused any voice in the matter'. The three societies I invited to advise me, were those that had previously written to my Father and myself on the subject of Stonehenge and its preservation. The Anthropological Society, of which I am ashamed to say I have never heard, wrote to me late in the day and said they would be glad to join, but I answered that the Committee had already been formed.[24]

In 1905, an action was brought against Antrobus, contending that he ought to allow free access at Stonehenge to the public. This appears to have been a test case supported by the Commons and Footpaths Preservation Society. Lockyer wrote forcibly in favour of Antrobus, and attended at court when the case was finally heard. In the event, he was not called as a witness, but the judge quoted what he had written in a summing up that came down firmly for Antrobus. Stonehenge remained in Antrobus' hands till his death in 1915. In 1918, its next owner presented it to the nation.

Lockyer's interest in Stonehenge did not end with his attempt to assign a date to it, nor with his vigorous defence of his dating procedure against attacks. He became intrigued by the fact that there appeared to be some remarkable geometrical relationships between Stonehenge and other landmarks in the neighbourhood. He pointed out that two straight lines could be drawn through Stonehenge which, when extended, passed through features which were probably, in origin, of great antiquity.[25] One line passed through Sidbury, Stonehenge, Grovely Castle, and Castle Ditches; the other through Stonehenge, Old Sarum, Salisbury Cathedral, and Clearbury. Moreover, the lines were so arranged that Stonehenge, Grovely Castle, and Old Sarum formed a perfect equilateral triangle. Having found these curious relationships for Stonehenge, as his interest in British stone monuments increased, Lockyer began to discern similar geometrical groupings elsewhere in Britain.

Lockyer's interest in Stonehenge naturally led him on to consider the even larger monument at Avebury, since suggestions that the two were in some way related had been made ever since John Aubrey first described Avebury in the mid-seventeenth century. John Lubbock, Lord Avebury, encouraged Lockyer to undertake an investigation of the circle, but Lockyer found it a distinctly harder problem than Stonehenge. He wrote to Lubbock, 'Avebury seems much more difficult to tackle. The circle seems to have been to the modern village what Memphis was to Cairo.'[26] Later, however, he decided that his observations showed Avebury to be at least a thousand years older than Stonehenge.

In 1903, Lockyer remarried. His second wife had many ties with the West Country, and this led him to turn his attention to that part of the world. In 1905, he went with his wife to examine some of the stone circles in Cornwall. He had already selected one group, called the 'Hurlers', as likely to be of especial interest, and an examination soon convinced him that they could be identified as an astronomically oriented temple. Indeed, he finally concluded that the Hurlers were a temple that had been in use for so long that it had been necessary to readjust its orientation during its existence, the same result that he had claimed a decade before for some Egyptian temples. He returned to Cornwall in 1906, this time to the 'Merry Maidens', a stone circle, owned by Lord Falmouth, which was situated near Penzance. In the following year, at Lockyer's instigation, a new society was established in Cornwall, with Lord Falmouth as its president, to investigate astronomical aspects of ancient stone monuments in the area. Also in 1906, Lockyer published a full description of his views on British stone circles and other monuments in *Stonehenge and other British Monuments Astronomically Considered*, which was very greatly expanded in a second edition, three years later.

Lockyer's archaeological investigations in Britain can be divided into three types : those he carried out purely from existing Ordnance Survey maps without visiting the site, those where he spent some time at the site himself, and those where the sites were examined and surveyed at his instigation but without his personal intervention. A surprising number fall into the second category, for Lockyer, though now in his seventies, remained very active. Whenever for any reason he visited a region of the British Isles where he knew ancient monuments to exist, he would always try to organise an expedition to inspect them. In 1906, he suggested to the Royal Society that they should set up a committee to supervise an astronomical survey of ancient monuments in Britain. After some discussion, the Royal Society agreed, and appointed, besides Lockyer himself, Sir George Darwin, Arthur Evans and H. F. Newall. They were to act as a supervisory body. Lockyer professed himself too old and too busy to undertake any of the survey work.

In the following year, 1907, Lockyer's interest turned to another part of the Celtic fringes, Wales. A Welsh clergyman, the Reverend John Griffiths, wrote to him, and drew his attention to the traditional Welsh assembly, the Gorsedd, and its associated lore, some of which seemed to refer to astronomy. What whetted Lockyer's appetite in particular was the belief that the Gorsedd was an assembly formerly held near one or other of the Welsh stone circles. In Lockyer's time, the term 'Gorsedd' was usually employed to describe meetings held in association with the Eisteddfod, the national bardic congress, which, after falling into abeyance in the eighteenth century, had been revived in the nineteenth century as a focus for the growing interest in the distinctive characteristics of Wales, and especially of Welsh literature. As a result of his now awakened interest in Welsh folklore and archaeological remains, Lockyer made a point of attending the 1907 Eisteddfod, which was held in Swansea. He was greeted with pleasure by the organisers, who conferred on him the title 'Gwyddon Prydain', Britain's Man of Science. Roscoe wrote to congratulate him, adding that Rücker had said Lockyer should now be addressed as 'Your Solar Prominence'.[27]

In his address to the assembly at Swansea, Lockyer remarked on the antiquity which he believed attached to such meetings, perhaps the original assemblies had been held at stone circles some 4000 years before. He further advocated that a society, similar to the one he had established in Cornwall, be set up in Wales, to study the astronomical implications of the ancient stone monuments to be found there. His suggestion was quickly acted on, and a parallel Welsh organisation was started with the Reverend John Griffiths as secretary. During the next year or two, Lockyer spent some time in Wales examining likely looking monuments.

In 1909, he was even called on to assist with the organisation of an Eisteddfod. It was decided to hold it in London that year, and Lockyer was approached by the organising committee for expert advice.

> His Majesty the King has been graciously pleased to permit us to hold the Gorsedd in Kensington Gardens on a piece of ground that slopes down to the Serpentine close to the Serpentine bridge. The Gorsedd Committee are anxious that everything should be done in the strictest order, and they are particularly desirous to have the stones outside the circle over which from the centre of the circle the rising sun would be seen on the Solstices and Equinoxes correctly placed. Knowing that you take great interest in the subject they are emboldened to ask whether you can assist them in fixing these points . . .[28]

When Lockyer ultimately came to sum up the progress he had made in his studies of ancient monuments in Britain, and, after the first decade, there were few further changes, he claimed to have established two general points. First that they could be arranged in an evolutionary order. The avenues and cromlechs formed an earlier, more primitive stage; the stone circles came later, and represented a more advanced stage of astronomical knowledge. Whereas an avenue could only be oriented towards the rising or setting of a single astronomical object, circles could be used with appropriate outliers to mark several different astronomical events. Lockyer's second main point was that the sunrise orientations normally used in the earlier stages were related to the vegetation year (i.e. the year associated with the growth and decay of plant life) which began in May. Later, the orientations used were associated with the solstitial year, beginning in June. He believed, for example, that there had been an earlier circle at Stonehenge, associated with May, before the present solstitial circle.

As in his discussion of Egyptian temples, Lockyer went far beyond the confines of orientation measurements in his investigation of British monuments. He also examined at some length traditional themes in British folklore which might be hoped to throw some light on their origin and use. For this he relied particularly on J. G. Frazer's *The Golden Bough* which had appeared in 1890. Lockyer corresponded with Frazer about his own ideas, and submitted his own notes on folklore to Frazer for comments and correction. In return, he offered to show Frazer round Stonehenge, which the latter apparently had never examined.[29] Lockyer came to believe that a wide variety of early religious activity, involving sacred fires, sacred trees and holy wells, could be connected with stone circles via the folklore tradition. In his final synthesis, therefore, he combined observation and folklore rather as he had for his Egyptian work. He believed, indeed, that there was an ultimate link between Britain and Egypt. He proposed that astronomer-priests, who were acquainted with Egyptian methods of orienting buildings, started constructing avenues in various parts of Britain about 3,600 B.C. This, he believed, followed on an actual influx of Semitic peoples into the British Isles. The druids, with whom Stonehenge was traditionally associated, were the lineal descendants of the original astronomer-priests.

The study of folklore had become academically respectable in the nineteenth century, in fact, the word itself was coined just before the middle of the century. It had achieved importance particularly because of its close relationship with the central nineteenth-century activity in philology, since folklore and philology were more often than not studied by the same people, for example, the brothers Grimm in Germany were experts in both areas. Both studies came to be used for the elucidation of pre-historical events.

> . . . every mythology, in the Aryan family at least, however puerile or absurd it might at first appear, was a fit subject for scientific investigation, and capable of yielding scientific results. The problem in each case was to trace the nursery tale to the legend, and the legend to the myth, and the myth to its earliest germ . . . In this way the history of a story, like the history of a word, was frequently found to be more interesting and instructive than the history of a campaign.[30]

Hence, the combination of mythology and archaeology was perfectly acceptable in the nineteenth century. Lubbock, for example, combined the two in his highly respected writings.

Lockyer's synthesis, however, came in for some heavy attack. There were several reasons, a growing caution concerning the use of folklore, a particular suspicion of his handling of it, doubt about the validity of orientation measurements, and so on, but one specific reason may be cited in conclusion. Lockyer, in strongly supporting Stonehenge as a druid temple, was going directly counter to the prevailing mood amongst archaeologists. John Aubrey had originally identified Stonehenge and Avebury as druid temples, and this had been enthusiastically backed in the eighteenth century by the antiquarian, William Stukeley. Proponents of the Romantic movement in the late eighteenth and early nineteenth century had pursued this cult of the druid vigorously. Subsequently the cult of pyramidology was extended to Stonehenge, and mystic significance assigned to its layout. By the early twentieth century, the extravagance of these various views had caused a necessary reaction amongst archaeologists. They were no longer very willing to listen to descriptions of Stonehenge which invoked either druids or Egyptian influence, and Lockyer's views conflicted on both counts.

EDUCATION AND NATIONAL PROGRESS

The Devonshire Commission recommendations had had disappointingly little immediate practical effect. The demands for scientific instruction therefore continued unabated. During the decade after the Commission's report the emphasis changed somewhat, and attention came to focus especially on the need for better teaching of technical rather than purely scientific subjects. The desirability of having more technical instruction available in Britain had been recognised, even in some official circles, from the 1860s onwards. When Forster introduced his Elementary Education Bill of 1870 into the House of Commons, he argued that better elementary education was a necessary preliminary to any increase of technical education in this country. But it was the depression of the latter part of the seventies, and the consequent greater fear of German commercial rivalry, that led to a new effort for improvement of the situation by concerned educationists.

The payment-by-results examinations were still increasing in scope, but they were regarded by many with mounting suspicion. Oddly enough, like the demand for more instruction in scientific and technical subjects, this suspicion was actually a consequence of admiration for the German model. The British examination system, in general, laid very little emphasis on practical work, relying very largely on written answers, but the German system based itself especially on research. Many scientists, Lockyer among them, therefore felt an aversion for the British system of written examinations. Lockyer believed, indeed, that the system harmed the examiners as well as the examinees.

> Go into a company of scientific men, and observe the most dogmatic, the most unfruitful and the least modest among them, you will find that this man is, as we may say, an examiner by profession. Speak to him of research or other kindred topics, he will smile at you—his time is far too precious to be wasted in discussing such trivialities; like his examinees, he finds they do not pay.[1]

The examination system nevertheless continued on its way, and whether despite it or because of it prospects for instruction in technical subjects gradually improved. While the Devonshire Commission was still sitting, a conference of the City Companies agreed to establish a major new centre for the teaching of technical subjects in the City of London. Progress was rather slow, and it was not until the end of the seventies that the City and Guilds Institute was finally inaugurated. Part of the delay was due to the search for a site. The 1851 Commissioners were prepared to offer ground at South Kensington, but the Guilds were for some time unhappy at the idea of establishing their institution outside the boundary of the City.

In 1879, Huxley took the chair for an address by S. P. Thompson on the need for increased technical education. Speaking afterwards, Huxley took the opportunity to warn the Guilds of public wrath unless they pressed ahead with support for technical teaching. This threat, which was prominently reported in *The Times*, led to a widespread discussion of the duties of the London Guilds. Whether as a result of this publicity or not, the City and Guilds Institute now rapidly undertook teaching commitments. At the beginning of the eighties, work was started on a technical college at Finsbury as well as a Central Institution at South Kensington. The latter opened in 1884, and offered a range of courses in engineering, applied physics and chemistry.

Meanwhile, in 1881, a new Royal Commission was set up, this time to study specifically technical instruction at home and abroad. The immediate impetus for the appointment of the Commission came from a relatively small group of Liberals who were intensely concerned with the need to increase the efficiency of British industry. When the members of the Commission were announced, it was found to consist almost entirely of this pressure group. Of the six men appointed, five either were already, or were to become, Liberal Members of Parliament. The chairman was Bernhard Samuelson, who, as we have seen, was an important figure in the agitation leading up to the appointment of the Devonshire Commission. Of the other commissioners, one was Roscoe and another was Philip Magnus, the first director of the City and Guilds Institute.

Lockyer was naturally prejudiced in favour of the Commission, since he agreed with most of the thinking of its members. Whilst it was still deliberating, he made a major speech on technical education at Coventry which evoked from Samuelson the comment that, '[It] has for the first time condensed into a few words, easily understood by all, the whole problem of the Education of the Industrial Classes in this country.'[2] In fact, the Samuelson Commission's report came to several of the same conclusions that had been reached a decade before by the Devonshire Commission. Continental countries, especially Germany, were ahead of us in the application of science; British employers did not appreciate the need for their employees to have a sound technical education; and the teaching of scientific and technical subjects should be considerably increased in our schools.

The main difference between the Devonshire and Samuelson Commissions was that several recommendations of the latter were actually acted on during the next decade or two. There was probably more than one reason for this effectiveness. One factor was the more favourable climate of opinion in the eighties. As a *Times* leader commented, 'There is a fashion in education, as in pictures and dress and china. Technical education is the fashion at present.'[3] But the main cause was almost certainly the continuing pressure exerted by members of the Commission. They stumped the country making speeches in favour of the Commission's recommendations, and, in 1887, formed a National Association for the Promotion of Technical Education to act as an organised pressure group for their ideas. The Association proceeded to present their case with as much vigour as possible, whilst behind the scenes Donnelly exerted what influence he could, and within a surprisingly short space of time they succeeded in their primary objective. In 1889, a Technical Instruction Act was passed through Parliament. Lockyer had helped with the agitation when opportunities presented themselves. His influence with *The Times* proved as useful in this respect as his editorship of *Nature*. Thus, Roscoe wrote to him, 'It is very important that we should get an article in the Times in favour of our Technical Education Bill. Can you do this for me?'[4]

It was something to have got a bill on the books that called on local authorities to make provision for technical education. It was equally important that in the following year money rather fortuitously became available to pay for the expansion of such education. Mainly as a result of pressure from the strong temperance interest in Parliament, sums of money derived from a tax on alcohol under the Local Taxation Act of 1890 were allocated to local authorities, to be used, if they wished, for subsidizing technical education. This formed the so-called 'whisky' money, and, if not all local authorities were initially willing to devote the funds to technical education, by the end of the century some three-quarters of a million pounds was being expended in this way.

Technical education at the lower levels thus only began to improve considerably in the last decade of the century. At the higher levels, too, the eighties were a period of slow progress. The opening of the Central College of the City and Guilds at South Kensington was, perhaps, the most important event. For Lockyer, it had the consequence that his friends Henry Armstrong and William Ayrton transferred from Finsbury to be professors of chemistry and physics, respectively, at the new institution. As they were situated just over the road from his observatory, he now saw a good deal more of them. But, during the eighties, the Central College attracted only a small number of students, and there were internal difficulties. The staff members, especially Armstrong, resented the rather strict control that the director, Philip Magnus, kept over them. Within a few years of the foundation of the College, Donnelly had come to the conclusion that it should be handed over to the control of his own Science and Art Department. Roscoe, speaking at the end of the eighties, proposed that it should be turned into a Government School of Applied Science to parallel the work of the Normal School of Science in pure science.

It was not, however, the possibility of a link with the Central College which most exercised Lockyer and his South Kensington colleagues in the sphere of academic politics. They were more concerned with the future of the University of London as a whole. The concept of a university had always been rather tenuous in London. The colleges had been formed first, and were the important institutions, the university acted purely as an administrative link and examining body. But in the latter area it exerted considerable influence. Thus the university had promulgated the first science degree in England. More importantly, it was the university that instituted the London external degree in 1858 by throwing the examinations open to anyone who wished to take them. The influence of this latter move on institutions outside London, for example, on the new university colleges, was undoubtedly beneficial. Equally certainly, the move was a source of considerable irritation to internal institutions of the university. The main problem was that the colleges had very little influence over the contents of the university examinations, which were as likely to be determined by external factors as by internal ones. Other irritants further exacerbated the differences between the colleges and the university, especially over the question who should have the most say in the government of the university, the teaching staff, the administrative staff, or the graduates.

The need for some reform of the system was already being discussed in the 1870s, but the first major attempt to force the issue came in the 1880s. In 1884, an Association for Promoting a Teaching University for London was formed with considerable backing, including that of Lockyer and many of his friends. As its title indicates, the main aim was to convert the university from being merely an examining body. The Senate of London University, thus urged, reviewed the position, but was unable to agree on any changes. The position was complicated at this time by an abortive move on the part of University College and King's College to form themselves into a new, and entirely separate, university. Ultimately, the Government was forced to intervene, and in 1888 a Royal Commission was established under Lord Selborne to inquire what kind of reorganisation, if any, was necessary to put the university on a sounder footing. (The members of the Commission included Stokes and Kelvin.) All the deliberations of the Commission proved ultimately to be in vain, for its proposals were rejected by Parliament. Similarly, subsequent proposals for reform by the Senate of London University came to nothing. In 1892, the Government found it necessary to set up yet another Commission. Its instructions were to investigate and comment upon a scheme for a new university in London, to be known as the Gresham University (after Gresham College, which was to be its centre).

At this point, a group of members of staff at the London colleges, led by Karl Pearson, formed an Association for Promoting a Professorial University for London. The significant change, as compared with the earlier association, was the substitution of 'professorial' for 'teaching'. Pearson wished the new University of London to be modelled in every way on the University of Berlin. Huxley was appointed President of the Association, but, in fact, was soon persuading members that the ideal of a university ruled entirely by professors was unattainable in England, a conclusion with which Lockyer was in full agreement. This change of front led to Pearson's resignation from the Association, but it became rapidly apparent that Huxley's more moderate position was much more likely to gain support. When the Gresham Commission reported, in 1894, its recommendations were found to parallel many of the points that the Association had been making. It proposed, in particular, that there should be a federal university in which the university teachers should have a strong, but not overwhelming, voice. The University of London should be transformed from a purely examining body, to a teaching and examining body. Although there was a fairly wide degree of agreement on this scheme, and although it was quite evident that no other type of reform would be likely to gain much support, it was not until 1898 that the necessary bill finally passed through Parliament. Much of the work involved in pressing the bill was undertaken by Lockyer's friend, Roscoe, who was at this time Vice-Chancellor of London University. He was also mainly responsible for the decision to rehouse the University administration at South Kensington, and in persuading Rücker to follow him in the position of Vice-Chancellor.

Although the ultimate result of the academic politics of the period was thus to involve South Kensington more closely in the interests of London University, the Royal College of Science had, in fact, stood rather aloof from the long debate. Members of the College had participated strictly as individuals. The leader of these, of course, up to his death in 1895 was Huxley, but Lockyer, too, was called on for much advice on matters where he was regarded as an expert. These ranged from the formulation of degree regulations, on the one hand, to the correct advertising rates to be charged by the *University Gazette* on the other. Despite their ready assistance, neither Huxley nor Lockyer, at this stage, thought that the Royal College of Science should itself become involved in the reformed University of London. They felt that its interests might be harmed by the predominantly non-scientific interests of that institution. Their chief, Donnelly, was of the same opinion.

During this period, as we have seen, Donnelly found himself faced with severe problems of his own, problems which not only affected the future of the institutions at South Kensington, but also expanded to affect the future of science teaching throughout the country. Suspicion of the Science and Art Department on the part of civil servants and politicians had been growing, and came to a head in the nineties. Until the middle of the decade, Donnelly's political superior, Sir Arthur Acland, was able to protect him, but Acland departed after the 1895 election, and an attack on Donnelly was soon mounted. When the latter retired in 1899, having reached the statutory age limit, the Science and Art Department he left behind was very much on the defensive.

With Donnelly out of the way, and the structure of London University placed on a rather more rational basis, the future of the teaching institutions at South Kensington became once more a matter for discussion. We have noted the great respect that British academics had for the German example in higher scientific and technological education. One especial target of their envy was the large and wealthy Charlottenburg Technical Institute of Berlin. In 1903, Lord Rosebery wrote to the Chairman of the London County Council proposing that an Imperial Technical College, based on this German model, should be set up in connection with London University. Rosebery had had a long interest in educational matters. He had been Prime Minister when the Gresham Commission reported, and was to speed the reform of London University, though his rapid fall from office prevented him from achieving much in this sphere. His proposal met with a considerable degree of sympathy, and a departmental committee of the Board of Education was formed to consider it in detail. Ultimately, R. B. Haldane, who had been a leading spirit in the movement for a technical university from the start, took charge of this committee. Another member was Sidney Webb, who had been Chairman of the L.C.C.'s Technical Education Board during the nineties, an activity that had brought him into contact with Lockyer.

Haldane had been considerably involved in the reorganization of London University during the previous decade, and was known to be a keen believer in the value of German educational methods. Under his guidance, the committee initially decided that a College of Applied Science should be established at South Kensington by forming a federal union of the Royal College of Science, the Royal School of Mines and the Central Technical College under a common administrative umbrella. On reconsideration, however, they recommended that these institutions should rather be combined into a unified School of Pure and Applied Science, and should not retain their separate identities. It was this latter recommendation that was accepted by the Board of Education, thus leading, in 1907, to the formation of the Imperial College. The Board of Education itself was happy to hand over the Royal College of Science and the Royal School of Mines, which, since the demise of the Science and Art Department, it had controlled, as rapidly as possible to the new Board of Governors, but the City and Guilds Institute was less happy with the proposals, and the final integration of the new college was delayed until 1910–11.

Although the establishment of Imperial College was thus agreed, there remained the problem of what its relationship should be to the reorganized University of London. Haldon was quite clear in his own mind that the major defect of higher technical education in Germany was that the polytechnics, which provided such education, were separated from the universities, where training in pure science was provided. He was correspondingly determined that Imperial College should abandon its isolationist attitude, and become associated with the University. Within the College itself opinion was divided, and there was a sustained attempt by some members of staff, such as Armstrong, to make it an independent institution. But Haldane won the day, and the South Kensington institutions moved within the orbit of the University. If this represented a loss of liberty, it was at least balanced by the influx of new funds. Haldane was highly successful in his attempts to solicit donations of money to the new foundation. This was urgently needed, for the accommodation had become quite inadequate. Roscoe told the Gresham Commission in the nineties that, because of the lack of funds, the state of the laboratories in the Royal College of Science was a national disgrace.

The Royal College of Science was not the only institution of higher education in financial difficulties during the latter years of the nineteenth century. The university colleges throughout England were all hard-pressed. Towards the end of the eighties, there was a considerable agitation, led by Playfair, for a Government grant to be given to these institutions. The Welsh colleges were already in receipt of grants. The former belief that State aid should not be given to education was coming under increasing attack, and, although the Conservatives, who were then in power, did not contain an effective scientific pressure group as did the Liberals, nevertheless their Prime Minister, Salisbury, was sympathetically inclined to the development of scientific and technical education, in which the university colleges generally specialised. By the end of the eighties, nine university colleges had been founded in the provinces, most of them in the preceding twenty years, and it was suggested that they should be given £4000 per annum each. The Conservative Government was a good deal less generous than this, it offered £15 000 a year to be distributed amongst all the colleges, but the important principle that these institutions should receive aid from the State had been established. The grant, which was initially for five years only, was renewed in 1894 by a Liberal Government, but in the following year, a delegation approached the Chancellor of the Exchequer asking that the amount should be increased. The Chancellor chose two representatives, one from Oxford and the other from Cambridge (the latter being Lockyer's old opponent, Liveing), to inspect the university colleges, and submit a report. Mainly as a result of their discussions, the grant was raised to £25 000, with slight increases in subsequent years.

Lockyer was an interested spectator of these various negotiations, but, although commenting on them in *Nature*, he took no active part himself. In 1903, however, he moved to the front of the stage, and became the main inspirer of the next developments. That year he was President of the British Association for its meeting at Southport in Lancashire, and took as the theme for his presidential address the need for the British nation as a whole to devote much more attention to science. He fully intended from the beginning that this speech should create a major stir which would quickly lead to some form of effective action. The time seemed suitable, as in the aftermath of the Boer War there was a general feeling that changes were necessary in many spheres. In the months preceding the British Association meeting, Lockyer circulated copies of his proposed address for comment and discussion. One person to whom he turned was S. P. Thompson, who had been campaigning for the advancement of scientific education over as long a period as Lockyer himself. In July 1903, Thompson wrote to Lockyer,

> Need I say how much I appreciate your courtesy in asking me to join you in talking over points in your forth-coming address, of which we expect and hope great things. Your pronouncement will come at a time that is ripe, and men will listen to facts as to the vital importance to the nation of science and education.
>
> One thing I hope you will make clear:— that the present greatest drawback to higher scientific training is not the lack of men to teach or institutions for teaching—though we be here sorely behind other rival nations—but in the pig-headed apathy of manufacturers who won't admit the scientific and trained assistant, nor encourage the existing training institutions by going to them for their best men wherewith to recruit their staffs. I had a talk not long ago with a Yorkshire engineer, a governor indeed of the Yorkshire College, who was *dead against* taking College trained students into his own works. I found that he took

premium pupils at £300 premium, and went off hunting 3 days a week. That is the kind of man who is responsible for England's want of science. The utter stupidity of our chemical manufacturers is notorious. Trade follows the brains:—and our manufacturers won't encourage the training of brains:— else why should —for instance—our Central Technical College have taken 18 years to fill up its Engineering Departments, and even now be almost empty in its Chemical Department? Again our old Universities have no policy except to foster the inept tutorial system in the interests of the dons, and are not even arranged in Faculties as all Universities in civilized countries are.

But there: you know all these things and can state them far better than I. More power to your elbow, and to your pen, and to your voice.[5]

The effort that Lockyer put into his address was well rewarded. It was widely commented on and quoted in speeches and in the press. Early in 1904, the Prince of Wales' private secretary wrote to Lockyer, saying,

I gave to the Prince of Wales the copy of your address to the British Association: having previously marked those passages which I thought would most interest HRH. He told me that he found it so very interesting that he read every word quite independent of my marks! Last night in replying to an address at the Battersea Polytechnic HRH quoted your forcible words urging the necessity of Brain as well as naval and military power in order to maintain our position among the Nations of the World.[6]

A major theme of Lockyer's address was the need for an immediate extension of university education. Education at lower levels had, he said, been appreciably improved in comparison with earlier decades, but institutions of university status were too few in number, and too inefficient, to provide the required amount of training at higher levels. Lockyer remarked that both Balfour, who had succeeded Salisbury as Prime Minister in the previous year, and Joseph Chamberlain, then on the point of resigning from the Government over the question of Imperial Preference, had expressed in their speeches the belief that the United Kingdom was deficient in higher education as compared with several of its competitors. Balfour, indeed, had said not long before,

The existing educational system of this country is chaotic, is ineffectual, is utterly behind the age, makes us the laughing-stock of every advanced nation in Europe and America, puts us behind, not only our American cousins, but the German and the Frenchman and the Italian.[7]

Lockyer concluded that the deficiencies of English education were well known, and explained that, in his opinion, 'Chief among the causes which have brought us to the terrible condition of inferiority as compared with other nations in which we find ourselves are our carelessness in the matter of education and our false notions of the limitations of State functions in relation to the conditions of modern civilisation.'[8] There were, he remarked, two possible sources of funds for universities. One was by private endowment. In this field, the negligible efforts made in Britain were most easily seen by a comparison with the United States. 'At present,' he observed, 'there are almost as many *professors and instructors* in the Universities and colleges of the United States as there are *day students* in the Universities and colleges of the United Kingdom.'[9] The other method of advancing the development of universities was by State action, as in Germany. It seemed to Lockyer that, since private action had evidently failed to produce the desired expansion, Britain must necessarily follow the German example, and depend on State subsidies to the universities and colleges, promoting some of the latter to university status. This concept was, needless to say, popular amongst those who were concerned with the problems of university finance. One of Lockyer's correspondents, writing to congratulate him on his address, added, 'It certainly ought to help much towards the proper endowment of the London University.'[10]

Lockyer was by no means the only significant agitator for the reform of science education at this time. Huggins, for example, took scientific education as the main theme for his Presidential address to the Royal Society in the same year. But Lockyer placed more emphasis on the importance of assistance at the university level than did most other speakers and writers on the subject, and he provided the call to action in the provision of State aid. Encouraged by the general response to his address, the British Association approached the universities and colleges to seek their support for a general appeal to the Government. Some time was spent in collecting data on the current state of these various institutions. Lockyer had already obtained a good deal of information on his own behalf, and had used it in a series of articles that appeared in *Nature* during 1903, comparing the statistics relevant to British universities with those from overseas. The main problem was in getting the different institutions to agree to a common policy of approach to the Government at all. This was particularly true of Ireland, where the religious tension considerably affected attitudes towards matters of university policy. There was, moreover, as Lockyer's correspondence indicates, some pessimism as to the outcome of any meeting with the Government, for the financial climate was unfavourable. Lyttelton, headmaster of Haileybury (and, in the following year, headmaster of Eton) was one of the supporters of an increased role for scientific education in Britain, but he wrote to Lockyer in May 1904, suggesting caution.

> If it is not too late I venture to make a suggestion with regard to the Memorial shortly to be sent to the Prime Minister. Sir Oliver Lodge's remark at the meeting last week reporting what the Chancellor of the Exchequer had said to him was significant. The debates last night show that the feeling in the House and in the country about the increase of expenditure is becoming a serious fact, and I feel that the cause which we have at heart will be furthered if we show in what we say that we are fully aware of the remarkable difficulties which

at present beset the Chancellor of the Exchequer. That is to say, I would suggest that the Government Grants should be asked for on a moderate scale and only after the authorities have been satisfied as to the urgent need of each individual University.[11]

If Lockyer was beset on the one side by councils of moderation, he also received counter-balancing suggestions of the need for radical changes in the approach to university education. Thomas Burt, a Liberal trade unionist M.P. representing the working class interest in a predominantly Conservative House of Commons, wrote to Lockyer in July assuring him of his best wishes, and continuing,

> The ideal aimed at should, I think, be to carry free education on from the Primary to the Secondary Schools and to the Universities. Every child who satisfies the teachers and authorities of his Primary School should have a right to proceed, without either examination or scholarship, to the Secondary School, if need be at the public expense. Such students, if they satisfy the teachers and authorities of the Secondary School, should have a right to go to the University without examination or scholarship. Further, they should have a right to all State positions on passing an examination relating to the work they will have to do, and not on what they may remember, at the moment, after they have been coached and crammed.
>
> The Civil Service Commission and much of our existing system of examination, which is often cumbrous and mischievous, would thus in time become unnecessary, the educational ladder, which now has many missing rungs, would be continuous, and every poor man's son who thirsts for education, and who has the capacity for receiving it, would have additional opportunities of serving the State.
>
> By thus widening the area of selection, and by opening the way for fit men in greater numbers to enter into the higher activities of the nation, the State itself would, beyond doubt, be an enormous gainer.
>
> Though in writing thus I am expressing my own personal opinions, I feel quite sure that large numbers of working men in all parts of the country would be in complete agreement with me.[12]

The discussions and arguments culminated in July, when Lockyer headed a large deputation which waited on Balfour and his Chancellor of the Exchequer, Austen Chamberlain, to urge the case for a much greater involvement of the State in the financing of universities and colleges. The delegates included some of Lockyer's old friends, such as Roscoe and William Ramsay, and also Joseph Chamberlain, as a representative of Birmingham University. Lockyer, in one section of his Presidential address, had attacked the current debate on Free Trade and Protection as irrelevant to the future of British commerce, asserting that the introduction of new science-based inventions was much more important. Although Lockyer had intended his rebuke to be even-handed, he found that his remarks had been taken in some quarters as supporting Joseph Chamberlain, and he had had to go to some pains to deny it. Nevertheless, Lockyer was certainly on amicable terms with Chamberlain. Geikie records that Lockyer and Chamberlain were both present at a small, specially arranged party at the House of Commons in 1905, and, as we shall see, Chamberlain supported Lockyer's educational effort in subsequent years. Lockyer, like many of his scientific and literary friends, seems to have been sympathetic to the Liberal Unionist position from the time of formation of that group in the 1880s.

The results of the meeting were only moderately satisfactory. Both Balfour and Austen Chamberlain were sympathetic, but were quite unprepared to increase the university grants to anywhere near the figure Lockyer had in mind. He wanted an immediate grant of £8 million plus £400 000 per annum. They offered, instead, simply to double the amount of the existing grant. Although from Lockyer's point of view the offer was disappointing, it must be judged against the fact that popular opinion would certainly have opposed any vast increase. In these terms, the agitation led by Lockyer probably produced as favourable a result as was possible at the time. More significantly, it had a longer term effect. The government grant was increased again in succeeding years by appreciable amounts, so that, when the First World War began, the income of the English universities and colleges from government and local sources combined had risen to about a quarter of a million pounds. The impetus of the 1903–4 debate on higher education also subsequently led to the formation of the present University Grants Committee to control the distribution of State aid.

Lockyer's British Association address led him to a brief involvement not only in national politics, but in local politics, too. The L.C.C. was at that time in the process of working out what reorganisation of educational facilities in London was required as a result of the Education Act of 1902. It was naturally proposed that the control of these facilities at all levels should be solely in the hands of members of the council, that is, representatives of the political groupings, but some leading figures in the London educational world felt that education should be treated on non-party lines. Lockyer was one of these, and, since his name was in the news by virtue of his recent address, he came to be regarded as a spokesman for this position. Witness, for example, the following letter written to him a few weeks before the L.C.C. elections in 1904.

> Is it not possible to do something to amend the scheme for the education committee of the County Council. The action of the Board of Education will depend on the County council elections on March 5th. If the 'progressive' scheme is allowed to go through, we may say goodbye for the next four years to any development either of secondary or higher education in London and it will be useless to attempt to get imperial funds for educational purposes while our education authority is squabbling over religious questions, playing at politics, or attempting to starve or confiscate voluntary schools . . .

> You have made a good start in waking up the country at large. Could you not get together an association of men for the purpose of waking up London to its needs, and to giving practical expression at the polls of its wishes?[13]

Lockyer was, in fact, approached by the Conservative and Liberal Unionist Party with the suggestion that he should serve as one of their aldermen on the L.C.C. with a special brief to cover education. Lockyer, however, although he proclaimed his support for the 1902 Education Act, which had been passed by the Conservatives and Liberal Unionists against Liberal opposition, remained true to his belief that education should be a non-party affair, and declined the honour.

Nothing typifies Lockyer's attitude to the national importance of education better than the title he gave to his British Association address, 'The influence of brain-power on history'. This was modelled on A. T. Mahan's book, *The Influence of Sea Power on History*, which had appeared in 1890. The latter had had a marked influence on the general public's approach to the question of sea power, and Lockyer had read it avidly. Lockyer, in the typical Victorian fashion, linked his firm belief in the virtue of imperialism to an equally firm belief in the importance of the part to be played by the Royal Navy. His British Association address was basically an attempt to point out that, just as the continuance of imperialism was linked to naval supremacy, so also it was linked with intellectual supremacy, especially in science. He pointed out that when Britain, at the end of the eighties, had found itself lagging behind in the expansion of its navy as compared with other countries, the result had been a massive injection of capital by the State into a crash programme of ship-building. Centuries before, such a programme might have been left to private enterprise, but this had long ago been found to be inefficient, and the government had been forced to intervene in order to protect the safety of the State. The same approach was now needed to higher education. Lockyer pushed his comparison of naval and brain power to the point of estimating what funds were necessary for higher education in terms of the cost of the corresponding number of battleships.

Although, as we have seen, Lockyer had always been interested in naval affairs, they were particularly in the forefront of his mind at this time, for he had just become involved in the question of what education was best fitted to the needs of naval officers. Sir John Fisher at the Admiralty had recently decided that the reform of naval education must form a vital part of his overall reorganisation of the navy. Lockyer meanwhile had been emphasizing that naval cadets should have considerably more scientific training in their curriculum. Fisher was impressed by Lockyer's arguments, and invited him to the Admiralty for discussions. As a result, when Fisher's revised education scheme appeared, it met with Lockyer's general approval. He had two main reservations. The new scheme of entry for naval cadets depended on nomination by senior naval personnel, and a limited, rather than open, examination. Lockyer, perhaps with his memories of the War Office in mind, stressed the obvious opportunities this offered for nepotism. He also objected to the system of specialization in naval education, which was being continued under the Fisher regime. In this, the naval officer who specialised, especially in navigation, was almost inevitably debarred from the highest ranks of his profession. Lockyer had a personal interest in this matter, for his son, Hughes Campbell Lockyer, was himself a trained navigation officer, and had written to his father with a tinge of bitterness whilst Fisher's reorganization was under way.

> . . . even if I had not had a row with friend Corry in the Hood I should not have expected to get promoted before 12 years, although as you know I had the best N.O. [Navigation Officer] record! And yet a Gunnery man who can teach an ord.[inary] seaman how to stick a sawdust man through with a cutlass gets his at $10\frac{1}{2}$ yrs.[14]

Despite these objections, Lockyer, much to Fisher's pleasure, praised the revised naval curriculum as being greatly in advance of anything the Army could offer.

Although Lockyer's emphasis on the expansion of higher education most rapidly involved him in action, the major proposal of his British Association address had actually lain elsewhere. He had mainly been concerned with the need for some national body to deal with the organisation and publicising of science, something which he had believed in since at least the days of the Devonshire Commission. He pointed out that the current general acceptance of Britain's technological deficiencies had resulted mainly from the warnings of the scientists. But though these warnings had been reiterated for some fifty years, still no really effective action had been taken. The reason, he explained, was because there existed no collective scientific voice to present the distinctive viewpoint of science to both the public and the Government. This lack was also responsible for the current misunderstanding of the role of science. Whereas the problem was generally considered to be how science could be applied to improve our technology, the real question was how scientific modes of thought could be applied to all the major problems facing the country.

Lockyer felt that his address to the British Association was an especially appropriate place to make this point, since one of the original objectives in founding the Association had been, 'to obtain a more general attention to the objects of science and a removal of any disadvantages of a public kind which impede its progress.'[15] But his acquaintance with the workings of the British Association over the previous forty years had led him to believe that this aspect of its work had become overshadowed by others. Nevertheless, at just about this time, the Association had begun to take a renewed interest in scientific education. In 1901, the Council, under pressure, had introduced a new section for education. The leading spirit behind this move had been Armstrong, but Lockyer's assistant at South Kensington, Gregory, had also been involved, and, as the section's work developed, it was he who became its chief organiser.

Gregory's early life seems in some ways to have been modelled on Lockyer's: astronomy at South Kensington, scientific journalism, especially on *Nature*, and a deep interest in the furtherance of scientific and technical education. He had even settled in Lockyer's first area of residence, Wimbledon. His contributions to *Nature* on questions of education were widely read, and in 1898 he helped launch a new educational monthly, *The School World*, which was particularly concerned with the problems of science teaching. Lockyer thus had, via Gregory, detailed information on the current educational interests of the British Association. He obviously felt there was a real possibility that the Association would respond to a call for it to become a scientific Parliament, exerting an influence both nationally and locally on behalf of science. He therefore proposed that a sub-committee should be set up to discuss the necessary procedural steps for such a development, and gently encouraged this with a barely concealed threat.

> Rest assured that sooner or later such a Guild will be formed, because it is needed. It is for you to say whether it shall be, or form part of, the British Association . . . Remember that the British Association will be as much weakened by the creation of a new body to do the work I have shown to be in the minds of its founders as I believe it will be strengthened by becoming completely effective in every one of the directions they indicated.[16]

As we have seen, the British Association did respond rapidly to one part of Lockyer's call : that requiring pressure on the Government to increase the efficiency of higher education, which was the first task he had allotted to his proposed scientific Parliament. But apart from this, the Association's response to his suggestions was disappointing. After Lockyer's address at Southport, a discussion developed there which ended with the passing of a resolution calling on the Council of the British Association to promote an organisation such as he had described. The Council made no definite response, and it soon became clear to Lockyer that any further progress would depend on his own initiative in forming a new association with the objectives he had outlined in his address.

Lockyer was by no means the first to propose the need for an influential organisation to debate questions of science policy and to exert political pressure on behalf of science. As long ago as 1870, Roscoe had proposed a National Science Union with much the same purpose. It is characteristic of Lockyer, however, that he was the first to turn the proposal into reality. During the early months of 1904, he consulted with his friends concerning the formation of a representative committee to consider the formation of the proposed new body. By June, he had received promises of help from many of his friends, amongst them Lubbock, Ayrton, Sir George Clarke, Meldola, Sir William Ramsay and S. P. Thompson, and was able to call the first meeting of the committee in the rooms of the Royal Society. The committee agreed with Lockyer that a new organisation, independent of the British Association, should be established. A public announcement soon followed. 'An organisation is being formed, under the name of the British Science Guild, with the object of insisting upon the importance of applying scientific method to every branch of the affairs of the nation.'[17]

The objectives of the new guild could be classified essentially under four headings—to bring together people throughout the Empire who were interested in the advancement of science; to act as a political pressure group for matters of science; to press for the greater application of scientific methods in industry; and to promote scientific education. In pursuing these aims, the committee of the British Science Guild, whilst not restricting membership, hoped especially to obtain adherents from people who could help in the achievement of its objectives, for example, from Members of Parliament. Although the original organising committee had only one M.P. amongst its eighteen members, others were subsequently persuaded to join the Guild once it had been launched, though never so many as Lockyer would have liked. There was, in fact, always a certain ambiguity about the aims of the Guild. Thus Lockyer decided that the annual subscription should be 2/6. Lubbock felt that this was much too low if the intention was to build a small influential body, and was initially reluctant to join the Guild. Lockyer eventually overcame his objections, and persuaded him, indeed, to become the honorary treasurer of the new organisation.

As a result of the executive committee's efforts, an inaugural meeting of the British Science Guild was organised for 30 October 1905. At the invitation of the Lord Mayor it was held in the Mansion House. The main aim of the meeting, apart from publicity, was to elect the chief officers of the Guild. It was intended that the new organisation should follow the common practice, and divide its interests between a variety of committees. Lockyer therefore took the key organisational post as Chairman of Committees. Numerous Vice-Presidents were elected from various walks of life. They included the Lord Mayor of London, the Bishop of Ripon, Alma-Tadema and Joseph Chamberlain, their main function being the provision of a good public image. But the most important post to fill, from the point of view of the Guild's future, was that of President. It required someone who was both a public figure, and also deeply concerned with the growth of scientific influence. Here Lockyer triumphed: he managed to persuade R. B. Haldane to accept the post. The Liberals were still in opposition, but within two months they were back in office. Haldane, now at the War Office, was an influential member of the Government throughout the early years of the Guild.

In his speech accepting the Presidency of the Guild, Haldane explained to his audience what he thought was his main qualification for the post. When his party had fallen from power nearly ten years before, he had, he said, decided that, 'he might as well turn his hand to the somewhat cobwebbed state of the higher education of this country.'[18] His studies of the problems facing Britain led him to take a similar stance to Lockyer's. What was needed was not the promotion of science in any narrow sense, but, 'the bringing of method, the bringing of thinking, into the modes of government which applied to our public affairs and which applied to our private industries alike.'[19] Haldane went on, however, to warn against too uncritical an admiration of the German example. Their scientific methods of organisation certainly led to greater efficiency, but were accompanied by a loss of individual liberty which would be considered intolerable in Britain. He urged, nevertheless, that this should not be taken as an argument against the real need for more scientific thinking in the realms of government.

It is evident that, despite the publicity given to the British Science Guild, there remained doubts in many minds as to its exact aims. For example, did scientific thinking applied to government mean what it said, or did it simply imply the application of logic to political problems? A letter from A. J. Balfour may be quoted in this context, written not long after he had lost office.

> I, as you well know, am most anxious to promote the cause of Science in every way in my power: but, frankly, I am not sure that I form a clear idea of the exact methods which your Guild proposes to adopt to attain that end. The words 'science' and 'scientific method' are, as no one knows better than yourself, loosely used; and, especially when applied to such subjects as Government or taxation, often mean little more than 'reasonable' or 'rational'. Every human being is desirous that the affairs of his country should be 'reasonably' or 'rationally' conducted: where mankind quarrel is over the means by which that excellent object can be secured. 'Science' gives no clear utterance on such subjects.[20]

Balfour, like his predecessor as Conservative party leader and Prime Minister, Lord Salisbury, was genuinely concerned with the encouragement of science. Indeed, Lord Rayleigh, who had been Salisbury's closest scientific confidant, was also Balfour's brother-in-law. But Balfour was a strong supporter of the British Association, and it may be that he feared that the formation of the British Science Guild would affect the Association's influence. The British Association was already in trouble before the formation of the Guild. In 1903, Balfour had written to Lockyer saying, 'I hear of the falling off of its members with some dismay. It would be a deplorable thing if anything should occur seriously to cripple its utility.'[21] The Association remained in a poor way throughout the first decade of the twentieth century, and there were continual arguments as to the type of reforms needed to save it from possible extinction.

Within a year or two of its foundation, the British Science Guild had produced some useful, though not very spectacular, results. The most important was due mainly to Haldane's personal intervention. He managed to have the government grant for the National Physical Laboratory, which had been formed in 1900 due to pressure from a group of scientists led by Rayleigh, doubled, from £5 000 a year to £10 000. The British Science Guild also tried to bring pressure to bear on the City Guilds to allow the unification of the institutions at South Kensington, this at a time when the attitude of the Guilds made progress seem unlikely.

Members of the British Science Guild also helped in the organisation of deputations to ministers over specific points of science policy. For example, the Government had been providing some of the funding for an international investigation of fishery problems in the North Sea, but it was rumoured in 1906 that these funds were to be cut off. One of the people immediately involved wrote to Lockyer and Haldane, asking for the Guild's assistance.[22] A deputation was organised to meet the Chancellor of the Exchequer, with Sir Michael Foster as the Guild's representative, and achieved some success. Marine zoological investigations were, indeed, grossly undersupported in Britain. Thus, in 1911, the Guild's assistance was again invoked in this field, in order that one research table should be kept open to British workers at the Naples Zoological Station, the main centre for marine zoology in Europe at that time. That same year the Germans paid for 22 tables there, and even Belgium and Holland had two each.[23]

In 1907, the British Science Guild was involved in another deputation, this time to the President of the Local Government Board, to request the introduction of measures for preventing the growing pollution of rivers and water supplies in Britain. Pollution was an early concern of the Guild. One of its first sub-committees, consisting of Ramsay, Meldola and Pedler, was set up to consider this problem. In the next year, 1908, the British Science Guild, with Lockyer in the lead, organised a major deputation of the various learned societies to meet the Postmaster General. The point at issue was the burden of postal charges on the small budgets of the societies. The cost of distributing journals could be larger than the cost of printing them, and some societies were even finding it cheaper to send journals to members in the London area by hand than to post them. They therefore asked that the literature produced by recognised learned societies should be allowed to go by post at reduced rates, but their plea fell on deaf ears.

During 1908, Lockyer and the Guild were both considerably involved with problems of time. One question was the possibility of introducing daylight saving. Lockyer was called on to give his opinion about this before a Select Committee of the House of Commons, and he indicated his opposition to the proposal. It was also opposed by the British Science Guild, but the Guild was more concerned at this time with the need for the synchronisation of clocks. Public clocks throughout London showed wide variations in time. According to the Guild's data the nearest approach to a standard-time public clock in the London area was Big Ben, for it had only twice been in error by more than three seconds during the previous year. The Guild tried to persuade the London County Council that something should be done to provide more uniform time standards, but the Council insisted that there was insufficient public interest to warrant the necessary legislation. Nevertheless, the pressure by the British Science Guild ultimately produced some effect. In 1912, the Postmaster General announced a new scheme for synchronising clocks.

In other, and, perhaps, more important areas, too, the British Science Guild was able to exert useful pressure in the years leading up to the outbreak of the First World War. It pressed continually for the greater application of science to agriculture, and was one of the bodies working behind the scenes for the passage of the Development Act of 1909, which provided aid for agricultural research. Again, the Guild never lost its enthusiasm, imbibed partly from Lockyer, for higher education, particularly for an integrated and systematic arrangement of education at all levels. By 1910, it was trying to persuade the Government that the universities should not be governed directly by the Treasury, but should be under the Board of Education. Within the next two years, this advice had been followed. The Board of Education set up an advisory committee that took over the distribution of the university grants from the Treasury.

The various deputations, petitions and memorials more often than not arose from the committees which were established within the British Science Guild. These, customarily, produced a considerable body of documentation on the questions investigated, as well as recommendations for action. For example, a major report on the problems associated with the standardization of time was produced in conjunction with the attempt to persuade the London County Council to do something in London on this issue. The London County Council, incidentally, was a convenient and near-at-hand target for the Guild's activities. There was, to give another example, a stout-hearted, though unsuccessful, attempt to get the Council's agreement to the renaming of some of London's streets after famous British scientists. The Guild regarded the lack of such names as shameful in view of the number of Parisian streets called after famous French scientists. Some of the reports of the committees contain important information which might otherwise be difficult to acquire. An example of this is the detailed analysis of the endowment of higher education in Britain, Germany and the United States, which Gregory carried out during the period round 1910. As time passed, some of the old committees were dissolved and new ones were formed. Perhaps the most interesting of these latter was the committee on the conservation of natural sources of energy, which began its investigations in 1910 under the chairmanship of Sir William Ramsay. The range of energy sources it studied ranged from coal and natural gas to solar power and atomic energy.

The British Science Guild thus certainly expended a good deal of time, paper, and talk on the aims for which it was established. To some extent, it received public recognition for its work. Some of the City livery companies gave small grants towards its operations, and it was sometimes officially called on to help with matters of scientific publicity. For example, Lockyer was made responsible for organising a science section in a Franco-British exhibition held in London during 1908. Another sign of success was that the Guild fairly quickly established branches elsewhere in the Empire. Yet, despite these genuine successes, there seems to have been a feeling amongst some members of the Guild of ultimate failure. The Guild was certainly helping in the production of small, piecemeal reforms, but a genuine official recognition of the importance of science and of scientific method, which it was committed to bring about, seemed no closer. The leaders of the British Science Guild asserted on several occasions that the main aim of the Guild was for gradual progress, but it was noticeable that they, too, returned more than once to the need for a National Scientific Council in Britain. In fact, the Guild's dreams of science in government were best realised not by its own efforts, but by the personal exertions of its President. Haldane, in pushing through his Army reforms, believed that he was following scientific principles. He subsequently pointed to the Committee of Imperial Defence as the prime example of the scientific approach to organisation.

In 1912, Haldane resigned as President, and was followed by Sir William Mather, the head of a well-known engineering firm, and a former Liberal M.P. Other new figures had come forward by this time. Sir Ernest Shackleton had come to play an important part in the Guild's affairs, whilst Lockyer's old assistant, Pedler, had become Secretary. Lockyer, himself, was playing a slightly less important role, partly, of course, because he was by this time in his mid-seventies, but also because of his involvement in a long-drawn-out controversy over the future of the Solar Physics Observatory. By 1914 Lockyer had decided to retire from active participation in the Guild, and tried, unsuccessfully, to persuade Ramsay to succeed him as the main driving force. Throughout the decade he had expended considerable effort, not only on questions of general science policy, but also, more personally, on behalf of colleagues who were in need. For example, a former colleague at South Kensington, J. W. Judd, solicited Lockyer's aid both for himself and for Ray Lankester over the question of pensions. The Treasury, he claimed, had found a legal loophole to refuse them superannuation, and he asked that something should be done to check the anti-scientific tendencies of the officials currently in the Treasury.[24]

From the beginning of the 1890s, Lockyer had been connected with another official body which was concerned with the promotion of scientific research, the 1851 Commissioners. In that decade he was co-opted to a committee which the Commissioners had set up to consider the establishment of scholarships for advanced work in science. The Commissioners had decided that, whereas they had hitherto been spending most of their income in London, it was now time for other areas to benefit. The chosen committee included many of the leading scientist-educators amongst its members—Playfair, Huxley, Roscoe—but it was Lockyer who was mainly responsible for mapping out the details of the scholarship scheme. In particular, he drafted the requirements that the awards should be given for research only, and not for other types of advanced study, and that it should not normally be permissible for the scholarship to be held at the scholar's home institution. He actually pressed for the awards to be limited to research in the physical sciences only.

Nearly all of Lockyer's proposals were accepted. The 1851 Commission scholarships were rapidly established and made available to university institutions throughout the Empire. The results were quickly recognised to have exceeded even the most sanguine expectations. Because so little assistance was then available for further research, the calibre of the students attracted was extremely high. The most famous of the early 1851 Exhibition scholars was certainly Ernest Rutherford, who came over from New Zealand to Cambridge in 1895, but many other lesser-known, but still important, scientists were aided in the same way. Lockyer remained on the scholarship committee for two decades, and was one of the members mainly responsible for the selection of scholars. He, himself, saw the entire scholarship scheme as a fulfilment of one part of the thinking which had motivated the Devonshire Commission years before. He noted, however,

> '. . . when I proposed this new form of the endowment of research it would not have surprised me if the suggestion had been declined. It was carried through by Lord Playfair's enthusiastic support. This system has been at work ever since, and the good that has been done by it is now generally conceded.[25]

Lockyer's views on science policy and education remained remarkably constant throughout his life. Yet his views were not, for the most part, unique. He was a member of a group of scientists whose overall approach to problems in this category was consistently similar. The leading member of this group, and one of Lockyer's heroes, was, of course, Huxley. Lockyer nevertheless brought his own particular ideas to modify in detail this general attitude. For example, he always stressed the imperial consequences if other countries, more particularly Germany and the United States, were ahead of Britain in scientific education. His high regard for the Prince Consort, another of his heroes, was stimulated in good part by Albert's foresight in comprehending the future importance of science in maintaining Britain's position in the world. Lockyer's support for the concept of Empire never wavered. In 1908, he was appointed an official of the British Empire League, and, under his guidance, the British Science Guild and the British Empire League frequently came into contact. For example, when, in 1911, there was a meeting of colonial prime ministers in London, the two bodies united to give a dinner in their honour.

The other noticeable characteristic of Lockyer's thinking on social and educational matters, which remained firm to the end, was his belief in the relevance of the concepts of Darwinian evolution. By the Edwardian era, Lockyer's full-blooded acceptance of the determinative importance of social Darwinism was already somewhat outmoded, but he remained convinced of its applicability. In 1910, the executive committee of the British Science Guild decided to set a prize essay, open to competition by the general public, in order to stimulate interest in the role of science in society. The subject chosen was, 'The best way of carrying on the struggle for existence and securing the survival of the fittest in national affairs'—a title almost certainly selected under Lockyer's guidance.

XI

THE FINAL PUSH

The Presidency of the British Association was not the only major event of Lockyer's life in 1903. A few months before the British Association meeting, he married Thomazine Mary Brodhurst, fifty at the time and sixteen years his junior. For his new wife, too, it was a second marriage; she was the widow of a surgeon. Lockyer had met her at South Kensington before her first marriage. Whilst away observing the 1882 solar eclipse in Egypt, he had written to her exultantly saying that the foreign observers present at the eclipse had been astounded by the accuracy of his predictions concerning the solar spectrum.[1] It seems possible that Lockyer, with a growing family on his hands, was already at that time looking around for a second wife. In the event Miss Browne, as she then was, married elsewhere, and he was forced to wait.

His second wife had as strong a character as his first, but was much less prepared to stay in the background. She enjoyed the glittering social round of the Edwardian era, and one suspects was not above a little head-hunting herself. She was certainly addicted to collecting the autographs of distinguished people. But at the same time, she had a keen social conscience. She was one of the women who associated with Octavia Hill in an effort to bring about housing reforms, and she had acted as a rent collector for one of the housing schemes established by Octavia Hill and John Ruskin. (It should be added that Octavia Hill and Lockyer had many friends in common, including Tom Hughes, Holman Hunt and Canon Barnett.) Mary Lockyer was, moreover, an ardent feminist. Her Unitarian beliefs emphasized the importance of intellect, and she was quite clear that, in Britain, the feminine intellect was being undervalued. Together with her elder sister and her mother, she was a leading protagonist in the creation of a hall of residence for the accommodation of women students at the University of London. This was during the early eighties, when the University had very little desire to see the numbers of its female students increased, and the Browne family faced an uphill struggle. Nevertheless, College Hall was ultimately established. Later, when a new building was erected for it in Malet Street during the thirties, Lady Lockyer was called on to lay the foundation stone in recognition of the part she had played in its origin.

With this background, it is hardly surprising that Lady Lockyer was a firm believer in the need for women's suffrage, and herself participated in the movement. Although she helped with the organization of the more spectacular marches (for example, she and Elizabeth Garrett Anderson acted as two of the advisers for a large procession through London in 1910) she believed that more lasting results were to be obtained by political persuasion rather than by the shock tactics of the more extreme suffragettes. It appears, indeed, from her correspondence that there was concern amongst her like-minded friends over the effect these shock tactics were having. One of them, Mrs Arnold-Forster, the wife of the Liberal Unionist M.P. who preceded Haldane at the War Office, wrote to her anxiously, 'I was amazed to find on talking to peers and their wives how much harm the suffragette demonstrations had done us. It seems so little to do with us.'[2] Yet Lady Lockyer remained on reasonably amicable terms with Hertha Ayrton who was one of the militants. But her own approach during the first decade of the century was to cultivate a close acquaintance with members of both the Commons and the Lords in an effort to promote relevant bills. Lockyer's numerous friends proved very useful to her here, and Lockyer himself became involved in her activities. She was, for example, one of the forces behind the bill to allow women to sit on County and Borough Councils which came before the House of Lords in 1907, and Lockyer assisted her efforts to put pressure on individual peers before the sittings.

Lockyer's second wife was a good deal less involved with his work than his first wife had been, although she accompanied him on some of his surveys of ancient monuments, and attended at least one eclipse expedition. She also participated in the activities of the British Science Guild, becoming its Assistant Treasurer. Lockyer's marriage, however, was in no sense one of convenience. This may be contrasted with his friend, Isaac Roberts, who was somewhat older than Lockyer and married in 1901. Lockyer invited Roberts to an 'at home' in October of that year, and his friend replied that he was sorry he would not be able to attend, but,

> . . . it happens that I shall be in attendance upon a wedding party from France from the 15th to the 17th inst. On the 17th. it is appointed that I shall be married to the eminent astronomer Miss Klumpke who has resigned her position at the Paris Observatory to be united with me as my wife in the furtherance of astronomical researches at my Observatory at Crowborough.[3]

Although the time was, perhaps, past when Lady Lockyer could have acted in any major way as a second mother to Lockyer's children, she gave them such assistance as she could. For example, H. C. Lockyer's promotion in the Navy seemed to her to be going too slowly. He had been promoted to Commander in 1902, but ten years later he had progressed no further. She therefore approached a friend in the Admiralty on his behalf. The friend was pessimistic, 'I can do nothing as I was connected with the old Board and Churchill hates my old chief Sir Arthur Wilson and has also quarrelled with Lord Fisher and my head would very soon be bitten off.'[4] Yet the year after, H. C. Lockyer was duly promoted to Captain.

Lockyer's compulsory retirement from the Royal College of Science, in 1901, necessitated a new look at the workings of the Solar Physics Observatory, for, unlike the College, its regulations specified no official retiring age. A committee, whose scientific representatives included Abney and Stokes, was therefore established to consider the future of the Observatory. It ultimately decided that the Observatory should continue with Lockyer as director, but that the links with the Royal College of Science should be severed. The Treasury, for once, proved amenable, and agreed to increase the Observatory's budget so as to compensate for the loss of support from the College. About this time, the chance arose for Lockyer to go out to India to take charge of the new solar physics observatory at Kodaikanal, but he finally decided to stay at South Kensington. He therefore continued at the Solar Physics Observatory, but Fowler left his staff to become Assistant Professor in the Royal College of Science, in order to continue the teaching of astronomical spectroscopy there.

During the first few years of the twentieth century, one or two of the old faces disappeared from the Solar Physics Committee, and new ones were added. Thus Stokes died in 1903, and Frank McClean, with whom Lockyer had become very friendly, initially as a result of their mutual interest in stellar spectroscopy, was elected. McClean was not a member for long, since he died in the same decade. But a connection between the McClean family and the Lockyers continued, for Jim Lockyer had become a flying enthusiast, a passion he shared with Francis McClean (one of Frank McClean's sons), who was also interested in astronomy. It is hardly surprising that the two were on excellent terms. In 1911, they travelled around the world together. Their ultimate object was to study a total solar eclipse from Tonga, but they took the opportunity to go via Dayton, Ohio, where they stayed as guests of Orville Wright. By this time their interests had turned definitely towards flying machines, though earlier they had been involved in the sudden British craze for ballooning.

Many of Lockyer's own friends in Britain were interested in the scientific aspects and uses of flight. Crookes, for example, was on the Council of the Aeronautical Society in the early 1900s, as was W. Napier Shaw, the Director of the Meteorological Office (who was also a member of the Solar Physics Committee). Generally speaking, the daring young men were more likely to be members of one of the aero clubs than of the Aeronautical Society, which was a little too staid for them. The leader of this younger group was the Hon. C. S. Rolls, and it was with him that Jim Lockyer made his first balloon flight. On the other hand, Jim Lockyer was also recognised as one of the early experts in photography from a balloon, and so had a foot in both camps. Members of both groups were therefore regular visitors at Lockyer's house. (Since Rolls was one of the great favourites of hostesses during this period, it is not surprising that Lady Lockyer was always particularly pleased when he put in an appearance. He usually arrived in his own Rolls-Royce, specially fitted with a container for a balloon.) Lockyer, in fact, made some attempt to bring the scientific and the sporting flyers together, in the hope that the latter might become more interested in the possibilities of scientific observation. It does not appear, however, that he had very great success. As Napier Shaw warned him,

> The difficulty is that without scientific experience they do not easily appreciate the horrible drudgery of scientific accuracy. With the very best of intentions scientific observations in a balloon want not only facilities but actual teaching —even the immortal Glaisher was (according to the Germans) not quite equal to the occasion in the upper air.[5]

The new developments in flight were, of course, followed and commented upon by *Nature*, but Lockyer was now somewhat less intimately involved in the running of the journal than he had been in previous decades. Much of the work was handed over to assistants, especially Gregory who, in 1905, became science editor to Macmillan. But Lockyer was still necessarily involved in major decisions or controversies. For example, there was criticism in *Nature* of an appointment in the Canadian Geological Survey, and the reply to this criticism was refused publication. Ernest Rutherford, on his transfer to McGill University from Cambridge, had rapidly established himself as a leading spokesman for science in Canada. He therefore was called on by his colleagues in Canada to write directly to Lockyer and resolve this dispute.[6] Fortunately, the two men were on very cordial terms, even though Lockyer was an old friend of Sir William Ramsay, a person whom Rutherford regarded with complete scorn, because of the inaccuracies he had detected in Ramsay's current work on radioactivity.

Lockyer, of course, was delighted with Rutherford's experimental demonstrations that one element could break down into another. His response here differed from that of some other members of his generation, who found it all rather difficult to digest. Thus Lord Kelvin refused to accept the idea of radioactive decay right up to his death in 1907. In

1903, Kelvin and Lockyer sat next to each other at a dinner, and during the speeches exchanged the following notes on the back of a menu.

Kelvin	I believe the atoms are durable, unchanged for ever—Don't you?
Lockyer	Do you mean the atoms of helium into which Radium is resolving itself?
Kelvin	I don't believe any atom resolves itself into others.
Lockyer	Both Rutherford and Ramsay have told me today that some of the 'rays' of radium are composed of helium.
Kelvin	I don't believe them.[7]

Lockyer, although excited by the new developments in the study of radioactive decay, never seems to have been tempted to try laboratory experiments in this field himself. He remained faithful to the spectroscopic approach to the end. He did, however, allow some of the results of his own spectroscopic work at South Kensington to lead him into other areas of research. Perhaps the most interesting of these, in the first decade of the twentieth century, was that stimulated by studies of the two elements, titanium and vanadium. These two were the main contributors to what Lockyer had formerly labelled the 'unknown' lines in sunspot spectra. He subsequently discovered that they also seemed to be found together in stellar spectra. This stimulated him to examine various terrestrial objects, for example volcanic dust, to see if these elements appeared in combination there, too. He found, as a result, that they seemed to occur together universally, even in plants. In view of this latter observation, he proceeded into a study of plant physiology, to try and discover how elements could be absorbed differentially from the soil through the roots.

But the main research topic at the Solar Physics Observatory during the first decade of the century was not something new. Instead, Lockyer concentrated with increased vigour on the attempt to find some relationship between terrestrial meteorology and the solar cycle. As we have seen in an earlier chapter, the measurements of the widening of sunspot lines, which had been carried on for several years at South Kensington, led Lockyer on to a new investigation of this area towards the end of the nineteenth century. Together with his son, Jim, he proceeded to correlate the changes in the lines with what he regarded as positive and negative heat pulses from the Sun. These, in turn, were related to climatic changes in India. Having derived a positive result for the latter part of the nineteenth century, for which sunspot spectra were available, they then extended their comparison backward to the beginning of the century, on the basis of sunspot numbers, and found again what seemed to be good correlation with the data on rainfall, and famine.

This result gave Lockyer perhaps more immediate pleasure than any other he had published. He had finally found the practical application of solar physics which he had sought for so long. The methods used for the Indian study were immediately applied elsewhere, especially to data from the Nile valley. Lockyer again thought he could discern a correlation between the solar heat pulses and the failures of the Nile. By 1901, Lockyer and Jim were ready to publish a definitive account of their investigations. Not long before it appeared, in the *Proceedings of the Royal Society*, Lockyer wrote confidently to Lord Salisbury.

> You may remember that in 1877 your kind and sympathetic action on a memorandum you had authorized me to send to you resulted in Solar Physics work being started in India.
>
> Owing to that work and other observations made in the Solar Physics Observatory at Kensington the riddle of the probable times of occurrence of Indian Famines has now been read, and they can be for the future accurately predicted, though not yet in the various regions. The Nile failures follow the same law. We may therefore hope that help is also in sight in relation to the droughts in our Colonies.[8]

Lockyer's friends showered him with congratulations. Sir John Eliot, who had supplied the Indian meteorological data (and who not long afterwards was coopted to the Solar Physics Committee), wrote to him enthusiastically. Langley, in the United States, remarked,

> It is an amazing thing to me that the enormous *utility* of such work as yours on the sun's connection with the conditions which bring famine or plenty, to India, for instance—the immense *utility* of like studies is lost sight of by almost all astronomers.[9]

Langley had reason to be pleased with Lockyer's results. He was at this time running into the same kind of criticism over the solar observatory he had established at the Smithsonian as Lockyer had experienced over the years regarding the observatory at South Kensington. In establishing his observatory, Langley, like Lockyer, had stressed its practical applications via solar-terrestrial studies, and the Federal authorities were now beginning to press him for some sign of applicable results. Lockyer's claim of a new breakthrough therefore came at a convenient time for Langley. In 1901, Langley wrote to Lockyer asking if he would draft a letter of support for the work of the Smithsonian Astrophysical Observatory, which could be shown to dubious members of the Senate.[10] Lockyer's reply, together with those of some other British and American scientists, may have helped swing the day in Langley's favour. At any rate, the Astrophysical Observatory survived.

Lockyer, meanwhile, was still seeking for possible additional methods of measuring changes in the heat emitted from the Sun besides the appearance of sunspot lines. One possibility he had in mind was to study variations in the number, or distribution, of solar prominences. A major criticism of Lockyer's correlations of sunspot changes with terrestrial weather had been that spots formed too small a part of the Sun's surface to produce the major results he ascribed to them. The introduction of the spectroheliograph by Hale and Deslandres during the 1890s had demonstrated, on the other hand, that prominences, or related structures, could cover a large proportion of the Sun's disc. Hence, they could not be criticized on this score. Moreover, the number and distribution of prominences round the solar limb had been mapped, especially by the Italians, since the 1870s; so there was plenty of material at hand for a statistical study.

The examination of the Indian meteorological data, mainly carried out by Jim, convinced Lockyer that atmospheric pressure might provide a better terrestrial reflection of solar change than rainfall, because, he believed, the pressure responded more rapidly to solar variations. Having studied sunspot numbers and rainfall in India, therefore, he next began a new correlation of solar prominence numbers and average atmospheric pressure changes over the same country. It soon became apparent that there was a prominence cycle which was distinct from the sunspot cycle. The number of prominences visible reached a peak at the same time as the spot maximum, but it also passed through another subsidiary maximum whilst the spot cycle was declining towards minimum. In fact, Lockyer decided he could trace a prominence cycle with a period of about one-third the sunspot cycle, and that this correlated very well with observed cyclic changes in atmospheric pressure over India.

Heartened by this result, Lockyer and his son looked around for other places on Earth for which long series of atmospheric pressure measurements were available, so that they could further test out their hypothesis. One place in Europe for which they had information was Cordoba in Spain, and an analysis of the data there seemed to indicate a similar cyclic pressure change. There was, however, an important difference : the pressure cycle in Spain appeared to have the opposite sense from that in India. When the atmospheric pressure was at a maximum over India, it was a minimum over Spain, and vice versa. During the early years of the present century, Lockyer and Jim extended their analysis of atmospheric pressure data to other stations throughout the world. In each case they found a cyclic variation, sometimes in phase with the Indian cycle, sometimes with the Spanish. They explained the existence of these two systems by what they called the barometric 'see-saw'. If the effect of the Sun was to create, say, a lower pressure at some point on the Earth's surface, this would necessarily be balanced by a higher pressure somewhere else. As the Sun's effect varied, so the atmospheric pressure at these two places would oscillate.

Although Lockyer thus regarded the pressure as a more useful yardstick than the rainfall, his main concern remained the possible practical applications of his results. He therefore spent some time on considering how pressure and rainfall fluctuations were related. Since there was considerable meteorological information available for the Thames basin, he and Jim examined the rainfall and pressure in this region over an extended period, and showed that the amount of rainfall varied inversely with the ambient atmospheric pressure. They also demonstrated that changes in the volume of water flowing down the Thames showed a five months' lag on the rainfall.

Along with the meteorological data on pressure and rainfall, information on terrestrial magnetism and aurorae were also being collected at South Kensington for comparison with the solar observations. In fact, the Solar Physics Observatory had now become a major centre for the collection of material relevant to the study of solar–terrestrial relationships. As a result, Lockyer was involved in a series of organisational problems that faced both meteorology and solar physics during the first decade of this century. In 1903, he was called to give evidence at a Government inquiry into the use the Meteorological Council was making of its Parliamentary grant. He insisted that the funds were being well used, and affirmed that the forecasts being provided by the Meteorological Office were as accurate as could reasonably be expected. 'The common distrust of the prophecies of the Meteorological Office is entirely baseless. I myself always take out my stick or my umbrella according to the Meteorological Office forecast when I go to work in the morning.'[11]

Lockyer naturally sympathised with the problems of the meteorologists in their efforts to extract money from the Government, more so since he still continued to have his own troubles with the Treasury at intervals. He may well have felt that Napier Shaw's description of life at the Meteorological Office fitted the history of the Solar Physics Observatory just as well.

> When one has gone on for years in the atmosphere that has prevailed at this Office where discouragements of all kinds have to be faced and generally speaking borne with a grin, one cannot avoid recognising even a small expansion as being something gained that may lead to more favourable conditions.[12]

Lockyer felt the real drawback to the Meteorological Office forecasts was that they were not being properly distributed. He advised Shaw to try and install signal stations for meteorological observation, which could communicate their results by the method of wireless telegraphy recently introduced by Marconi. Marconi had set up his Wireless Telegraph Company in London in the late nineties, but had first really hit the headlines in 1901, when he managed to transmit a message from Cornwall to Newfoundland. Lockyer rapidly developed a high regard for Marconi, whose work, he thought, was being under-rated by most contemporary scientists, and the two men were soon on cordial terms.

In 1903, there was a session of the International Meteorological Committee at Southport, in conjunction with the meeting of the British Association. Lockyer was asked to address the meteorologists concerning recent work at South Kensington on solar–terrestrial relations. As a result of the ensuing discussion, it was decided to establish a special commission to consider the relationship of terrestrial and solar phenomena. This commission met the following year, at the British Association meeting in Cambridge, and appointed Lockyer as its President. It had been decided that a data centre was needed which would tabulate and compare relevant information on solar–terrestrial phenomena, to be sent in by the various members of the international body. Almost inevitably, this task fell to the lot of South Kensington, which thus became the recognised world centre for this type of data. One result of the consequent influx of data to South Kensington was that, in the latter part of the decade, Lockyer initiated a new series of studies of changes in the general atmospheric conditions over the Earth as a whole.

It is clear that meteorologists were considerably interested in Lockyer's ideas on the influence of the Sun on the Earth's atmosphere, and, indeed, there had been a growing interest in the existence of short-term climatic cycles since at least the 1890s. This is not to say, however, that the scientific world in general accepted his views. On the contrary, despite his efforts, the overall attitude at the end of the decade was as lukewarm as it had been at the beginning. The opinion, in 1901, was that, 'it cannot be said that any very definite result has been reached, or that there is any accepted belief among men of science as to the validity of the hypothesis.'[13] Similarly, in 1911, Shaw, speaking as an expert witness, said, 'I keep a perfectly open mind . . . but I should say that no definite relation has been established.'[14]

In understanding this cautious attitude, it must be remarked that agreement with the South Kensington results depended on a willingness to accept their statistical probability, since there were always greater or smaller deviations from true coincidence. This uncertainty was increased by the fact that the climatological data could be processed in more than one way, so producing slightly different results. Moreover, it seemed suspicious that the number of periodicities found in the data grew, the longer the data were studied. Thus, besides finding a different length for the prominence cycle as compared with the sunspot cycle, the Lockyers also noted different periods superimposed on the normal 11-year cycle for sunspots. Furthermore, their analysis of atmospheric pressure changes suggested a long-period variation of about 19 years, which did not appear to be directly correlated with any solar phenomenon. On the other hand, though there were general reservations, few scientists were prepared to say outright that there was no connection betwen solar changes and terrestrial weather, and some meteorologists were definitely prepared to accept it as a working hypothesis. Lockyer, therefore, ended the first decade of the century reasonably satisfied with the developments in solar–terrestrial relations.

But he grew increasingly dissatisfied during the decade with the amount of solar physics that could be done at South Kensington. His unhappiness was fed by the rapid growth of the subject in the United States, more especially, by the spectacular advances initiated by his young friend, G. E. Hale. These had started back in 1890, when Hale, still a student at Massachusetts Institute of Technology, had worked out the principle of the spectroheliograph. In one sense, this instrument extended the idea that Lockyer had helped introduce into astronomy, the observation of the Sun in the light of a single spectral line. However, whereas Lockyer's method had permitted the visual examination of the chromosphere and prominences only round the solar limb, Hale's more complex approach made it possible to record both of these photographically across the disc of the Sun. The principle of the device had been suggested long before by Janssen in France, and, almost simultaneously with Hale, another French solar physicist, Deslandres, managed to construct a similar, though not identical, instrument. The resulting prolonged priority dispute between Hale and Deslandres was somewhat embarrassing to Lockyer since he was on good terms with both, and so found himself in the rather unusual position of acting as mediator.

From 1896 onwards, Lockyer was busy seeing whether a spectroheliograph could be installed at South Kensington, and in 1901 actually managed to persuade the Treasury to grant £1000 to pay the costs of building such an instrument. Hale complimented Lockyer on his new apparatus, but it is evident that the poor meteorological conditions at the Solar Physics Observatory site limited the results that could be obtained. Jim Lockyer emphasized to Hale that the quality of the South Kensington pictures was appreciably lower than the standard being attained in the United States.[15]

During the 1890s, Hale branched out in other directions. First he founded a new journal for the study of astrophysics, which rapidly came to be the main source of information on developments in this field. Then, on his appointment as the first Director of the new Yerkes Observatory in Wisconsin, he used his spectroheliograph to begin the first ever regular survey of the entire solar chromosphere. (C. T. Yerkes, who provided the money for the observatory, was an American millionaire who subsequently became involved in financing underground railways in London. During the first decade of the present century, he bought up both the District line and the Brompton and Piccadilly Circus Railway, thus owning all the underground access routes to South Kensington.) But Hale was unhappy both with the instrumental facilities—he thought that a new type of 'stationary' telescope would be more convenient for solar work—and with the meteorological conditions at Yerkes. He was therefore soon looking around for a new site. In 1902, he was invited to become a member of an advisory committee on astronomy being established by the recently created Carnegie Institution of Washington. His work on this committee soon convinced him that the Carnegie Institution might be prepared to pay for a new observatory devoted to solar and stellar astrophysics. Early in 1903, he began to write round to a few of the leading astrophysicists in the world, outlining his plans for a new observatory, and requesting their comments and advice. One of those he approached was, inevitably, Lockyer, who wrote back cordially welcoming and approving of his scheme.

One can hardly blame Lockyer if there was an occasional tinge of envy in his comments on the new solar observatory which Hale now proceeded to set up on Mount Wilson in California, and which, within a few years of its founding, came to be regarded as the leading observatory of its type in the world. The two men held several discussions together at just the time that the work at Mount Wilson was getting under way, for, in the autumn of 1905, there was a meeting of the International Union for Co-operation in Solar Research* at Oxford, which both Hale and Lockyer attended. The Union was in many ways another of Hale's achievements, for he had engineered its first meeting in the United States the year before. Lockyer became an immediate, and keen, adherent.

One topic that certainly cropped up in these conversations was the question of selecting sites for new observatories. Observational conditions at South Kensington were becoming noticeably worse as time passed. Not only were the lights and smoke of London increasing, but the growth of trees and the appearance of new buildings on the South Kensington site were appreciably affecting the extent of the observatory's skyline. These developments made it increasingly obvious that, if the Solar Physics Observatory was to continue its observational programme, it would have to find a new site. The immediate problem seemed to be to convince the authorities of the need for such a move. In the event, however, it was the

* This subsequently developed into the present International Astronomical Union.

Board of Education who moved first. In 1906, the Board advised the Solar Physics Committee that it was considering the erection of new buildings for the Royal College of Science, as a result of the reorganisations then under way. The only suitable space was provided by the site of the Observatory. The Board therefore asked the Committee to consider the possibility of transferring the Observatory away from South Kensington altogether.

To the Committee, this seemed a heaven-sent opportunity for placing the Observatory on a more reasonable site. Lockyer, too, was well pleased with the new situation, though he also immediately stressed the limitations which would have to be imposed when selecting the new position. One point he had emphasized to Hale when the latter was considering sites in the United States was that an observatory should not be too far away from a large town, 'as modern work requires the investigation to be in touch and personal contact with scientific men for purposes of mutual assistance and advice.'[16] On the other hand, the site should be at an elevation of at least 250 feet, and away from smoke and river valleys.

Lockyer believed that it should be possible to find a place satisfying these conditions within striking distance of London, where most of his scientific contacts were. At about this time, there were several sites round London owned by various Government departments, especially the War Office, which had become surplus to their requirements. Jim was dispatched to investigate whether any of them might prove suitable for a solar observatory. Simultaneously, the solar researches at South Kensington were curtailed pending the expected move. Jim wrote to Fowler in 1906,

> After consultation with Sir Norman Lockyer it has been decided, in consequence of the probable removal of this observatory to some locality outside London in the near future, that no new work regarding sunspot spectra will be commenced at present.[17]

This decision was a blow to Fowler. He had just made the important discovery that some of the lines in sunspot spectra were due to the presence of molecules, a discovery that Hale made almost simultaneously in California. This result needed to be urgently followed up by comparisons of solar and laboratory spectra. Hale had facilities for both, but Fowler, with the Solar Physics Observatory not cooperating, was in difficulties. (Identification of molecular lines in sunspot spectra was, incidentally, the final and conclusive piece of evidence that sunspots were cooler than the general solar surface—not hotter, as Lockyer had supposed years before.)

By January 1907, Jim had inspected all the available sites. Lockyer, going over his results, concluded that a piece of land owned by the War Office at Fosterdown, near Caterham in Surrey, offered the best prospects for a solar observatory. Jim recorded the caretaker there as saying that he had, 'never seen or smelt London smoke'.[18] The Solar Physics Committee agreed with Lockyer's choice, and recommended to the Board of Education that the observatory should be transferred from South Kensington to Fosterdown.

Here the matter rested for over two years. Some progress was made in this time. Haldane, the person ultimately responsible for the disposal of War Office land, was very happy to help Lockyer, and signified his willingness to hand over the land concerned for the erection of an observatory. By June 1909, the Solar Physics Committee felt that official prevarication had gone on for long enough, and they reminded the Board of Education of the 1907 decision to re-site the observatory. The Board's answer, though it came as a bombshell, at least to Lockyer, indicated clearly enough why the tempo of development had been so slow. The Committee was told that,

> It has been decided that the appliances in the Observatory should be handed over, with an appropriate annual grant-in-aid for maintenance, salaries, etc., to some responsible body or institution whose other interests and activities are sufficiently akin to those of the Observatory to afford a guarantee that the grant would be expended to advantage in the special field for which it is intended.
>
> The University of Cambridge, where a Professorship of Astro-physics has recently been established, appears to possess exceptional opportunities of discharging to the public advantage such a trust as is here contemplated.[19]

How was it that Cambridge suddenly came to be specified as the most suitable place for a solar physics observatory? The story really begins with the great Newall refractor which Lockyer had used years before at Gateshead. In 1889, R. S. Newall had written to the Vice-Chancellor of Cambridge University offering the telescope with dome and other accessories as a gift to the University. He noted in doing so that a part of the reason was the unfavourable weather conditions at his own site near Gateshead, and part was his wish that astrophysics should be cultivated at Cambridge. A University committee appointed to report on the offer recommended acceptance, but the University decided against this as there was no source of money for staffing, or maintaining, the telescope. Meanwhile, R. S. Newall had died. His son, H. F. Newall, who was then working under J. J. Thomson in the Cavendish, came forward, and offered both to provide money for the maintenance of the telescope, and to act as unpaid observer, himself, for five years. The University naturally agreed to this generous proposal, and, in 1891, the Newall refractor was set up in Cambridge. H. F. Newall, with his background in physics, proceeded to promote astrophysics at Cambridge as his father had hoped. He was personally successful in this endeavour, being appointed Assistant Director of the Cambridge Observatory in 1904 (still without stipend), and finally Professor of Astrophysics in June 1909. Thus by the end of of the first decade of the present century, Cambridge had established itself as a possible centre for astrophysical research.

The development of astrophysics at Cambridge does not in itself, however, explain the sudden change of mind on the part of the Board of Education. Here, it is almost certainly not accidental that the new decision by the Board coincided in time with the creation of Newall's Chair at Cambridge. It seems that one or two interested scientists had been intriguing both in Cambridge and at the Board of Education. They wished astrophysics to be developed at Cambridge, and the transferral there of the Solar Physics Observatory offered an obvious way of achieving this. The leader in these manoeuvres was probably J. Larmor, Stokes' successor as Lucasian professor of mathematics at Cambridge, and a member of the Solar Physics Committee. Newall, who was also on the Solar Physics Committee, does not seem to have been so involved in the early stages, but his diary for 1909 shows Larmor visiting him in the winter of that year, and urging him to do everything possible to ensure the transfer of the Solar Physics Observatory to Cambridge.[20]

Lockyer was utterly taken aback by the new proposal. He had built up the Solar Physics Observatory as a forerunner of the sort of Government-aided research institution he had always held to be vital for scientific development in Britain. The idea of it passing into private hands dismayed him. Although the Board of Education's suggestion obviously found favour with some members of the Solar Physics Committee, Lockyer managed to persuade the Committee as a whole to return a cautious response.

> The Committee recognise that great advantage both as to expert direction and as to scientific life would be secured for the Observatory by the proposed transference to Cambridge, but they feel that the resulting gain in the work of the Observatory would depend upon questions of organisation, as to which further information is necessary before they could give a confident opinion on the matter.[21]

They went on to point to a possible alternative, the re-siting of the Observatory at Fosterdown, as previously planned, but now with its control transferred to the Meteorological Committee. The Observatory was currently working so closely with the Meteorological Office—Lockyer had one of his computing assistants more or less permanently sited there—that such a transfer made quite reasonable sense. Rather than pressing this alternative, however, the Solar Physics Committee suggested that a small impartial committee should be formed to investigate the comparative advantages offered by the sites at Fosterdown and Cambridge.

The President of the Board of Education agreed to this proposal, and set up a committee under the chairmanship of Sir Thomas Heath, Assistant Secretary to the Treasury (though better remembered as one of the great authorities on Greek mathematics). The other members were F. W. Dyson, the new Astronomer Royal (he took over from Christie in 1910), R. T. Glazebrook, Director of the National Physical Laboratory, and Arthur Schuster, with F. G. Ogilivie (who was appointed Director of the Science Museum at South Kensington in 1911) as Secretary. Lockyer might reasonably have complained that this hardly made up an impartial committee. All four of its members were Cambridge men—even Schuster, though educated elsewhere, had spent five years working in the Cavendish. But Lockyer's attention was otherwise engaged.

The committee took an unconscionable time to get together, and, towards the end of 1910, Lockyer heard that, unless some immediate action was taken, the issue would be prejudiced. The War Office, presumably irritated by the long delay, had decided to put up the site at Fosterdown for public auction. He proved equal to this new emergency, getting the British Science Guild, led by Sir David Gill, to organise a memorial to Prime Minister Asquith from amongst his numerous friends, both scientific and political. They called for the land to be withheld from sale until such time as the Board of Education's committee reported. This request was agreed to, and the committee finally met in the spring of 1911.

Only three witnesses were examined, Lockyer, Newall and Napier Shaw. Lockyer's comments were devoted mainly to the importance of continuity in both observing and in the reduction of observations. But he also stressed the importance of close contact with the Meteorological Office, and the need for an observatory to be on as high ground as possible. He finally remarked that he thought it invidious to single out a particular university to be given an entire observatory together with all the running costs from Government funds. All these factors, of course, militated against Cambridge as a possible site, and it was no surprise when he concluded firmly that Fosterdown was the only reasonable choice.

There was one rather unusual point in Lockyer's evidence. He insisted strongly that the Solar Physics Observatory did not need a director, like himself, in charge of all aspects of the work. It could function satisfactorily with an administrative director under the control of a board of scientific specialists. This idea was not well received by the members of the committee examining him. They remarked that this was essentially the system tried in the early days of the Solar Physics Observatory, but it had not then proved very successful.

The point behind Lockyer's remarks was almost certainly concern for the future of his son, Jim. Jim had been his Chief Assistant at South Kensington since the nineties, and it seems to have been in Lockyer's mind that he would retire when the Observatory transferred to Fosterdown, allowing Jim to take his place. Lockyer had, in fact, been trying to help his son's advancement in the scientific world during the previous decade, perhaps with the Directorship ultimately in view. A major disappointment a few years before had been his failure to have Jim elected into the Royal Society. This was relevant to the Board of Education's investigation, for two of the members of the committee had been involved. Lockyer had written to various of his friends who were Fellows of the Royal Society, and whose interests lay in the field of astrophysics, asking them to back his son's candidature. Some, such as David Gill and George Darwin, agreed. Others, including Dyson, refused, and Schuster, although he finally agreed, evidently did so against his better judgement. In the event, Jim was not elected to the Royal Society, either then or later. A part of the reluctance to back Jim was evidently due to the belief that he had not carried out any major piece of research that qualified him for election. But there was also clearly a feeling that he was too closely connected with his father's researches, which were themselves under something of a cloud.

Dyson had, moreover, inherited a dispute with Lockyer from his predecessor, Christie. Both the Solar Physics Observatory and the Greenwich Observatory included, as part of their duties, measurements of sunspots, terrestrial magnetism and meteorology. From the beginning, therefore, there had been a possibility of an overlap in their functions. During the latter part of the nineteenth century, Lockyer's astrophysical interests had differed appreciably from the main concerns at Greenwich, so that no difficulties resulted. But when, in the late nineties, Lockyer's attention refocused on solar-terrestrial problems, and, in particular, on the collection of routine data concerning them, a clash with the Greenwich Observatory soon developed. It seemed to the staff at Greenwich that Lockyer was both using their results in his own compilations without acknowledgement, and also obtaining money from the Treasury for work that ought properly to have been done at Greenwich. Relations between Lockyer and Christie soon degenerated. In 1909, Christie wrote to the Secretary of the Solar Physics Committee, saying that he could not approve of the draft report drawn up by Lockyer.

> The Report is the basis of an application to the Treasury for an increased grant on account of the work done by Solar Physics Observatory and I should be putting myself in a false position with the Treasury if I endorsed claims which are made on behalf of the Solar Physics Observatory for work which is carried out at the Royal Observatory.[22]

Shortly afterwards, he wrote again, in even stronger terms, '. . . many of the Members of the Committee are, I believe, imperfectly informed. The Reports of 1889 and 1901 are calculated to mislead them as well as the Treasury.'[23] It is evident that, during the period when the future of the Solar Physics Observatory was being decided, the Astronomer Royal was most unhappy with the way in which it was currently being run.

The next witness to face the Board of Education's committee after Lockyer was Newall. He was evidently determined that, if the Observatory moved to Cambridge, it should be on his own terms. These included, in particular, a move away from routine and 'useful' research towards something more purely academic. He added that although the horizon at Cambridge was moderately clear for solar observation in the early morning, it might not necessarily continue to be so in the future (with the implication that, if that occurred, it might be necessary to curtail some of the solar observing). He also made it clear that he could not guarantee the jobs of any of the current members of the Solar Physics Observatory, and that, whereas Lockyer had estimated the cost of the move to Fosterdown at about £3000, the move to Cambridge would certainly not cost less than £5000.

The final witness was Shaw, who had been invited along to give the viewpoint of the meteorologists. He was not, he explained, particularly concerned whether the Observatory moved to Fosterdown or Cambridge. He simply wished to ensure that there should continue to be a governmental voice in its administration, so that the work of value to the Meteorological Office would be continued. He added, however, that he was not at all sure whether this would be possible if the Observatory became part of a university.

The Committee now deliberated for some six weeks before issuing its report. Part of this time was spent in trying to collect more comparative meteorological data on the two proposed sites. Schuster had asserted that, when on a visit to Fosterdown, he had been told that it was frequently misty there. But a part of the delay was due to disagreements both within the committee and outside. When the recommendations finally appeared in June 1911, it was found that not all the members of the committee had signed them. The majority report, signed by Heath, Dyson and Schuster, recommended that the Solar Physics Observatory should be transferred to Cambridge, and that the Professor of Astrophysics there should become its new Director. They pointed out that solar observing conditions at the two sites appeared to be comparable, and therefore the advantages of associating with a university seemed decisive in favour of Cambridge. To meet Shaw's reservations, it was recommended that the observatory should be required to undertake a definite amount of routine investigation. To ensure that this was done satisfactorily, Cambridge University should be required to set up an advisory committee, of which both the Astronomer Royal and the Director of the Meteorological Office should be *ex officio* members. This majority report was followed by a minority report, signed by Glazebrook, which declared briefly that the evidence he had heard indicated to him that a transfer to the Fosterdown site would be preferable.

During the next few months after the appearance of the report, Newall was busy negotiating the acceptance of its proposals with the University. In December 1911, Cambridge University formally accepted the Board of Education's offer, and the fate of the Solar Physics Observatory was finally sealed. Newall was soon arranging to visit South Kensington, both to make an inventory of the instrumentation there, and to look over the staff. Apart from Lockyer and his son, this consisted of two assistants and four computers who, despite their title, also acted as assistant observers. The assistants, F. E. Baxendall and C. P. Butler, had been at the Observatory for nearly 25 and 20 years respectively, the computers had been engaged during the previous decade. Newall ultimately decided to offer positions at Cambridge to the two assistants and the two senior computers. All accepted, in at least one case with a feeling of relief. Newall subsequently noted that the man 'speaks with evident feeling about Lockyer's theories'.[24]

As a sop for the meteorologists, it was also decided to establish a new post when the Solar Physics Observatory moved to Cambridge, that of Observer in Meteorological Physics, and C. T. R. Wilson was appointed to it. Wilson's main research interests were in atmospheric electricity and cloud physics. It was his work in the latter area that led to the development of the cloud chamber, a device vital for the growth of nuclear physics at Cambridge under Rutherford. Shaw, however, was by no means satisfied. The Solar Physics Observatory under Lockyer had become a centre for collecting information on world-wide weather changes. This activity was dropped on the transfer to Cambridge, and Shaw hardly considered the appointment of Wilson as adequate compensation. Indeed, there were various complaints during the first few years of the Observatory's existence at Cambridge. Not so much because the work became more academic, as Newall had foreshadowed, but because the Observatory became noticeably less productive after its transfer. In 1920, A. L. Cortie at Stonyhurst College wrote to Jim Lockyer.

> It seems to me that they have not done very much at Cambridge since the Solar Physics Observatory was placed there. But Cambridge strives to monopolise natural science, and no one is of much importance in scientific circles unless he has the Cambridge mark upon him.[25]

As was noted above, Lockyer had probably already decided before the question of Cambridge arose to relinquish the Directorship of the Observatory when it was transferred. But the problem of Jim Lockyer's future remained. Newall had made no commitment to take Jim on at Cambridge. He did, in fact, enquire whether Jim was interested in transferring, and was told that he would be prepared to come to Cambridge at the same salary he was earning at South Kensington (£450 p.a.). Whether, or not, Newall would have offered Jim a post is unknown, for in July 1912, Lockyer senior came back forcibly into the picture.

Lockyer had, of course, been both distressed and angered by what he regarded as the betrayal of his life's work in the assignment of the Solar Physics Observatory to Cambridge. Oddly enough, one of his oldest colleagues at South Kensington was placed in a very similar position at just the same time. When, in 1910, the link between the new Imperial College and the City and Guilds Institute was finally agreed, there was an immediate move to rationalise the course structure. It was decided that the chemistry course, lovingly devised over the years by Henry Armstrong, should be abolished. Consequently, in 1911, Armstrong was given six months' notice, with no pension rights. The difference in reaction of the two men is utterly typical of Lockyer's approach to life. Armstrong had gone into retirement. Lockyer, now aged 76, decided to fight back.

To begin with, there was just a hope that it might be possible to alter the Government's decision, if enough influence could be brought to bear. Several of his friends agreed with Glazebrook's minority report. Sir David Gill, another member of the Solar Physics Committee, wrote to Lockyer at the end of 1911.

> I feel that you and your work have been treated most unfairly—that the conclusion of the Committee is contrary to such evidence as has been collected —for I entirely concur in Glazebrook's view of it. Evidence on a much broader basis and of a very much more conclusive character was required before the organisation which you founded and carried on so successfully for so many years was ruthlessly upset.[26]

But Gill also advised the angry Lockyer not to be too virulent in his protests against the committee's recommendations, 'on the principle that you will catch more flies with sugar than with vinegar.'[27]

Before the committee reported, Lockyer had already been gathering support abroad for the transfer to Fosterdown, his leading supporter being Deslandres in France. Deslandres told him, however, that it would be difficult for foreign scientists to interfere directly in English affairs, and so there was little chance of support for a specific attack on Schuster, Dyson or Newall.[28] Rather he should try to get the leading astrophysicists from a number of different countries to sign a general protest. After the final decision of the committee had been handed down, Deslandres wrote again, strongly sympathizing with Lockyer, and adding that he believed that the result had been influenced by prejudice against Lockyer.

> In 1910 I had occasion to attend a conference on solar research at the Royal Institution, and, in conversing with various scientists, I was struck by the hostility which you evoked amongst them. The following day, I encountered Sir William Crookes at lunch and asked him about this unfortunate attitude; he replied: the strong man always has enemies.[29]

Deslandres attempted to console Lockyer further by telling him that the French universities were also trying to take over independent scientific establishments, though the main observatories at Paris and Meudon were managing to resist engulfment.[30]

One of the people in Britain to whom Lockyer turned for sympathy and advice was F. K. McClean. In April 1911, McClean had gone on a solar eclipse expedition to Tonga in the South Pacific organised by Lockyer. (He was also busy during the year training naval officers in the gentle art of flying.) In one sense, the McClean family had been partly responsible for the development of solar physics at Cambridge. A bequest to Cambridge University by Frank McClean in 1904 had been used for a new solar telescope and spectrograph. Newall's work with this, and the good results he claimed were obtainable with it on the Cambridge site, played a significant part in the final choice of Cambridge for the Solar Physics Observatory. Now, in conversation with F. K. McClean, Lockyer's thoughts turned away from trying to alter the Board of Education's decision towards the idea of founding a new observatory, to make up for the loss (as Lockyer saw it) of the observatory at South Kensington. A major objection to Cambridge had been its low elevation. Lockyer therefore began to consider the possibility of building a hill observatory in England for astrophysical research. (The term, 'hill observatory', had been introduced in India to denote an observatory at some height above the plains.)

For such an observatory, the site selected itself. Lady Lockyer had for many years been a visitor to Sidmouth in Devon. Her sister had suffered from poor health as a girl, and they had both spent long holidays in Sidmouth with grandparents who lived there. After their marriage, Lockyer and she spent several of their vacations in Sidmouth, and eventually decided to build themselves a house on Salcombe Hill, overlooking Sidmouth, on some land owned by Lady Lockyer. Lockyer decided that this would do very well for the site of a hill observatory. Although it was only about 550 feet above sea level, this was considerably better than either South Kensington, which was at an elevation of only 25 feet above sea level, or Cambridge.

In July 1912, Lockyer organised a meeting of his supporters at his London house, and there (over the inevitable clay pipes) they agreed to launch an appeal for funds for the proposed new observatory. Many of Lockyer's oldest friends, such as Abney, Crookes and Ramsay, were present, and the McClean family was represented by W. N. McClean (the brother of F. K. McClean). He, together with Gill, took the leading part in drawing up and circulating the appeal. Nevertheless, one can surely hear Lockyer's voice in places, for example, in the sweeping first paragraph of the appeal.

> Astrophysics, as an exact science, may be said to have had its origin in England—and yet at the present moment there is no observatory in this country that is adequately equipped and at the same time favourably situated for its pursuit.[31]

The appeal goes on to note that a considerable amount of money has already been raised. Lockyer had donated £4000, Lady Lockyer £1000, and F. K. McClean £9000.

The question that had to be resolved first was whether sufficient instrumentation would be available to equip the observatory. Lockyer still possessed the 30-inch reflector he had used at Westgate-on-Sea, but the largest telescope in use at South Kensington was a 36-inch reflector. This latter telescope did not belong to the Solar Physics Observatory, but was on loan from the Science Museum. It had not been of great use in the nineties, as its driving mechanism had continually malfunctioned. Since then it had been working better, and Lockyer evidently hoped that, since the avowed intention of the move to Cambridge was to improve solar observation, the telescope, which was employed for stellar spectroscopy, would be available for him to borrow and take to Sidmouth. Newall, however, had no intention of losing such an instrument, and took over the 36-inch telescope with the other equipment of the Solar Physics Observatory. But he agreed to let Lockyer have the 10-inch refractor, and this, together with Lockyer's own instrumentation and some more donated by F. K. McClean, provided a sufficient foundation for a new observatory. One great disappointment was that most of the collection of stellar spectra that Lockyer had built up at South Kensington went to Cambridge. As it turned out, Lockyer was still a good deal better off than Fowler, who had been using some of the equipment of the Solar Physics Observatory for his own spectroscopic studies in the laboratory. After the spoils had been shared out, Fowler found himself left with very little major instrumentation, and had to start building up again almost from scratch.

The appeal for funds was going ahead quite well (at least so far as publicity was concerned). After some hesitation, Prince Arthur of Connaught agreed to become President of the Appeals Committee, which was officially set up early in 1913. At the same time, many of Lockyer's friends in various fields were writing to add their names to the list of sympathisers, Lord Avebury, Lauder Brunton, Philip Magnus and Fowler amongst others. Abroad, other friends were also rallying to Lockyer's support, Hale and E. C. Pickering in the United States and Deslandres in France (Janssen had died a few years before). Work therefore went ahead on the Hill Observatory, as it was now named, at Sidmouth. Meanwhile, dismantlement of the old Solar Physics Laboratory at South Kensington continued during the spring and summer of 1913, and the last meeting of the Solar Physics Committee was held. By this time, a new Committee to supervise the work at Cambridge had been appointed, though it actually contained many of the members of the old committee.

Arrangements now had to be made for the running of the Hill Observatory. Lockyer, reinvigorated, was to be its Director. Jim, all thought of a transfer to Cambridge thrust aside, was to continue as his chief assistant. It was decided that one assistant observer should be recruited immediately. At first it seemed that one of the computers at South Kensington, Rolston, would prefer a job at Sidmouth to one in Cambridge. (Newall noted: 'He [Rolston] is evidently fearful of East wind at Cambridge for his children.'[32]) But, in the end, Rolston decided to go to Cambridge, and Goodson, one of the junior computers who had not been taken on at Cambridge (Newall described him as: 'Nervous, excitable'[33]) was engaged. Limited observations were already being made in 1913, and, in the following year, it was felt necessary to appoint an additional assistant.

It had been decided that the Hill Observatory should be run as a corporation, and preparations were under way in 1914 to apply to the Board of Trade for incorporation. These plans were scotched by the outbreak of the first World War, which also led fairly rapidly to the effective cessation of the observing programme. In 1915, Johnson, the junior assistant, left to join the Royal Flying Corps, and Jim was commissioned into the Royal Naval Air Service (as was his friend, F. K. McClean). The following year, Goodson departed to work on the manufacture of munitions, and Lockyer was left in isolation at the observatory. Observations had, in fact, been drastically cut back before the first winter of the war, as the result of a request from the Chief Constable of Devon.

I understand that you have an observatory at your house for the purpose of studying the stars. Would it be asking too much my asking you to kindly put off your photographic study of the stars till after the war is over, as it is creating a certain amount of alarm in Sidmouth, as it is a seacoast town?[34]

Lockyer's patriotic fervour was as strong as ever. Since star-gazing was out, he turned to the problems of coast-watching, and spent some of the war years working out how coast-watchers might most easily communicate with each other.

Another interlude in the war years was provided by Rider Haggard, who wrote to Lockyer in 1916, posing him a problem that had arisen during the writing of a new novel. The question was how a person who had slept for a very long period of time could check the length of that period from observations of the stars.

Could it be done in this way? My Superman, Oro by name, before he puts himself to sleep 250,000 years ago, after he has caused the deluge, makes a simple map of the portion of the heavens, or rather of the constellations necessary to his purpose as they were at that date—say on a sheet of metal. He also makes a map on another sheet of metal of the same constellations as his astronomical knowledge tells him they should be 250,000 years later—that is at the present time.

When Oro wakes up, he is naturally anxious to learn whether he has slept the exact 250,000 years . . . He produces his two maps and compares No. 2 with the existing positions of the chosen constellations, or constellation if only one is necessary. Then he triumphantly points out to his English discoverer, Arbuthnot, that he has hit off exactly.[35]

Lockyer turned to Arthur Eddington, who, as a member of the Society of Friends, was sitting out the war, almost alone, at the Cambridge Observatory, and asked for his cooperation. Eddington proposed a series of observations, starting from the motions of the stars, and working down to the relative positions of the Sun, Jupiter and Saturn, which would be capable of fixing the period of 250,000 years to within a month or two from astronomical observations only. Lockyer concurred, and Rider Haggard gratefully accepted their suggestions.

As was the case with several of his friends, Ramsay and Lankester, for example, Lockyer's attitude towards Germany, as a model to be imitated by England, underwent a drastic change with the coming of the war. The previous complimentary comparison between the organisation of German scientific and military effort, which had been so common in the latter part of the nineteenth century, now seemed to redound to the discredit of German science.

The greatest advances in scientific thought have not been made by members of the German race; nor have the earlier applications of science had Germany for their origin. So far as we can see at present, the restriction of the Teutons will relieve the world from a deluge of mediocrity. Much of their previous reputation has been due to Hebrews resident among them . . .[36]

Ironically, Lockyer's old mentor, R. B. Haldane, was subsequently to fall from power at the War Office because of his alleged partiality to things Germanic. Yet he had been one of the first to warn the scientific public, through the British Science Guild, of the dangers of too slavishly copying the German example. He had done this at a time when Lockyer and most of his friends were still sympathetic to Germany.

The war acted as a stimulant to the British Science Guild. As the points it had been hammering home during the previous ten years became more apparent to the informed public, its influence grew. Lockyer's age and poor health prevented him from playing an important part in the new activities of the Guild, but his ideas were still very much in the minds of members. As the President remarked in his annual address in 1915,

> I venture to believe that we could claim for Sir Norman Lockyer the character of a prophet, for foreseeing, as he appears to have done, the movements of the world which have come to pass since, and more especially the great need, in regard to English culture and education generally, for more thorough scientific training.[37]

The Guild was particularly active in pointing out the need to allow for the cessation of certain vital imports from Germany which would result from the war. In 1913, it was already emphasizing that the availability of nitrates for the explosive industry should be considered as a matter of urgency. Subsequently, the Guild called the Board of Trade's attention to the problems facing the optical industry, since the supply of high-quality glass had been cut off. Significantly, the Board of Trade responded by asking the Guild for its advice on what to do. When the Guild suggested that the National Physical Laboratory should be put in charge of supervising the manufacture of such glass, the Board of Trade ordered accordingly. There was, indeed, a new willingness on the part of Government and Civil Service to listen to the spokesmen of science. This, though it was frequently based on no very deep conviction as to the efficacy of science in wartime activities, was a considerable advance. It led, in fact, to what was certainly one of the greatest triumphs of the British Science Guild, the decision by the Government in 1915, to set up an Advisory Council for industrial and scientific research. This body, the forerunner of the Department of Scientific and Industrial Research, represented the first admission by the Government that it had a general duty, and need, to support the advancement of research and to take the advice of scientists. Of course, the Guild had not been the only source of agitation for the formation of such a body. Nevertheless, it was acknowledged at the time that it had been one of the most influential.

In the United States, Hale was preparing American scientists similarly for what he saw as the inevitable war with Germany. (He told Lockyer that the German ambassador should have been sent home after the sinking of the *Lusitania* in 1915, but the prevailing pacificism in the U.S.A. had prevented it.[38]) He had thrown himself with characteristic energy into the task of organising a National Research Council to co-ordinate the scientific resources of the country. In the summer of 1916, when he was in England in connection with these activities, he wrote to Lockyer, saying,

> Prompt action on the part of President Wilson enabled me to complete the organization of our National Research Council somewhat sooner than I expected, and I have come over here to learn what I can about the actual needs of a country under war conditions. Of course I have followed with great interest your leading articles in 'Nature'—they served me well, as I sent the standing advertisement, giving the full list of them, to President Wilson and also to Justice Hughes, the Republican candidate for the Presidency, to impress them with the importance of science to a nation.[39]

Hale remarked to Lockyer that he had been unable to keep up with advances in astronomical spectroscopy during the previous few years, but, he added, H. N. Russell's work at Princeton University seemed to indicate that the old idea of a single-branched evolutionary development of stars was unlikely to survive.[40] He was reflecting here the considerable change of attitude that took place during the early years of the twentieth century towards Lockyer's idea of a double-branched evolutionary curve. In 1905, the Danish astronomer, E. Hertzsprung, following up a spectroscopic clue in the Henry Draper catalogue, had provided evidence, independently of Lockyer, for a division between intrinsically bright and intrinsically faint stars of the same colour. His work was not published in an astronomical journal, and attracted less attention than it deserved. The major swing of opinion towards Lockyer's view did not take place until after 1912, when Russell, one of the rising theoretical astrophysicists in the U.S.A., pointed out that a sufficient number of stellar distances had been determined to show that yellow and red stars were either faint ('dwarfs') or bright ('giants').

Lockyer was naturally delighted by this support—though it followed a different line from his own approach—and when Russell was over in England, in 1913, the two held a long discussion on the evidence for a two-branched curve. Within the next few years most of the more influential astrophysicists in both the United States and Britain had accepted the idea. We have seen Hale's comments in 1916. In the same year, Eddington wrote to Lockyer, 'Although I find some difficulties in the theory of ascending and descending temperature stars, I believe that in the main it must be right.'[41] The observations now seemed to support the view of stars which Lockyer had propounded for decades. Despite this, there was no tendency amongst astronomers to revive belief in his meteoritic hypothesis. It had been too thoroughly disproved. Theoretical attempts to explain the double-branched curve were rather based on ideas of fluid spheres.

Lockyer, in fact, spent part of the war trying to salvage his system of stellar classification, since more recent spectroscopic work had shown various of his earlier suppositions to be incorrect. Not long before the war began, he received a letter from Fowler telling him that studies at South Kensington had definitely disproved the claim that some red stars showed bright carbon flutings in their spectra. Fowler had discovered, indeed, that there were no bright flutings at all. Lockyer had been looking at bits of the bright continuous spectrum which appeared between dark flutings due to titanium oxide. Lockyer was very unhappy with this result for it played havoc with his meteoritic hypothesis, but Fowler's evidence was entirely convincing.[42]

Lockyer's preoccupation with his usual scientific interests was not disturbed by the death of any of his family in the war, though two members, Jim and Hughes Lockyer, were intimately involved in it. (One son—Edmund—had died earlier at sea on his way back to Britain from Canada.) Jim was flying naval scout planes, and was placed in command of various anti-Zeppelin units in Britain, ending the war as a major. Hughes made an even greater mark in the wartime Navy. The battleship which he commanded played an important role in the landings at Gallipoli in 1915, and in the subsequent New Year's Honours List he was created a C.B.

Lockyer was by this time in touch with a very wide family circle, mainly as a result of his Presidency of the British Association earlier in the century. It was customary at the meeting to fly a flag bearing the family arms of that year's President. In ensuring that his family were correctly represented, Lockyer had become greatly interested in tracking down his own family tree, and had therefore written far and wide to distant, and hitherto unknown, branches of his family. Some of the results were unexpected. He heard during the war that a distant relative on his mother's side of the family, the son of a cousin who had emigrated to the United States, was no longer in touch because he had been ambushed and killed by train bandits. Another result of Lockyer's enhanced interest in his family history was that he also fostered a connection during the war years with one or two schools in Dorset which had been founded by his great-grandfather.

If the war years had little effect on Lockyer's immediate family circle, they saw the death of some of his oldest friends. Lauder Brunton and Ramsay, for example, both died in 1916. Other old acquaintances ceased to be in close touch, mainly, of course, because of old age. In at least one case there was a more romantic reason. Kropotkin, the Russian revolutionary, had been an exile in London since the seventies, and during that period he had undertaken various assignments for *Nature.* In 1917, he wrote to Lockyer bidding him an affectionate farewell; he was about to return to Russia following the downfall of the Tsar. Some of the people who died had been leading supporters of the Hill Observatory when it was formed. The observatory had finally been incorporated in 1916, but as the war came to an end, it was clear that the council controlling the observatory would have to be completely reconstructed. Lockyer's old assistant at South Kensington, Richard Gregory, knighted in 1919, agreed to take charge of this work, and was appointed the new chairman. Indeed, Gregory was taking over Lockyer's role in various directions at this time. As one of Lockyer's Vice-Chairmen of Committees, he already had a major hand in the running of the British Science Guild. At the end of 1919, he also took over from Lockyer as editor of *Nature.* He had, of course, effectively been editor for some time, but Lockyer decided to resign officially on the fiftieth anniversary of the founding of the journal.

Work at the Hill Observatory resumed rather slowly. In May 1919, a new assistant, D. L. Edwards, was appointed on Fowler's advice. Jim was not demobbed until August of that year, spending the intervening period on the East coast as a meterological officer in the newly constituted R.A.F. His return to Sidmouth evidently depressed him. Not only was the financial state of the observatory insecure, but also the freedom from parental surveillance, which he had obviously enjoyed during the war years, was over. Early in 1920, W. N. McClean was still trying to console him, 'I wonder whether the Director really thinks you waste time at the R.A.S.! Perhaps he does not fancy the free lance!!'[43]

By this time, Lockyer had withdrawn from virtually all his other activities except that of Director of the Observatory. He was still mentally alert, and took a continuing interest in new developments in science, for example, in the attempts to obtain observational confirmation of Einstein's theory of general relativity at the solar eclipse of 1919. But his physical movement was now greatly restricted. He passed the summer of 1920 sitting in the garden or the house at Sidmouth, and there on 16 August he died peacefully. He was buried at the church of St Peter and St Mary in Salcombe Regis.

Lockyer's was a complex and often contradictory character, but his contemporaries were in doubt as to the impact of his ideas, regardless of whether they were accepted or rejected. The various obituaries, notices and articles concerning him which appeared in the months following his death bear evidence to this.[44] But perhaps the tribute which Lockyer, himself, would have liked best came in a personal letter from Hale to Lady Lockyer.

I remember with what fascination I read for the first time his 'Spectrum Analysis', 'Solar Physics', 'Chemistry of the Sun', and other books, to which I find myself repeatedly going back for fresh inspiration. He painted for me a great picture of the possibilities of astrophysical research, and stirred my enthusiasm as a boy, when I first began to dream of the role of the spectroscope in astronomy. Only recently I have been taking quotations from his 'Solar Physics' bearing on the nature of sun-spots, and I cannot open these books without a thrill of the old excitement that they brought to me many years ago. . . .

I have also shared Sir Norman's belief in the unlimited possibilities of science and research in the world's development, and consequently I followed with keen interest and much advantage the work he did through 'Nature' and the British Science Guild. I have often said that if I could take but one scientific journal it would be 'Nature', and I have admired in its pages the wide vision and editorial skill of its founder and dominant spirit.

His loss will be felt far outside the ranks of science by all those who have felt the influence he exerted on advancement in many fields.[45]

EPILOGUE

After Lockyer's death, Jim finally attained the position of director of the Hill Observatory, and shortly afterwards, perhaps in celebration, he married. The observatory itself was soon renamed 'The Norman Lockyer Observatory', but its finances continued to be too limited to permit the kind of expansion that was needed. When Jim Lockyer died in 1934, the observatory was kept going on similar lines for the next twenty years or so by his assistant D. L. Edwards. In 1948, the running of the observatory was handed over to the University College of the South West (now the University of Exeter), and though it has subsequently ceased functioning as an astronomical observatory, it is actually busier than before, since the University is using it as a geophysical out-station.

Five of Lockyer's other children, besides Jim, survived him. Norman died fairly soon after him in 1922, Ormonde in 1935, Hughes in 1941, and Rosaline in 1957. Winifred remained at Sidmouth looking after Lady Lockyer and Jim until her death in 1934. Lady Lockyer, herself, survived until the Second World War, dying at the age of 91. She remained a notable figure in the locality, continuing to have a say in the governance of the Norman Lockyer Observatory, and in Unitarian circles in the county, up to the end.

Lockyer's main scientific interests have fared rather variously. The idea of a double-branched curve of stellar evolution was generally accepted during the twenties, though the meteoritic hypothesis was never revived. But other attempts at a comprehensive theory of stars also failed until the 1940s and 1950s. During these latter years the nuclear sources of stellar energy were finally defined, thus establishing the basis for our present understanding of stellar evolution. Despite this entire eclipse of the meteoritic hypothesis, it is worth remembering that Lockyer was one of the pioneers of meteoritic studies in Britain. Some of the problems he investigated have only been taken up again in detail during the past two decades.

Lockyer's final version of the dissociation hypothesis (as modified by Fowler's work) also triumphed during the twenties. Shortly after Lockyer's death, an Indian physicist, M. N. Saha, came to work under Fowler at Imperial College. The paper he wrote during this visit, 'On a physical theory of stellar spectra' showed how the spectra of stars could be understood in terms of the new quantum theory of the atom together with the dissociation hypothesis. After some initial opposition, his results were rapidly accepted. The theory showed that both temperature and pressure affected the dissociation of atoms in stellar atmospheres. So both Lockyer and his opponents had been partly right. It is only fair to Lockyer to add that the influence of temperature on stellar spectra is much more marked than that of pressure.

Other aspects of Lockyer's work—as on the structure of the solar atmosphere, or on the classification of stellar spectra—were superseded and forgotten almost by the time of his death. Thus Saha had originally intended to call his paper, 'On the Harvard classification of stars'. Fowler had to point out to him that some of the pioneering research in this area had been carried out at South Kensington.

Two other areas in which Lockyer was interested—the orientation of ancient monuments and the influence of the Sun on terrestrial weather—have remained controversial. During the decades after his death, scientific interest in these waned, but during the last decade there has been a revival of activity, and articles on both problems have, for example, been appearing again in *Nature*. The trouble in both cases is that interpretation of the measurements involved seems to depend partly on subjective opinion. Even when there is agreement on the data, there need be no agreement on their interpretation. Nevertheless, Lockyer's work in both these areas might now bear re-examination.

The British Science Guild continued its operations during the twenties under the guidance particularly of Gregory. In 1925, an annual 'Norman Lockyer Lecture' was established (to be followed in 1929 by another annual lecture named after Alexander Pedler). But the Guild's impetus gradually declined, and this decline was hastened by a reviving interest in questions of science policy on the part of the British Association. The Guild, as we have seen, was started because the B.A. did not wish to involve itself in such questions. When, during the thirties, the B.A. did become involved in matters of science policy, the *raison d'être* of the Guild disappeared, and, in 1936, it amalgamated with the British Association.

Finally, and most importantly, there was *Nature*. As all scientists throughout the world are aware, this has, if anything, increased in importance and prestige since Lockyer's death, and now stands as the last great monument to his life's work.

SELECT BIBLIOGRAPHY

This brief bibliography consists mainly of books; where possible, ones that are moderately easily available. Unless otherwise indicated, they are published in London. Several of the titles mentioned are relevant to more than one chapter. This is especially true of the biographies and autobiographies of the various scientists. For further background material, reference may be made to *Nature* itself from 1869 onwards, and also to the annual British Association *Reports.*

Chapter I

THE MILITANT CIVIL SERVANT

Further details of William James can be found in: E. M. S. P[aine], *The Two James's and the Two Stephensons* (G. Phipps, 1861). For the War Office and Civil Service background, see: G. Cousins, *The Defenders: a History of the British Volunteer* (Muller, 1968); L. Wolf, *Life of the First Marquess of Ripon* (Murray, 2 vols., 1921); M. Wright, *Treasury Control of the Civil Service: 1854–1874* (Clarendon Press, Oxford, 1969). For further relevant material on the Christian Socialists, see: E. C. Mack and W. H. G. Armytage, *Thomas Hughes* (Benn, 1952).

Chapter II

THE MAN OF LETTERS

Additional details on Alexander Macmillan and the Macmillan company can be found in: C. L. Graves, *Life and Letters of Alexander Macmillan* (Macmillan, 1910); C. Morgan, *The House of Macmillan: 1843–1943* (Macmillan, 1943). The history of the *Reader* is treated by J. F. Bryne, *The 'Reader': a Review of Literature, Science and the Arts, 1863–1867* (Northwestern University, U.S.A., unpublished Ph.D. thesis, 1964); and the events leading up to the foundation of *Nature*, together with the general publishing background of this venture, by R. M. McLeod, *Nature* Vol. 224, pp. 423–40 (1969). For T. H. Huxley, see: L. Huxley, *Life and Letters of Thomas Henry Huxley* (Macmillan, 2 vols., 1900). Some of Lockyer's contemporaries who were involved in his early publishing activities are described in the following biographies or autobiographies: F. Galton, *Memories of my Life* (Methuen, 1909); A. Geikie, *A Long Life's Work* (Macmillan, 1924); L. Huxley, *Life and Letters of Sir Joseph Dalton Hooker* (Murray, 2 vols., 1918); H. Spencer, *An Autobiography* (Williams and Norgate, 1904); C. G. Knott, *Life and Scientific Work of P. G. Tait* (Cambridge University Press, 1911).

Chapter III

THE MAN OF SCIENCE

For more extensive information on the study of the Sun during the nineteenth, and early twentieth, centuries, the following books may be consulted: A. C. Clerke, *A Popular History of Astronomy during the Nineteenth Century* (A. and C. Black, 1885 and subsequent editions); A. J. Meadows, *Early Solar Physics* (Pergamon, 1970); S. A. Mitchell, *Eclipses of the Sun* (Columbia University Press, New York, 1923; C. A. Young, *The Sun* (Kegan, Paul and Trench, 1882 and subsequent editions). A brief description of the developments in the Royal Society during Lockyer's lifetime is given in: D. Stimson, *Scientists and Amateurs: a History of the Royal Society* (Henry Schuman, New York, 1948). Biographies and autobiographies of influential scientists with whom Lockyer came into contact early in his research career include: *Sketches from the Life of Edward Frankland* (Spottiswoode, 1902); W. Airy (Ed.), *Autobiography of Sir George Biddell Airy* (Cambridge University Press, 1896); H. E. Roscoe, *The Life and Experiences of Sir Henry Enfield Roscoe* (Macmillan, 1906); S. P. Thompson, *The Life of William Thomson* (Macmillan, 2 vols., 1910). There is no full-scale biography of Huggins, but reference may be made to: E. W. Maunder, *Sir William Huggins and Spectroscopic Astronomy* (The People's Books, 1912).

Chapter IV

THE DEVONSHIRE COMMISSION

The main source of information is, of course, the series of eight reports published by the Royal Commission on Scientific Instruction and the Advancement of Science between 1872 and 1875. For a general discussion of the developments in scientific and technical education during Lockyer's lifetime, see : M. Argles, *South Kensington to Robbins* (Longmans, 1964); D. S. L. Cardwell, *The Organisation of Science in England: a Retrospect* (Heinemann, 1957). The parts played by the Prince Consort, The Duke of Devonshire, Playfair and Strange in the development of science policy and education are described in : J. G. Crowther, *Statesmen of Science* (Cresset, 1965). For Playfair, see also : W. Reid, *Memoirs and Correspondence of Lyon Playfair* (Cassell, 1900). Appleton, who was Lockyer's colleague in agitating for an extension of science teaching, and, in some ways, his competitor as an editor, is briefly described in : J. H. Appleton and A. H. Sayce, *Dr. Appleton: his Life and Literary Relics* (Trubner, 1881). A member of the Devonshire Commission with whom Lockyer was in contact for many years was Lubbock. There is a record of his life in : H. G. Hutchinson, *Life of Sir John Lubbock, Lord Avebury* (Macmillan, 2 vols., 1914). The vicissitudes of the Royal Astronomical Society during Lockyer's lifetime are, to some extent, described in : J. L. E. Dreyer and H. H. Turner, *History of the Royal Astronomical Society, 1820–1920* (Royal Astronomical Society, 1923). The quarrels during, and after, the Devonshire Commission are frequently reflected in the pages of the *Astronomical Register* and the *English Mechanic*. Proctor's main contribution to the debate has recently been reprinted : R. A. Proctor, *Wages and Wants of Science-Workers* (Frank Cess, 1970).

Chapter V

SOUTH KENSINGTON AND METEOROLOGY

Cole's activities at South Kensington are recorded in : H. Cole, *Fifty Years of Public Work* (Bell, 1884); and Donnelly's are briefly described by W. H. G. Armytage in : *Vocational Aspect*, Vol. 2, pp. 6–21 (1950). Huxley's educational interests, both at South Kensington and elsewhere, are discussed by C. Bibby, *T. H. Huxley: Scientist, Humanist and Educator* (Watts, 1959). One of the constituent parts of the South Kensington complex, the Royal School of Mines, has been described by M. Reeks, *Register of the Associates and Old Students of the Royal School of Mines and History of the Royal School of Mines* (Royal School of Mines, 1920). There is a brief review of the history of the Science Museum at South Kensington in *The Science Museum: the First Hundred Years* (H.M.S.O., 1957). The life of one of the most important members of the Solar Physics Committee, G. G. Stokes, is recorded in : J. Larmor, *Memoir and Scientific Correspondence of the late Sir George Gabriel Stokes, Bart.* (Cambridge University Press, 2 vols., 1907). There is a general description of developments in meteorology during Lockyer's lifetime in: W. Napier Shaw, *Manual of Meteorology*, Vol. I (Cambridge University Press, 1932).

Chapter VI

WHAT IS AN ATOM?

A general analysis of the concept of atoms during the nineteenth century is to be found in: D. M. Knight, *Atoms and Elements* (Hutchinson, 1967). A description of the development of spectroscopy and its relevance to nineteenth-century atomic ideas is given in: W. McGucken, *Nineteenth-Century Spectroscopy* (Johns Hopkins, 1969). The attitude of chemists to Lockyer's first dissociation hypothesis has been discussed by W. H. Brock in: *Ambix*, Vol. 16, pp. 81–99 (1969). Of the scientists who were in some way involved with Lockyer's atomic concepts, the following four may be mentioned: J. J. Thomson, *Recollections and Reflections* (Macmillan, Toronto, 1937); J. V. Eyre, *Henry Edward Armstrong, 1848–1937* (Butterworth, 1958); M. W. Travers, *A Life of Sir William Ramsay* (Arnold, 1956); E. E. F. d'Albe, *The Life of Sir William Crookes* (Fisher Unwin, 1923). Schuster recorded the earlier part of his life in some detail in *Biographical Fragments* (Macmillan, 1932). A briefer account, but covering all of his life, can be found in: J. G. Crowther, *Scientific Types* (Cresset, 1969). This also contains an account of Dewar.

Chapter VII

THE PHILOSOPHER'S STONE

For Rayleigh's life (and, in particular, his work on the noble gases) see: R. J. Strutt, *Life of John William Strutt, Third Baron Rayleigh* (University of Wisconsin Press, 1968). There is a description of the developments in astronomical spectroscopy at Harvard and of contacts between astronomers there and their British colleagues in: B. Z. Jones and L. G. Boyd, *The Harvard College Observatory (1839–1919)* (Harvard University Press, Cambridge, Mass., 1971).

Chapter VIII

FAMILY AND FRIENDS

Some of Lockyer's friends of long standing are described in the following: H. O. Barnett, *Canon Barnett* (Murray, 1921); E. Romanes,

The Life and Letters of George John Romanes (Longman, 1898); E. Sharp, *Hertha Ayrton, 1854–1923* (Arnold, 1926). There are brief descriptions of two clubs that Lockyer belonged to in : H. Ward, *History of the Athenaeum, 1824–1925* (London, 1926); H. H. Turner, *Records of the R.A.S. Club, 1820–1903* (Oxford, 1904). A discussion of the history of *Nature* during its first fifty years is given by R. M. McLeod in *Nature,* Vol. 224, pp. 441–61 (1969). For the development of anthropology in the nineteenth century, and for ideas on social Darwinism, see : J. W. Burrow, *Evolution and Society* (Cambridge University Press, 1966). Langley, and the astronomical work at the Smithsonian, are described in : B. Z. Jones, *Lighthouse of the Skies: the Smithsonian Astrophysical Observatory* (Smithsonian Institution, Washington, 1965).

Chapter IX

A NEW ORIENTATION

For a discussion of investigations into the orientation of Stonehenge, including recent attempts at interpretation, see : G. S. Hawkins, *Stonehenge Decoded* (Souvenir Press, 1966). Some recent investigations of the astronomical significance of megalithic monuments are contained in : A. Thom, *Megalithic Lunar Observatories* (Oxford University Press, 1971). Lubbock's interest in archaeology (and anthropology) is described in : A. G. Duff (Ed.), *The Life-Work of Lord Avebury 1834–1913* (Watts, 1924). This also contains some general material on the background to archaeology during Lubbock's lifetime.

Chapter X

EDUCATION AND NATIONAL PROGRESS

The main source of information on the activities of the British Science Guild are its reports, which were published annually from 1907 onwards. The general development of universities and university colleges during Lockyer's lifetime (with emphasis on science teaching) is described in : W. H. G. Armytage, *Civic Universities* (Benn, 1955). Some of the problems of reforming London University are outlined by T. L. Humberstone, *University Reform in London* (Allen & Unwin, 1926). Three of the people with whom Lockyer was involved in his later endeavours in science policy and education are described in the following books : W. H. G.

Armytage, *Sir Richard Gregory: his Life and Work* (Macmillan, 1957); R. B. Haldane, *Autobiography* (Hodder, 1929); J. S. and H. G. Thompson, *Silvanus Phillips Thompson: his Life and Letters* (Fisher Unwin, 1920).

Chapter XI

THE FINAL PUSH

For further details of Hale, see : H. Wright, *Explorer of the Universe: a Biography of George Ellery Hale* (Dutton, New York, 1966). The life of one of Lockyer's leading supporters in the Solar Physics Observatory dispute, Gill, is recorded in : G. Forbes, *David Gill: Man and Astronomer* (Murray, 1916). For a summary of ideas concerning variations in terrestrial meteorology with the solar cycle, see : E. Huntington, *Earth and Sun: an Hypothesis of Weather and Sunspots* (Yale University Press, 1923). The establishment of the Solar Physics Observatory at Cambridge is described by F. J. M. Stratton, *Annals of the Solar Physics Observatory, Cambridge,* Vol. I (1949).

REFERENCES

Books are published in London, unless otherwise specified. All unannotated manuscript material is taken from the collection at the Norman Lockyer Observatory, Sidmouth, Devon. [IC-H] indicates that an item derives from the Huxley archives at Imperial College; [Capt. L] that it belongs to Captain H. C. Lockyer, R.N., of Helston, Cornwall; [Camb.] that it is in the archives of the Cambridge Observatories; [RS] that it is in the possession of the Royal Society; and [RGO] that it is in the archives of the Royal Greenwich Observatory.

Chapter I

THE MILITANT CIVIL SERVANT

1. R. C. Adams to Lord Leigh, 31 March 1856.
2. A. W. à Beckett, *Recollections of a Humorist*, Pitman (1907) pp. 25, 28–31.
3. M. Wright, *Treasury Control of the Civil Service: 1854–1874*, Clarendon Press, Oxford (1969), p. 214.
4. War Office Circular Memorandum No. 11, 7 December 1865.
5. J. S. Vacher to Lockyer, 31 May 1866.
6. Often spelt 'Trebinshon'.
7. It has proved difficult to establish this with complete certainty, but all the evidence points this way.
8. E. M. S. P[aine], *The Two James's and the Two Stephensons*, G. Phipps (1861), p. 103.
9. W. Huggins to Lockyer, 20 June 1864.
10. A. R. Wallace to Lockyer, 17 October 1865.
11. O. Lodge to Lockyer, 22 February 1907.

Chapter II

THE MAN OF LETTERS

1. M. Wright, *Treasury Control of the Civil Service: 1854–1874*, Clarendon Press, Oxford (1969), p. 208.
2. Lockyer to T. H. Huxley, 14 February 18–[?] [IC-H].
3. T. Hughes to T. H. Huxley, 22 November 1864 [IC-H].
4. T. H. Huxley to Lockyer, 25 November 1864 [IC-H].
5. T. Hood to Lockyer [undated].
6. F. C. Penrose to Lockyer, 23 November 1864.
7. L. Huxley, *Life and Letters of Sir Joseph Dalton Hooker*, Murray (1918) Vol. 1, p. 541.
8. F. Galton, *Memories of my Life,* Methuen (1909), p. 168.
9. T. H. Huxley to Lockyer, 22 August 1865 [IC-H].
10. Statement handed to Sir John Pakington by Lockyer in May 1868.

11. Lord Farrer to Lockyer, 12 December 1868.
12. G. B. Airy to Lockyer, 15 June 1868.
13. Charles Morgan, *The House of Macmillan* (1943), p. 71.
14. Charles Morgan, *op cit.*, p. 69.
15. A. Geikie to Lockyer [undated].
16. J. N. Lockyer to T. H. Huxley, 14 February 18–[?] [IC-H].
17. J. D. Hooker to A. Macmillan, 27 July 1869. (The appearance of this letter amongst Lockyer's correspondence indicates that Macmillan passed it on to him.)
18. M. Foster to Lockyer, 4 August 1869.
19. T. H. Huxley to Lockyer, 16 July [1869] [IC-H].
20. This point is made by R. M. McLeod: *Nature*, **224**, 439 (1969).
21. T. M. and W. L. Lockyer (Eds.), *Life and Work of Sir Norman Lockyer*, Macmillan (1928), p. 48.
22. J. J. Sylvester to Lockyer, 15 October 1869.
23. C. L. Graves, *Life and Letters of Alexander Macmillan*, Macmillan (1910), pp. 302–3.
24. J. D. Hooker to A. Macmillan, 27 July 1869.
25. M. Foster to Lockyer, 4 August 1869.
26. T. M. and W. L. Lockyer, op. cit., pp. 46–7.
27. C. Kingsley to Lockyer, 8 November 1869.
28. C. Kingsley to Lockyer, 8 November 1872.
29. A. Macmillan to Lockyer, 10 November 1871.
30. Lockyer, *Nature*, **2**, 1 (1870).
31. R. M. McLeod, op. cit., p. 443.
32. T. H. Huxley to Lockyer, 23 May 1873 [IC-H].
33. J. Brett to Lockyer, 20 May 1878.
34. F. Galton to Lockyer, 5 April 1878.
35. E. L. Youmans to Lockyer, 16 July 1872.
36. T. H. Huxley to Lockyer, 28 December 1872 [IC-H].
37. J. D. Hooker to T. H. Huxley, 19 November 1872 [IC-H].
38. J. D. Hooker to T. H. Huxley, 11 November 1872 [IC-H].
39. J. D. Hooker to T. H. Huxley, 13 November 1872 [IC-H].
40. J. D. Hooker to T. H. Huxley, 16 November 1872 [IC-H].
41. T. H. Huxley to Lockyer, 24 November 1872 [IC-H].
42. B. A. Gould to Lockyer, 8 December 1871.
43. R. H. Scott to Lockyer, 10 March 1874.
44. A. Geikie to Lockyer, 19 March 1877.
45. D. M. Horne to Lockyer, 7 February 1879.
46. P. G. Tait to Lockyer, 12 July 1875.
47. N. Pole to Lockyer, 1 May 1876.
48. H. L. F. von Helmholtz to Lockyer, 7 June 1874.
49. *Nature*, **8**, 399 (1873).
50. Ibid.
51. P. G. Tait to Lockyer, 26 September 1873.
52. T. H. Huxley to Lockyer, 8 October 1870 [IC-H].
53. H. C. Bastian to Lockyer, 18 January 1876.
54. H. Spencer to Lockyer, 19 May 1874.
55. Quoted by H. Dingle, *Nature*, **224**, 829–30 (1969).
56. C. W. Siemens to Lockyer, 24 February 1879.
57. T. A. Edison to Lockyer, 1 November 1878.
58. W. Jack to Lockyer, 24 July 1876.
59. W. Jack to Lockyer, 29 August 1877. (Quoted in: R. M. McLeod, op. cit., p. 454.)
60. A. Macmillan to Lockyer, 21 March 1877.
61. J. Chenery to Lockyer, 10 October 1878.

Chapter III

THE MAN OF SCIENCE

1. Admiral W. H. Smyth, *Sidereal Chromatics*, London (privately printed), (1864), p. 90.
2. J. N. Lockyer, *Contributions to Solar Physics*, Macmillan (1874), p. xi.
3. W. R. Birt to Lockyer, 31 May 1861.
4. W. R. Birt to Lockyer, 21 May 1861.
5. J. Phillips, *British Association Report*, Part II, p. 15 (1853).
6. W. Huggins to Lockyer, 20 June 1864.
7. J. N. Lockyer, *Proceedings of the Royal Society*, **15**, 256 (1866).
8. J. N. Lockyer, *Philosophical Transactions*, **159**, 425 (1869).
9. B. Stewart, *Nature*, **7**, 301 (1873).
10. J. N. Lockyer, *Proceedings of the Royal Society*, **17**, 128 (1868).
11. W. De la Rue to G. G. Stokes, 17 December 1868.
12. E. Frankland to Lockyer, 7 April 1869.
13. E. Frankland to Lockyer, 9 September 1872.
14. J. N. Lockyer, *Contributions to Solar Physics*, Macmillan (1874), p. 645.
15. J. N. Lockyer, *Proceedings of the Royal Society*, **17**, 350 (1869).
16. G. G. Stokes to Lockyer, 1 October 1869.
17. J. N. Lockyer, *Philosophical Magazine*, **39**, 63 (1870).
18. J. N. Lockyer to A. Secchi, 24 January [1873?].
19. L. Respighi to Lockyer, 13 June 1872.
20. H. E. Roscoe to Lockyer, 12 January 1870.
21. C. A. Young to Lockyer, 21 January 1871.
22. W. K. Clifford, *The Commonsense of the Exact Sciences*, Knopf, New York (1946 Reprint), p. xxii.
23. G. B. Airy to Lockyer, 7 July 1871.
24. C. A. Young to Lockyer, 29 September 1871.
25. J. N. Lockyer, *Contributions to Solar Physics*, Macmillan (1874), p. 341.
26. J. N. Lockyer, ibid., p. 344.
27. C. Pritchard to Lockyer, 29 December 1865.
28. Ibid.
29. B. Stewart to G. B. Airy, 6 January 1869 [RGO].
30. G. B. Airy to B. Stewart, 7 January 1869 [RGO].
31. E. Frankland to Lockyer, 23 March 1871.
32. W. De la Rue to Lockyer, 30 April 1870.
33. There is some uncertainty as to the authorship of these lines (and they appear in various guises), but Clerk Maxwell seems to be the probable author. *See:* A. L. Cortie, *Astrophysical Journal*, **53**, 241 (1921).

Chapter IV

THE DEVONSHIRE COMMISSION

1. T .H. Huxley, *Science and Education*, Macmillan (1899), p. 105.
2. M. Reeks, *Register of the Associates and Old Students of the Royal School of Mines and History of the Royal School of Mines*, Royal School of Mines (Old Students' Association), (1920), p. 98.
3. M. Arnold, *Higher Schools and Universities in Germany*, Macmillan (1874), p. 212.

4. A. Strange to Lockyer, 25 March 1869.
5. C. Appleton to Lockyer, 18 November [1869].
6. C. Pritchard to Lockyer, 11 December 1869.
7. The Royal Commission on Scientific Instruction and the Advancement of Science. (This will be referred to below as 'Devonshire Commission'.) First Report (9 March 1871).
British Sessional Papers, Vol. XXV, p. 3.
8. Ibid., p. 6.
9. Ibid., p. 9.
10. Ibid., p. 23.
11. Devonshire Commission. Third Report (1 August 1873).
British Sessional Papers, Vol. XXVIII, p. 694.
12. Devonshire Commission. Sixth Report (18 June 1875).
British Sessional Papers, Vol. XXVIII, p. 65.
13. Ibid.
14. Ibid.
15. Devonshire Commission. Eighth Report (18 June 1875).
British Sessional Papers, Vol. XXVIII, p. 425.
16. Ibid.
17. Ibid., pp. 426–7.
18. Ibid., p. 429.
19. Ibid., p. 449.
20. Quoted in: J. G. Crowther, *Statesmen of Science,* Cresset Press (1965), pp. 252–3.
21. Devonshire Commission. Eighth Report (18 June 1875), p. 430.
22. Ibid., p. 431.
23. Ibid.
24. Ibid., p. 437.
25. Ibid., p. 508.
26. Royal Astronomical Society Circular, 21 June 1872.
27. R. A. Proctor to G. B. Airy, 26 June 1872 [RGO].
28. J. N. Lockyer, *British Association Report,* p. 503 (1873).
29. G. B. Airy to Lockyer, 6 November 1872.
30. J. N. Lockyer to G. B. Airy, 7 November 1872 [RGO].
31. Balfour Stewart to Lockyer, 31 March 1873.
32. R. A. Proctor to G. B. Airy, 9 November 1872 [RGO].
33. *English Mechanic,* **16**, 528 (1873).
34. *Astronomical Register,* **11**, 66 (1873).
35. Circular dated 25 March 1873 [RGO].
36. Circular dated 29 March 1873 [RGO].
37. C. Pritchard to Lockyer, 18 November 1873.
38. Copy of document circulated at the University of Virginia, 17 January 1871
39. J. Henry to Lockyer, 4 August 1874.
40. H. E. Roscoe to Lockyer, 30 July 1875.
41. G. G. Stokes, Balfour Stewart and General Strachey, *Report on Observations in Astronomical Physics* [to Viscount Sandon] (1877). p. 4.
42. Ibid., p. 6.
43. Ibid., p. 6.
44. J. L. E. Dreyer and H. H. Turner (Eds.), *History of the Royal Astronomical Society: 1820–1920,* Royal Astronomical Society (1923), p. 209.
45. Ibid., pp. 209–10.
46. G. B. Airy, *English Mechanic,* **32**, 586 (1881).
47. J. W. L. Glaisher (Ed.), *The Collected Mathematical Papers of Henry John Stephen Smith,* Clarendon Press, Oxford (1894), p. xliv.
48. W. T. Thiselton-Dyer, *Nature,* **106**, 22 (1920).

Chapter V

SOUTH KENSINGTON AND METEOROLOGY

1. B. H. Becker, *Scientific London*, H. S. King (1874), p. 150.
2. H. A. Rowland to Lockyer, 7 January 1885.
3. S. P. Langley to Lockyer, 3 May 1879.
4. T. M. and W. L. Lockyer (Eds.), *Life and Work of Sir Norman Lockyer*, Macmillan (1928), p. 455.
5. P.R.O. Works, 17.20/5, 167–201 [IC—Mrs. Pingree].
6. J. F. D. Donnelly to Lockyer, 3 September 1891.
7. L. Playfair to Lockyer, 19 December 1889.
8. H. H. Turner to Lockyer, 23 December 1889.
9. Napier Shaw, *Manual of Meteorology*, Vol. I, Cambridge University Press (1932), p. 308.
10. Report of the Director of the Solar Physics Observatory for the period 1889–1909. H.M.S.O. (1911), p. 16 [Cd. 5923].
11. Ibid., p. 29.
12. H. G. Wells 'Filmer' in: *The Complete Short Stories of H. G. Wells*, Ernest Benn (1927), p. 830.
13. T. M. and W. L. Lockyer, op. cit., p. 457.
14. Meldola MS Diary (1875–6) [IC].
15. H. Dingle, *Obituary Notices of Fellows of the Royal Society*, **3**, 484 (1941).
16. Meldola, op. cit.
17. T. M. and W. L. Lockyer, op. cit., p. 461.
18. R. A. Proctor, *English Mechanic*, **15**, 318 (1872).
19. J. Larmor, *Memoir and Scientific Correspondence of the late Sir George Gabriel Stokes, Bart.*, Cambridge University Press, Vol. 1 (1907), p. 350.
20. T. M. and W. L. Lockyer, op. cit., p. 462.

Chapter VI

WHAT IS AN ATOM?

1. C. A. Young, *Nature*, **7**, 17 (1872).
2. J. N. Lockyer: Laboratory Notebooks (1872–4 and 1877–84). In the possession of the Norman Lockyer Observatory, Sidmouth.
3. H. St.-C. Deville, *Comptes Rendus*, **76**, 1175 (1873); quoted by W. H. Brock in *Ambix*, **16**, 84 (1969).
4. J. N. Lockyer, *Chemistry of the Sun*, Macmillan (1887), p. xi.
5. J. B. A. Dumas to Lockyer, 11 October 1873.
6. *Chemical News*, **28**, 176 (1873).
7. T. Andrews to Lockyer, 9 January 1874.
8. A. Crum Brown to Lockyer, 12 November 1873.
9. H. E. Roscoe to Lockyer, 24 May 1874.
10. W. Crookes to Lockyer, 16 December 1878.
11. J. P. Cooke, Jnr. to Lockyer, 9 February 1879.
12. H. E. Roscoe to Lockyer, 26 May 1874.
13. H. A. Jevons, *Letters and Journal of W. Stanley Jevons*, Macmillan (1886), p. 393; quoted by W. H. Brock, op. cit., p. 96.
14. H. E. Roscoe to Lockyer, 17 November [?] 1878.
15. W. Crookes to Lockyer, 16 December 1878.
16. H. E. Roscoe to Lockyer, 10 December 1878.

17. J. Larmor, *Memoir and Scientific Correspondence of the late Sir George Gabriel Stokes, Bart.*, Cambridge University Press, Vol. I (1907), pp. 406–7.
18. B. C. Brodie to Lockyer, 8 July 1879.
19. W. Crookes to Lockyer, 31 May and 7 June 1879.
20. H. E. Roscoe to Lockyer, 11 April 1880.
21. H. E. Roscoe to Lockyer, 17 November [?] 1878.
22. J. Larmor, op. cit., p. 430.
23. W. Roberts-Austen to Lockyer, 15 October 1898.
24. J. N. Lockyer to the Royal Society, 23 March 1878.
25. J. N. Lockyer to J. Dewar, 9 April 1879.
26. W. C. Roberts to Lockyer, 10 April [1879].
27. W. Abney to Lockyer, 1 September 1881.
28. C. A. Young to Lockyer, 24 November 1879.
29. C. A. Young to Lockyer, 25 April 1879.
30. N. Reingold (Ed.), *Science in Nineteenth-Century America*, Macmillan (1966), p. 274.
31. See: J. N. Lockyer, *Inorganic Evolution*, Macmillan (1900), p. 103.
32. J. N. Lockyer, *The Chemistry of the Sun*, Macmillan (1887), p. 371.
33. Lord Kelvin to Lockyer, 25 January 1896.
34. See: J. N. Lockyer, *Inorganic Evolution*, p. 117.
35. Ibid.
36. H. E. Armstrong to Lockyer, 1 September 1885.
37. C. A. Young, *Trans. New York Acad. Sciences*, **5**, 234 (1886).
38. J. N. Lockyer to J. D. Hooker. Undated; but it evidently refers to the 1875 eclipse preparations.
39. A. Schuster to Lockyer, 11 June 1874.
40. A. Schuster to Lockyer, 18 September 1876.
41. J. N. Lockyer, *The Chemistry of the Sun*, p. 303.
42. J. Larmor, op. cit., p. 426.
43. J. N. Lockyer, *The Chemistry of the Sun*, p. 445.
44. A. L. Cortie, *Monthly Notices Roy. Astron. Soc.*, **58**, 370 (1898).
45. G. E. Hale to H. Goodwin, 21 August 1891. Hale Centennial Exhibition Catalogue 1968. The original is in the Huntington Library.
46. E. B. Frost, *An Astronomer's Life*, Houghton Mifflin, New York (1933), p. 45.
47. T. M. and W. L. Lockyer (Eds.), *Life and Work of Sir Norman Lockyer*, Macmillan (1928), p. 458.
48. H. Dingle, *Journal of the British Society for the History of Science*, **1**, 210 (1962–3).
49. H. Kayser to Lockyer, 11 November 1899.
50. J. J. Thomson, *Philosophical Magazine*, **44**, 311 (1897).
51. J. J. Thomson to Lockyer, 15 November 1899.
52. T. Preston, *Nature*, **60**, 178 (1899).
53. J. N. Lockyer, *Inorganic Evolution*, p. 157.
54. W. C. Roberts-Austen, *Proceedings of the Royal Institution*, **14**, 497 (1895).
55. J. N. Lockyer, *Inorganic Evolution*, p. 174.

Chapter VII

THE PHILOSOPHER'S STONE

1. W. Huggins, *Phil. Trans.*, **154**, 437 (1864).
2. W. Huggins to Lockyer, 6 May 1889.
3. J. N. Lockyer to W. Huggins, 7 May 1889.
4. W. Huggins to Lockyer, 7 May 1889.
5. J. N. Lockyer, *The Sun's Place in Nature*, Macmillan (1897), p. 98.

6. W. Huggins, *Report of the British Association* (1891), Part I, p. 9.
7. J. N. Lockyer, op. cit., p. 105.
8. Royal Society Archives [RR. 10.360], 27 May 1890.
9. W. Huggins to Christie, 25 October 1890 [RGO].
10. G. G. Stokes to Lockyer, 7 October 1887.
11. J. N. Lockyer, *Meteoritic Hypothesis*, p. 333.
12. J. N. Lockyer, ibid., p. 336.
13. G. H. Darwin to Lockyer, 18 March 1888.
14. I. Roberts to Lockyer, 29 December 1888.
15. W. Huggins, *Proceedings of the Royal Society*, **14**, 42 (1865).
16. J. N. Lockyer, *The Sun's Place in Nature*, p. 84.
17. H. C. Vogel, *Astronomische Nachrichten*, **84**, No. 2000 (1874).
18. W. Ramsay, *Proceedings of the Royal Society*, **58**, 81 (1895).
19. W. Ramsay, *Nature*, **53**, 366 (1896).
20. T. M. and W. L. Lockyer (Eds.), *Life and Work of Sir Norman Lockyer*, Macmillan (1928), p. 160.
21. C. Runge to Lockyer, 20 October 1895.
22. T. G. Bonney to Lockyer, 16 February 1899.
23. J. N. Lockyer, *Meteoritic Hypothesis*, p. 466.
24. I. Roberts to Lockyer, 2 October 1889.
25. J. N. Lockyer, *The Sun's Place in Nature*, pp. 232–3.
26. J. N. Lockyer, ibid, p. 248.
27. Lockyer to E. C. Pickering, 3 May 1887 [IC-HCO]
28. Lockyer to E. C. Pickering, 21 January 1899 [IC-HCO].
29. G. B. Airy to Lockyer, 11 February 1888.
30. A. Tennyson, 'God and the Universe' (1892).
31. H. Macpherson (Jnr.), *Astronomers of Today*, Gall and Inglis (1905), p. 67.
32. *Punch*, **92**, 252 (1887).

Chapter VIII

FAMILY AND FRIENDS

1. N. Reingold (Ed.), *Science in Nineteenth-Century America*, Macmillan (1966), p. 257.
2. E. Romanes, *The Life and Letters of George John Romanes*, Longmans (1898), p. 133.
3. H. O. Gray, *The Life of Sir William Quiller Orchardson*, Hutchinson (1930), p. 94.
4. H. F. Blandford to Lockyer, 24 June 1884.
5. I. Roberts to Lockyer, 2 October 1889.
6. *Nature*, **36**, 218 (1887).
7. R. H. Bacon to Lockyer, 29 August [1890].
8. E. B. Frost, *An Astronomer's Life*, Houghton Mifflin, New York (1933), pp. 45–6.
9. W. J. S. Lockyer to Lockyer, 30 October 1891 [Capt. L].
10. W. J. S. Lockyer to Lockyer, 8 November 1891 [Capt. L].
11. W. J. S. Lockyer to Lockyer, 5 April 1899 [Capt. L).
12. W. Crookes to Lockyer, 20 August 1895.
13. Report of Huxley's speech in the *British Medical Journal*, p. 1262 (1894).
14. See: R. M. McLeod, *Nature*, **224**, 456 (1969).
15. For a discussion of anthropology during this period, see: J. W. Burrow, *Evolution and Society*, Cambridge Univ. Press (1966) (especially Chapter 4, Section 3).
16. A. H. Pitt Rivers to Lockyer, 14 February 188[4?].

17. F. M. Müller to Lockyer, 19 January 1885.
18. J. Buchanan to Lockyer, 31 December 1895.
19. E. R. Lankester to Lockyer, 25 September [?].
20. T. H. Huxley to Lockyer, [10?] July 1887 [IC-H].
21. W. Thiselton Dyer to Lockyer, 6 January 1890.
22. S. P. Langley to Lockyer, 22 May 1888.
23. S. P. Langley to Lockyer, 13 May 1896.
24. J. J. Sylvester to Lockyer, 5 August 1888.
25. W. H. Preece to Lockyer, 9 July 1884.
26. For details, see: R. M. McLeod, *Nature*, **224**, 449 (1969).
27. R. M. McLeod, ibid.
28. Willard Fiske to Lockyer, 30 November 1891.
29. E. Gosse to Lockyer, 28 November 1889.
30. F. Harris to Lockyer, 11 February 1892.
31. E. G. Anderson to Lockyer, 4 March 1899.
32. J. N. Lockyer to A. Rücker, 28 June 1899 [RS].
33. Article on 'Golf' in the *Encyclopaedia Britannica* (Ninth Edition; 1879).
34. P. G. Tait to Lockyer, 17 August 1893.
35. P. G. Tait to Lockyer, 22 August 1893.
36. H. H. Turner to Lockyer, 14 August 1890.
37. C. Pritchard to Lockyer, 23 March 1893.
38. H. H. Turner, *Records of the R.A.S. Club: 1820–1903,* Oxford (1904), p. xxxvi.
39. T. H. Huxley to Lockyer, 6 November 1887 [IC-H].
40. T. H. Huxley to Lockyer, 10 November 1887 [IC-H].
41. J. N. Lockyer to T. H. Huxley, 14 November 1887 [IC-H].
42. J. W. Gregory to Lockyer, 15 February 1901.
43. J. W. Gregory to Lockyer, 6 May 1901.
44. *Nature*, **12**, 509 (1875).
45. *Nature*, **67**, 104 (1902).
46. J. Evans to Lockyer, 5 December 1902.
47. T. M. and W. L. Lockyer (Eds.), *Life and Work of Sir Norman Lockyer,* Macmillan (1928), p. 69.
48. H. Ayrton to Lockyer, 16 January 1911.
49. J. N. Lockyer, *Report on Indian Observatories and their Organization,* India Office (1898), p. 36.
50. J. F. Tennant to Lockyer, 25 July 1875.
51. G. S. Clarke to Lockyer, 11 June 1908.
52. Sir Norman Lockyer and W. L. Lockyer, *Tennyson as a Student and Poet of Nature,* Macmillan (1910), p. 2.
53. A. Tennyson to Lockyer, 17 November 1890.
54. J. N. Lockyer to T. H. Huxley, 14 February 1888 [IC-H].
55. T. H. Huxley to Lockyer, 15 February 1888 [IC-H].
56. T. M. and W. L. Lockyer, op. cit., p. 196.
57. J. F. D. Donnelly to T. H. Huxley, 2 January 1893 [IC-H].
58. Lord Salisbury to Lockyer, 16 June 1897.
59. See: Lawrence Badash, *Rutherford and Boltwood,* Yale University Press (1969), p. 266.

Chapter IX

A NEW ORIENTATION

1. J. N. Lockyer, *The Dawn of Astronomy,* Cassell (1894), Preface.
2. W. M. Flinders Petrie to Lockyer, 10 October 1890.
3. J. N. Lockyer, op. cit., p. 179.

4. W. E. Garstin to Lockyer, 17 March 1894.
5. H. G. Lyons to Lockyer, 13 February 1895.
6. H. G. Lyons to Lockyer, 25 November 1895.
7. J. N. Lockyer, op cit., p. 2.
8. Ibid., p. 3.
9. G. A. Macmillan to Lockyer, 22 May 1895.
10. J. N. Lockyer, op cit., p. 416.
11. E. A. Wallis Budge to Lockyer, 3 April 1895.
12. E. A. Wallis Budge to Lockyer, 6 April 1899.
13. E. A. Wallis Budge to Lockyer, 7 March 1893.
14. T. M. and W. L. Lockyer, *Life and Work of Sir Norman Lockyer*, Macmillan (1928), pp. 151–2.
15. *Edinburgh Review*, **180**, 418 (1894).
16. T. M. and W. L. Lockyer, op. cit., p. 404.
17. J. N. Lockyer, op. cit., p 12
18. Piazzi Smyth to Lockyer, 24 April 1890.
19. J. N. Lockyer, *Stonehenge and other British Stone Monuments Astronomically Considered*, Macmillan (1906), Preface to First Edition.
20. H. G. Hutchinson, *Life of Sir John Lubbock, Lord Avebury*, Macmillan, Vol. II (1914), p. 137.
21. Sir John Lubbock to Lockyer, 14 October 1906.
22. J. N. Lockyer, *Stonehenge* (Second Edition; 1909), pp. 46–7.
23. R. H. Caird to Lockyer, 26 January 1908.
24. Sir Edmund Antrobus to Lockyer, 13 October [?].
25. J. N. Lockyer, *Stonehenge* (1909), p. 412.
26. H. G. Hutchinson, op. cit., p. 137.
27. T. M. and W. L. Lockyer, op. cit., p. 206.
28. W. E. Davies to Lockyer, 5 June 1909.
29. J. G. Frazer to Lockyer, 19 March 1905; 2 February 1906.
30. Article on 'Folklore', *Encyclopaedia Britannica* (Ninth Edition), 1879.

Chapter X

EDUCATION AND NATIONAL PROGRESS

1. J. N. Lockyer, *Education and National Progress*, Macmillan (1906), p. 16.
2. Ibid., p. 18.
3. *Times*, 16 May 1884.
4. H. E. Roscoe to Lockyer, 22 June 1889.
5. S. P. Thompson to Lockyer, 28 July 1903.
6. A. Bigge to Lockyer, 25 February 1904.
7. *Times*, 15 October 1902.
8. J. N. Lockyer, op. cit., p. 190.
9. Ibid, p. 194.
10. P. L. Sclater to Lockyer, 11 October 1903.
11. E. Lyttelton to Lockyer, 18 May 1904.
12. T. Burt to Lockyer, 12 July 1904.
13. E. H. Starling to Lockyer, 5 February 1904.
14. H. C. Lockyer to Lockyer, 23 May 1903 [Capt. L].
15. J. N. Lockyer, op. cit., p. 183.
16. J. N. Lockyer, op. cit., pp. 186–7.
17. *Nature*, **70**, 343 (1904).
18. *Nature*, **73**, 12 (1905).
19. Ibid.
20. A. J. Balfour to Lockyer, 10 August 1906.

21. A. J. Balfour to Lockyer, 3 November 1903.
22. A. E. Shipley to Lockyer, 17 November 1906.
23. S. J. Hickson to Lockyer, 17 March 1911.
24. J. W. Judd to Lockyer, 15 August 1906.
25. J. N. Lockyer, op. cit., p. 208.

Chapter XI

THE FINAL PUSH

1. J. N. Lockyer to Miss T. M. Browne, 18 May 1882 [Capt. L].
2. Mrs M. Arnold-Forster to Lady Lockyer, 13 June [?] [Capt. L].
3. I. Roberts to Lockyer, 5 October 1901.
4. H. F. Oliver to Lady Lockyer, 14 May 1912 [Capt. L].
5. W. N. Shaw to Lockyer, 11 January 1906.
6. E. Rutherford to Lockyer, 21 October 1905 [?]
7. Written on a menu for the Allied Universities Dinner, 10 July 1903.
8. J. N. Lockyer to Lord Salisbury, 23 October 1900.
9. Report by the Director of the Solar Physics Observatory [Cd. 5923] (1911), p. 17.
10. S. P. Langley to Lockyer, 30 November 1901.
11. T. M. and W. L. Lockyer, *Life and Work of Sir Norman Lockyer*, Macmillan (1928), pp. 184–5.
12. W. N. Shaw to Lockyer, 29 January 1903.
13. *Observatory*, **24**, 68 (1901).
14. Report of the Departmental Committee on the Solar Physics Observatory [Cd. 5924] (1911), p. 37.
15. W. J. S. Lockyer to G. E. Hale, 15 March 1909 [S.P.O. Letter Book—Camb.].
16. J. N. Lockyer to G. E. Hale, 23[?] February 1903 [S.P.O. Letter Book—Camb.].
17. W. J. S. Lockyer to A. Fowler, 11 October 1906 [S.P.O. Letter Book—Camb.].
18. J. N. Lockyer to F. G. Ogilvie, 22 January 1907 [S.P.O. Letter Book—Camb.].
19. Report of the Departmental Committee on the Solar Physics Observatory [Cd. 5924] (1911), p. 5.
20. H. F. Newall, 5 December 1909 [MS. Diary—Camb.].
21. Report of the Departmental Committee on the Solar Physics Observatory [Cd. 5924] (1911), p. 6.
22. W. H. M. Christie to F. R. Fowke, 8 May 1909 [RGO].
23. W. H. M. Christie to F. R. Fowke, 9 June 1909 [RGO].
24. H. F. Newall, 31 January 1913 [MS. Notebook—Camb.].
25. A. L. Cortie to W. J. S. Lockyer, 17 December 1920 [Capt. L].
26. G. Forbes, *David Gill: Man and Astronomer*, Murray (1916), p. 351.
27. Ibid.
28. H. Deslandres to Lockyer, 2 January 1911.
29. H. Deslandres to Lockyer, 27 January 1912. (The original letter is in French.)
30. H. Deslandres to Lockyer, 7 February 1912.
31. Copy of the *Appeal*, dated 16 July 1912.
32. H. F. Newall, 31 January 1913 [MS. Notebook—Camb.].
33. H. F. Newall [Undated] [MS. Notebook—Camb.].
34. Chief Constable of Devon to Lockyer, 28 September 1914.
35. H. Rider Haggard to Lockyer, 11 November 1916.
36. W. Ramsay, *Nature*, **104**, 138 (1914).
37. Sir William Mather, *Nature*, **105**, 520 (1915).
38. G. E. Hale to Lockyer, 9 April 1916.
39. G. E. Hale to Lockyer, 14 August 1916.

40. G. E. Hale to Lockyer, 9 April 1916.
41. A. S. Eddington to Lockyer, 6 March 1916.
42. A. Fowler to Lockyer, 4 June 1913; 27 October 1914.
43. W. N. McClean to W. J. S. Lockyer, 20 March 1920 [Capt. L].
44. See, for example:
    W. W. Campbell, *Publications of the Astronomical Society of the Pacific*, **32**, 265 (1920).
    H. Deslandres, *Comptes Rendus*, **171**, 591; 1252 (1920).
    A. Fowler, *Nature*, **105**, 781; 831 (1920).
    A. Geikie, *et al., Nature*, **106**, 20 (1920).
    W. E. Rolston, *Observatory*, **43**, 358 (1920).
    A. L. Cortie, *Astrophysical Journal*, **53**, 233 (1921).
    H. E. Armstrong, *Nature*, **122**, 870 (1928).
45. G. E. Hale to Lady Lockyer, 27 August 1920 [Capt. L].

INDEX

Abney, W. 108–9, 116, 154, 152, 166, 231, 282, 300
Acland, Sir Arthur 237, 263
Adams, J. C. 45, 55, 102, 107, 246
Admiralty 66, 91, 214, 249, 270, 282
Airy, G. B. 24, 51, 58, 66, 68, 72, 94–5, 96–7, 98, 99, 100, 102, 109, 110–11, 122, 159, 206
Alma-Tadema, L. 211, 273
Anderson, Elizabeth, G. 223, 281
Andromeda nebula 188–9
Ångstrom, A. J. 48, 65, 139, 143
Anthropological Society 22, 218, 253
Anthropology 19, 217–18, 229
Antrobus, Sir Edmund 252–3
Appleton, C. E. 27, 83, 94
Archaeology 229, 238–57
Armstrong, H. E. 121, 152, 154, 157, 228, 260, 261, 264, 272, 298
Army 1, 6, 9
Army Regulations 8, 9–10, 23
Aswan dam 240–1
Athenaeum 19, 23, 25
Athenaeum 37, 210, 214, 235
Atoms 135–74, 199
Aviation 220–1, 282–3, 305
Avogadro, A. 136, 137
Ayrton, Hertha 231, 281
Ayrton, W. E. 121, 231, 250, 273

Bacon, R. H. 214–15
Ballooning 18, 122, 283
Balfour, A. J. 266, 268–9, 274–5
Barnett, S. 212, 280
Basic lines 144, 154, 155, 168, 169
Bastin, H. C. 35–6
Beer, W. 41, 43–5
Birt, W. R. 41–2
Blandford, H. F. 127, 172
British Association 19, 24, 26, 42–3, 45, 60, 69, 71, 78, 81–2, 89, 93, 98, 99, 106, 123, 147, 152, 154, 157, 164, 185, 265, 267, 270–3, 275, 280, 288, 305, 309
British Astronomical Association 224–5
British Museum 2, 6, 31, 87, 89, 177, 247–8
British Science Guild 273–9, 281, 294, 303, 306, 307, 309
Brodie, B. 138, 146, 148, 152
Browning, J. 52, 55, 95–6
Brunton, Lauder 38, 106, 223, 235, 301, 305
Budge, E. A. T. Wallis 230, 247
Bunsen, R. 48, 60, 138, 159

Caesium 60, 148
Calcium 141, 160, 194
Cambridge 26, 73, 77, 82, 87, 88, 89, 96, 106, 215, 226, 233, 246, 265, 278, 292–4, 296–8, 299, 301, 302
Carbon 154, 179, 180, 189, 193, 200, 206
Cardwell, E. 76, 214
Chamberlain, J. 266, 268–9, 273
Chemical News 19, 217
Chemistry of the Sun, The 155, 166, 167, 169, 194, 307
Chief nebular line 183–6
Chlorine 148–9
Christian Socialists 8, 9, 14, 16, 18, 19, 213, 222, 246
Christie, W. H. M. 186, 231, 295
Chromosphere 51, 57–8, 59, 61–2, 64, 65, 70, 72, 140, 172–3, 289
City and Guilds 121–2, 259, 260–1, 264, 266, 275, 298
Civil Service 6, 7, 8, 16, 90–1, 236
Clarke, Sir George 233, 241, 273
Clifford, W. K. 67, 174
Cole, H. 76, 79, 81, 85–6, 113
Comets 175, 177–81, 186–7, 193
Commissioners for the 1851 Exhibition 119, 121, 259, 278–9
Common, A. A. 110, 213, 224
Conservatives 12, 164, 265, 268
Cooke, T. 6, 15, 40–1, 52, 79, 177
Corona 49, 51, 65–6, 68, 70–1, 73, 96, 160, 165–6, 172–3, 176, 184
Cortie, A. L. 167, 297
Crimean War 1, 8, 113
Crookes, W. 19, 26, 43, 60, 67, 76, 148, 149, 151, 152, 157, 158, 217, 228, 237, 282, 299, 300

Daily News 25, 210, 211
Darwin, C. 9, 21, 173–4, 219
Darwin, G. H. 84, 188, 190, 254, 295
Dawes, W. R. 42, 44, 45, 46, 49
Dawn of Astronomy, The 242, 246, 250
de Grey, Earl 9–10
De la Rue, W. 42–3, 46–7, 49, 53–4, 57–8, 73, 76, 93, 94, 95–102, 109

Denison, E. 99, 101, 111
Deslandres, H. 286, 289, 298–9, 301
Devonshire Commission 32, 75–112, 115, 132, 146, 150, 159, 258, 259, 279
Devonshire, Duke of 82–3, 105
Dewar, J. 34, 153–6
Diffraction gratings 118, 155, 156
Disraeli, B. 105, 108
Dissociation hypothesis 128, 135, 146–74, 194, 215
Donnelly, J. 81, 86, 113, 116, 117, 119, 120, 121, 129, 210, 235, 260, 261, 263
Doppler effect 61–2, 144, 171, 203
Draper, H. 153, 204, 210
Druids 256–7
Dumas, J. B. 137, 145, 147, 148
Dunkin, E. 95, 108
Dyer, T. 220, 228
Dyson, F. W. 294, 295, 296

Eclipse expeditions 66–70, 88, 127, 132, 159–60, 162–5, 171–2, 210, 211, 280, 281, 299
Eclipses of Sun 29, 50–1, 54, 65, 68, 70–1, 160, 162–3, 172, 179
Eddington, A. S. 302, 304
Egypt 162, 211, 238–46, 256
Eisteddfod 254–5
Elementary Lessons in Astronomy 23, 24
Eliot, J. 172, 232, 285
English Mechanic 101, 111
Enhanced lines 168–9, 171, 172, 198–9, 206
Evans, A. 230, 254

Faye, H. 46, 50, 63, 94
Fisher, Sir John 270–1, 282
Flash spectrum 162, 163, 164–5, 171–3
Foster, M. 26, 27, 31, 231, 275
Fowler, A. 129–34, 163, 168, 169, 171, 172, 189, 217, 251, 282, 291, 300, 301, 305, 306, 309
France 1, 79, 86, 89, 218
Frankland, E. 20, 58–60, 73, 77, 81, 85, 88, 93, 95, 96, 105, 113, 121, 131, 132, 140, 141, 145, 160, 194, 210, 237
Fraunhofer lines 47–8, 50, 51, 59, 64, 65, 68, 70, 140, 142, 153, 161, 162, 172–3, 194
Friswell, R. 132, 146, 148

Galton, F. 21, 30
Garstin, Sir William 240–1
Geikie, A. 24–5, 33, 37, 221, 268
Germany 28, 79, 86, 88, 156–7, 196, 215–16, 239, 258, 263, 264, 267, 274, 302–4
Gill, D. 294, 295, 298, 300
Gladstone, W. E. 8, 9, 31, 67, 82, 92, 219, 237
Glaisher, James 18, 122, 123–4, 283
Glazebrook, R. T. 294, 297, 298
Gold Medal of R.A.S. 95–6, 99–100, 102, 108, 225
Golf 223–4
Great Exhibition of 1851 75–6, 78
Greece 238, 246
Gregory, J. W. 227–8
Gregory, R. 133, 272, 283, 306
Guthrie, F. 113, 119

Haggard, Rider 233, 302
Haldane, R. B. 263–4, 274–5, 277, 292, 303
Hale, G. E. 167–8, 197, 222, 286, 289–90, 291, 301, 303–4
Hampstead 13–15, 42
Harvard College Observatory 203–5
Harvard spectral classification 204–6, 304, 308–9
Heath, Sir Thomas 294, 296
Helium 58–60, 158, 194–8, 284
Helmholtz, H. 34, 189
Henry, J. 103–4
Herbert, S. 2, 9
Herschel, A. S. 84, 178
Herschel, Sir John 42, 46, 49, 64, 249
Herschel, Sir William 39, 42, 49, 124, 182–3
Hooker, J. D. 21, 25, 27, 31–2, 87, 159, 228
House of Commons 31, 81, 120, 122, 226–7, 258, 268, 269, 276, 281
House of Lords 82, 281
Huggins, W. 13, 35, 40, 43, 45–6, 50, 55–6, 58, 60–1, 62, 67, 68, 81, 93, 96, 97, 98, 99, 102, 108, 110, 145, 146, 150, 152, 157, 175, 179, 181, 182, 183–6, 189, 200–1, 202, 203, 206, 210, 222, 236, 267
Hughes, T. 8–10, 14, 16–17, 19, 20, 211, 212, 280
Hunt, W. H. 16, 211, 234–5, 280
Huxley, T. H. 16–22, 24, 25, 26, 28, 29, 31–2, 35–6, 77, 81, 83, 85, 86, 87, 98, 113–16, 121, 173, 210, 217, 219, 226–7, 230, 235, 236, 237, 259, 262–3, 278, 279
Hydrogen 58–61, 64, 66, 70, 136, 144,

146, 152–3, 157, 158, 170, 175, 183, 194

Imperial College 264, 298
India 68–70, 80, 94, 97, 107–8, 124–8, 132, 165, 171–2, 232–3, 282, 284–5, 286
Inorganic Evolution 169, 173–4
Iron 65–6, 144, 168, 169, 172, 198–9, 206

James, W. 12–13
Janssen, P. J. C. 53–5, 56, 67, 70, 99, 111, 118, 147, 289, 301
Judd, J. W. 116, 119, 252, 278
Jupiter 44, 46

Kayser, H. 158, 169
Kelvin, Lord, *see* W. Thomson
Kenilworth 3, 5, 10
Kew 31–2, 46, 50, 87, 109, 122–3, 220
Kingsley, C. 16, 19, 28–9, 246
Kirchhoff, G. 48–50, 51, 53, 60, 64, 65, 70, 138, 139, 142, 143, 150, 159, 172

Laboratory spectra 138–43, 148–57, 168, 176–80, 195–9
Langley, S. P. 118, 126, 220, 285
Lankester, R. 219, 228, 278, 302
Laplace, P. S. 182, 190
Leigh, Lord 3, 5, 7
Leverrier, U. J. J. 33, 45, 55, 147, 234
Liberals 12, 83, 92, 226, 259, 264, 268, 274, 278
Lick Observatory 104, 185
Lindsay, Lord 30, 181
Lithium 131, 152
Liveing, G. D. 153–6, 186, 265
Liverpool and Manchester Railway 12–13
Lockyer, A. E. 209, 305
Lockyer, F. E. 210
Lockyer, H. C. 209, 214, 249, 271, 281–2, 305, 308
Lockyer, Joseph Hooley 3–5
Lockyer, N. J. 209, 308
Lockyer, O. H. 209, 308
Lockyer, Rosaline A. 209, 308
Lockyer, Thomazine M. 254, 280–3, 299–300, 308
Lockyer, W. J. S. 128, 133, 163, 171, 209, 215–17, 232, 245, 282–3, 284–7, 289, 291–2, 295, 298, 301, 305, 306, 308
Lockyer, Winifred 12–14, 29, 53, 67, 209–10
Lockyer, Winifred L. 209, 234, 308
Lodge, O. 14, 267
London County Council 263, 269–70, 276, 277
London Review 17, 21, 42
Lubbock, J. 21, 31, 218, 226, 251, 252, 253, 256, 273, 301
Ludlow, J. M. 8–9, 16–17, 19, 21, 222, 246
Lyons, H. G. 240, 241

McClean, F. K. 282, 299–300, 301
McClean, W. N. 300, 306
McLeod, H. 131–2
Macmillan, A. 14, 16, 17, 24–9, 33, 37, 38, 89, 106, 217
Macmillan's Magazine 16, 19, 54
Mädler, J. H. 41, 43–5
Magnesium 61, 140, 145, 152, 160, 179, 183, 206
Magnus, P. 259, 261, 301
Manchester 77, 150, 159
Mariette, A. 238, 243
Marine biology 219, 275
Markham, Sir Clements 227–8, 235
Mars 44–5
Maspero, G. 247, 248
Masson, D. 16, 19, 235
Maxwell, J. C. 73, 107, 137
Meldola, R. 131, 132, 138, 159, 231, 273, 275
Meteorites 176–81, 194
Meteoritic hypothesis 176–94, 198–208, 211, 215
Meteoritic Hypothesis, The 187, 234
Meteorological Office 33, 94, 282, 287, 293, 294, 296, 297
Meteorology 18, 33, 96, 98, 107, 122–9, 132, 172, 284–9, 309
Meteors 122, 175–6, 178
Moon 40–4
Müller, M. 218, 247, 250
Murchison, R. 22, 79, 87
Mythology 244, 245, 256

Nasmyth, J. 42, 46
Natural History Review 18, 25
Nature 25–38, 65, 87, 98, 101, 111, 119, 124, 133, 138, 144, 178, 196, 211, 214, 216, 217–22, 229, 230, 231, 239, 248, 249, 260, 265, 267, 272, 283, 306, 307, 309
Navy 7, 214, 270–1, 281, 305
Nebulae 61, 181–3, 185, 187–91, 200

Newall, H. F. 254, 292–4, 296–9, 301
Newall, R. S. 177, 292
Nitrogen 60–1, 183, 195
Noble, W. 101, 109, 111
Normal School of Science 86, 115, 116, 117, 121–2
Novae 175, 181, 186–7, 201–4

Orchardson, W. 211–12
Orientation of monuments 238–57, 309
Origin of Species 18, 21, 24
Orion nebula 184, 187, 188, 194, 198
Owen, R. 31–2, 87
Oxford 3, 42, 77, 81, 83, 87, 88, 89, 125, 138, 218, 225, 233, 247, 265, 290
Oxygen 148, 153

Palmerston, Viscount 1, 9
Paschen, F. 197–8
Payment-by-results examinations 79–80, 84–5, 258
Pedler, A. 132, 171–2, 275, 278, 309
Peirce, B. 66–7
Penrose, F. C. 246–7, 248, 251
Percy, J. 87, 119, 142
Petrie, W. M. Flinders 239, 251, 253
Phillips, J. 42–3, 44, 45
Philology 218, 256
Photography 42, 108, 110, 116, 162–5, 168, 181, 188–9, 193, 197, 201, 213, 283, 289
Photosphere 49, 57, 64, 68, 140, 143, 144, 161, 162, 172–3
Physiography 116, 227
Pickering, E. C. 158, 204–6, 301
Pitt Rivers, A. H. 218, 229
Planetary nebulae 181, 182, 188
Playfair, L. 76, 80, 81, 113, 121, 226, 264, 278
Pollock, G. 6, 8, 40
Popular Science Monthly 30–1
Prince Consort 75–6, 279
Pritchard, C. 43, 71, 83, 95, 99, 102, 103, 109, 163, 225
Proceedings of the Royal Society 50, 53, 63, 169
Proctor, R. A. 45, 55, 96, 98–103, 108, 110, 133, 159, 220
Prominences 51–8, 61, 63, 144, 285–6, 288
Prout, W. 136, 152
Pyramids 239, 249–50

Ramsay, W. 195–7, 200, 268, 273, 275, 277, 278, 283, 302, 305
Ranyard, A. C. 30, 67, 108–9, 225
Rayleigh, Lord 195–6, 232, 274–5
Reader 17–23, 25, 33, 79
Respighi, L. 62–3, 64, 70, 99
Reversing layer 68, 161, 162
Roberts, I. 188, 200, 213, 281
Roberts-Austen, W. C. 142, 148, 153, 154, 174
Romanes, G. J. 174, 210, 212, 219, 221, 228
Roscoe, H. 24, 64, 77, 98, 104, 107, 126, 138, 139, 147, 148, 149, 150, 151, 159, 226, 255, 259, 260, 261, 262, 264, 268, 272, 278
Rosse, Lord 42, 182
Rowland, H. A. 118, 155, 156
Royal Astronomical Society 29–30, 44, 50, 54, 66, 68–9, 71, 72, 78, 80, 95–103, 108–12, 122, 163, 189, 224–6, 230, 246
Royal College of Chemistry 58–9, 76, 84–7, 89, 105, 113, 132
Royal College of Science 115, 120, 129, 262–4, 282, 291
Royal Engineers 6, 85, 107, 108
Royal Indian Engineering College 132, 233
Royal Institution 58, 71, 146, 153
Royal Mint 142, 154, 174
Royal Observatory Greenwich 93, 94, 97, 109, 117, 122, 129, 295
Royal School of Mines 76, 84–7, 89, 113, 115, 264
Royal Society 6, 14, 18, 52, 54, 62, 66, 68–9, 72, 79, 80, 98, 109, 118, 123, 146, 149, 151, 152, 154, 157, 159, 163, 186, 187, 211, 212, 213, 225, 226–32, 241, 246, 254, 267, 273, 295
Royal Society catalogue 231–2
Royal Society Government Grant 52, 91, 111, 213
Rücker, A. 119, 121, 122, 186, 223, 235, 255, 262
Rugby 4–5, 9, 57, 90, 127
Runge, C. D. 197–8
Rutherford, E. 278, 283–4, 297
Rutherfurd, L. 118, 191

Sabine, E. 6, 18, 123
Salisbury, Lord 79, 92, 108, 119–20, 132, 264, 266, 274, 285
Samuelson, B. 81, 259
Saturday Review 18, 23
Schiaparelli, G. V. 96, 175
Schuster, A. 154, 159, 162, 166, 179, 294, 295, 296

Science and Art Department 76, 79–81, 84–5, 88, 105–8, 113, 116, 117, 120, 129, 236, 240, 261, 263, 264
Scott, R. F. 227–8
Seabroke, G. M. 38, 56–7, 89–90, 127
Secchi, A. 43, 45, 49, 63–4, 191
Sharpey, W. 26, 50, 53, 57, 83
Shaw, W. N. 282–3, 287, 288, 294, 296, 297
Sidmouth 299–302, 306, 308
Smith, H. J. S. 93, 111–12
Smithsonian Institution 103, 220, 285
Smyth, Piazzi 125, 194, 249–50
Society of Arts 79, 80, 81
Sodium 47–8, 58, 61, 145, 152–3
Solar Physics 38, 307
Solar Physics Committee 109–12, 116, 117, 118, 122, 125–7, 129, 163, 167, 186, 282, 285, 291–3, 301
Solar Physics Observatory 117, 126, 127, 129, 133, 164, 216, 235–6, 278, 282, 284, 285, 287, 289, 290–300, 301
Sorby, H. C. 60, 177
South Kensington 78, 79, 84–5, 87, 105, 107–9, 113–22, 129, 132, 142, 143, 149, 150, 163, 165, 168, 171, 172, 185, 197, 198, 200, 204–6, 213, 215, 216, 221, 235–6, 259, 262–4, 275, 282, 287–91, 297–8, 300
South Kensington Museum 78, 89, 105–6, 114, 116, 119, 134, 218, 240, 294, 300
Spectator 17, 23, 100, 102
Spectroheliograph 286, 289
Spectroscope and its Applications, The 25, 37
Spectroscopes 46, 50–8, 70, 140, 175
Spencer, H. 16, 21, 30, 31, 36–7, 235
Stellar colours 145, 190–2
Stellar spectra 190–4, 198–206
Stewart, Balfour 23, 24, 46–7, 53, 55, 72, 81, 94, 98, 99, 107, 109, 123, 125–6, 127, 148, 149, 150, 159
Stokes, G. G. 57, 63, 81, 83, 107, 109, 118, 134, 140, 151, 153, 161, 187, 226–7, 262, 282
Stonehenge 250–3, 256, 257
Strange, A. 80–2, 91–2, 93, 94, 96–102
Sunspot cycle 124–8, 160, 166, 172, 284, 288
Sunspots 6, 46–7, 49, 50, 124, 127–8, 144, 160, 165, 166–7, 176, 291, 305
Surgeon-apothecaries 3–4
Switzerland 3, 10
Sylvester, J. J. 26, 27, 73, 221

Tait, P. G. 34–6, 177, 183, 224
Telescopes 40, 44, 46, 213, 300
Tennyson, A. 1, 14, 16, 66, 81, 206, 221, 234
Terrestrial magnetism 6, 123–5, 249, 287, 295
Thalén, T. R. 143, 150
Thallium 148, 151, 179
Thompson, S. P. 259, 265, 273
Thomson, J. J. 169–70, 232, 292
Thomson, W. 34, 60, 69, 81, 93, 94, 98, 107, 186, 189, 237, 262, 283–4
Times, The 1, 7, 38, 100, 102, 228, 253, 259, 260
Treasury 9, 33, 68–9, 103, 105, 117, 119, 120, 129, 276, 278, 282, 289, 295–6
Turner, H. H. 122, 163, 224–5, 226
Tyndall, J. 21, 29, 31, 34–7, 67, 81, 214

Underground railways 121–2, 290
Unitarians 280, 308
University colleges 264–5, 267, 268–9.
University of London 76, 77, 86, 89, 261–4, 267, 280
U.S.A. 30, 65, 66–7, 103, 118, 127, 167, 185, 210, 215, 267, 289, 303–4.

Vanadium 148, 167, 284
Vogel, H. 191–2, 215
Volunteers 1–2, 9, 211

Wales 12, 254–5, 264
Wallace, A. R. 14, 116
War Office 2–3, 6–12, 16, 23, 57, 58, 83, 103, 104, 105, 214, 274, 281, 291, 292, 294, 303
Warwickshire 3, 5, 12
Westgate-on-Sea 211–12, 213, 223
Wimbledon 2, 5–6, 8–9, 13–14, 16, 19, 246, 272
Woolner, T. 14, 16, 211

X Club 20, 31, 58

Yerkes Observatory 289–90
Youmans, E. L. 30–1
Young, C. A. 60, 61, 63, 66, 68, 69, 144, 154–5, 158, 160, 162, 180, 194

Zöllner, J. C. F. 56, 63